THE METAPHYSICS
AND NATURAL PHILOSOPHY
OF JOHN BURIDAN

MEDIEVAL AND EARLY MODERN SCIENCE

Editors

JOHANNES M. M. H. THIJSSEN
University of Nijmegen

CHRISTOPH LÜTHY
University of Nijmegen

Editorial Consultants

JOËL BIARD, University of Tours
SIMO KNUUTTILA, University of Helsinki
JOHN E. MURDOCH, Harvard University
JÜRGEN RENN, Max-Planck-Institute for the History of Science
THEO VERBEEK, University of Utrecht

VOLUME 2

THE METAPHYSICS AND NATURAL PHILOSOPHY OF JOHN BURIDAN

EDITED BY

J.M.M.H. THIJSSEN AND JACK ZUPKO

BRILL
LEIDEN · BOSTON · KÖLN
2001

On the cover: Wood print showing an astronomer contemplating the skies with the help of an armillary sphere and geometrical tools. From Johannes Angelus: *Astrolabium planum in tabulis ascendens continens qualibet hora atque minuto equationes domorum celi* (Venice: Iohannes Emericus de Spira, 1494), *f.* 1*v*. Reproduction by courtesy of Biblioteca Trivulziana, Milan. On the back: After a manuscript drawing of a regular solid by Leonardo da Vinci.

This book is printed on acid-free paper.

Library of Congress Cataloging-in-Publication Data

The metaphysics and natural philosophy of John Buridan / edited by
    J.M.M.H. Thijssen and Jack Zupko.
       p.   cm. — (Medieval and early modern science ; ISSN 1567-8393 v. 2)
    ISBN 9004115145 (alk. paper)
    1. Buridan, Jean, 1300-1358—Contributions in metaphysics.
    2. Metaphysics.   3. Buridan, Jean, 1300-1358—Contributions in physics.
    4. Physics.   5. Philosophy, Medieval.   I. Thijssen, J.M.M.H.
II. Zupko, Jack.   III. Series.
B765.B844 M48   2000
189'.4—dc21   00–064178
                                                                                          CIP

Die Deutsche Bibliothek - CIP-Einheitsaufnahme

The metaphysics and natural philosophy of John Buridan / ed. by J. M. M. H. Thijssen and Jack Zupko. - Leiden ; Boston ; Köln : Brill, 2000
  (Medieval and early modern science ; Vol. 2)
  ISBN 90-04-11514-5

ISSN   1567-8393
ISBN 90 04 11514 5

© *Copyright 2001 by Koninklijke Brill NV, Leiden, The Netherlands*

All rights reserved. No part of this publication may be reproduced, translated, stored in a retrieval system, or transmitted in any form or by any means, electronic, mechanical, photocopying, recording or otherwise, without prior written permission from the publisher.

Authorization to photocopy items for internal or personal use is granted by Brill provided that the appropriate fees are paid directly to The Copyright Clearance Center, 222 Rosewood Drive, Suite 910 Danvers MA 01923, USA. Fees are subject to change.

PRINTED IN THE NETHERLANDS

TABLE OF CONTENTS

Preface vii
JOHANNES M. M. H. THIJSSEN & JACK ZUPKO, John
    Buridan, Metaphysician and Natural Philosopher.
    An Introductory Survey ix

PETER KING, John Buridan's Solution to the Problem of
    Universals   1
GYULA KLIMA, Buridan's Theory of Definitions in his
    Scientific Practice 29
OLAF PLUTA, Buridan's Theory of Identity 49
SIMO KNUUTTILA, Necessities in Buridan's Natural
    Philosophy 65
JOËL BIARD, The Natural Order in John Buridan 77
GERHARD KRIEGER, *Naturaliter principiis assentimus*:
    Naturalism as the Foundation of Human Knowledge? . . 97
JOHN E. MURDOCH & JOHANNES M. M. H. THIJSSEN,
    John Buridan on Infinity 127
DIRK-JAN DEKKER, Buridan's Concept of Time. Time,
    Motion and the Soul in John Buridan's Questions
    on Aristotle's *Physics* 151
JACK ZUPKO, On Certitude 165
PETER G. SOBOL, Sensations, Intentions, Memories,
    and Dreams 183
FABIENNE PIRONET, The Notion of "non velle" in
    Buridan's Ethics 199
EDITH DUDLEY SYLLA, *Ideo quasi mendicare oportet intellectum
    humanum*: The Role of Theology in John Buridan's
    Natural Philosophy 221
PAUL J. J. M. BAKKER, Aristotelian Metaphysics and
    Eucharistic Theology: John Buridan and Marsilius of
    Inghen on the Ontological Status of Accidental Being . . 247
ROLF SCHÖNBERGER, Philosophical Theology in John
    Buridan 265

Bibliography 283
Index of Names 299
Index of Manuscripts 301

PREFACE

That John Buridan's brilliance as a logician is now widely recognized is due in no small part to the 1976 publication of a collection of essays, *The Logic of John Buridan*. Edited by the late Jan Pinborg, that volume brought together some of the best scholars in the field to explore – and, in many cases, to reveal for the first time – Buridan's distinctive contribution to medieval logic and semantics.

The idea for the present volume came about some two decades later. As more and more scholars were becoming interested in Buridan's non-logical corpus, the idea of trying to assemble some of them for a conference on Buridan's metaphysics and natural philosophy began to be feasible. The conference, organized by the editors, was held at Emory University on October 23–25, 1998. The success of that conference, and especially the many fruitful, collaborative discussions to which it gave rise, led directly to the present volume. Almost all of the essays collected here began as informal presentations at the conference, though most of them have been revised by their authors on the basis of exchanges with the other participants. None of the essays has been published before.

Our title, *The Metaphysics and Natural Philosophy of John Buridan*, is deliberate insofar as we have tried to create a kind of companion or successor volume to *The Logic of John Buridan*. Our intent has not been to hitch our own project to the success of the latter volume – though all Buridan scholars begin by treading in Pinborg's footsteps, of course – but rather to follow its basic investigative assumptions into the fields of metaphysics and natural philosophy. We are one with Pinborg in our determination to bring to the whole range of Buridan's thought the attention it deserves.

For their generous support of the conference on which this volume is based, the editors would like to thank Dean Steven Sanderson of the College of Arts and Sciences at Emory University, the Department of Philosophy at Emory, and the Center for Medieval and Renaissance Natural Philosophy at Nijmegen University (NWO-grant 200-22-295).

This book was typeset by *Typographica Academica Traiectina* in Utrecht. We should like to thank Johannes Rustenburg and Ivo Geradts for the wonderful job they did, and Kim van Gennip for her help with preparing the manuscript and the index.

JOHN BURIDAN,
METAPHYSICIAN AND NATURAL PHILOSOPHER.
AN INTRODUCTORY SURVEY

JOHANNES M. M. H. THIJSSEN &
JACK ZUPKO

It is now widely acknowledged that John Buridan was the most important philosopher at the University of Paris in the fourteenth century. As a native of Béthune, Buridan belonged to the Picard Nation. He held several important administrative posts, such as rector, delegate of the Picard Nation, and negotiator of that nation in its conflict with the English Nation. He died in 1361.[1]

For reasons which are not known, Buridan never moved on to study theology, but spent his entire career in the arts faculty. In this sense, he is one of the few bona fide philosophers of the later Middle Ages. He left a large body of writings which had a considerable impact on his contemporaries and on thinkers of later centuries. With the exception of the *Summulae de dialectica* and a few polemical works, which probably originated as disputations within the faculty of arts, Buridan's entire philosophical corpus is in the form of commentaries on the works of Aristotle,[2] covering logic, metaphysics, natural philosophy, psychology, and ethics.

Although Buridan's contribution to medieval philosophy is now recognized, it is still often omitted in general surveys of medieval philosophy. This is partly due to the fact that it is only in the last few years that his works have started to become available in modern editions, making it possible for a larger number of scholars and students to gain a first-hand knowledge of his thought. Until recently, the study of Buridan was a specialized endeavour, requiring expertise in the reading of medieval manuscripts and early printed editions.

In our own time, the appreciation of Buridan's work has undergone a few interesting shifts. In the first half of this century, Buridan was mainly praised for his contributions in the field of mechanics.

---

[1] See Michael, *Johannes Buridan*, for the details of Buridan's academic career. We know that he died no later than 1361 because one of his benefices was given to another person in that year.

[2] An almost complete survey of Buridan's works and of the manuscripts and early editions in wich they have been preserved is given in vol. 2 of Michael, *Johannes Buridan*.

These were first brought to light by Pierre Duhem, who drew attention to the notion of impetus, an impressed force which Buridan had introduced to explain the continuation of projectile motion and the acceleration of falling bodies.[3] Duhem claimed that Buridan's impetus theory had anticipated Galileo's law of inertia. The ensuing controversy over Duhem's claims did much to uncover certain aspects of Buridan's natural philosophy. Most notably, Anneliese Maier and Marshall Clagett started studying specific passages of Buridan's commentaries on the *Physics* from the manuscript sources, and placed them in their fourteenth-century context.[4] Buridan's impetus theory even came to be absorbed in more recent discussions of the growth of scientific knowledge by, for instance, Thomas Kuhn and Paul Feyerabend.[5]

A second major factor in the growth of interest in Buridan's philosophy occurred in 1976, when Jan Pinborg edited and published a volume of essays by various scholars on Buridan's logic and semantics.[6] The volume proved to be seminal not only in that it channeled subsequent research into precisely this area of Buridan's thought, but also in that it helped clear the way for a critical edition of Buridan's major work in logic, his *Summulae de dialectica*.[7]

During the last decade or so, scholarly attention has turned to other aspects of Buridan's thought, in particular to his natural philosophy and metaphysics.[8] Most of the resulting literature appears in the bibliography of this volume, and most of its authors are featured in the ensuing chapters.

The essays are organized around five themes, which roughly follow the ordination of the speculative sciences in the late medieval arts curriculum. Beginning with metaphysics (Chapters 1–3), we pro-

---

[3] Duhem, *Études*, esp. vol. 3.
[4] Maier, especially *Die Vorlaüfer*, pp. 132–154, and *Zwei Grundprobleme*, pp. 201–236, and Clagett, *The Science of Mechanics*, esp. pp. 505–682.
[5] Kuhn, *The Structure of Scientific Revolutions*, pp. 119–120, and Feyerabend, *Realism, Rationalism and Scientific Method*, pp. 44–97.
[6] Pinborg, *The Logic*.
[7] The edition of the *Summulae* is a joint international effort directed by prof. Sten Ebbesen of Copenhagen University, The *Summulae* consists of eight treatises, four of which have now been edited and published by Ingenium Publishers in Nijmegen.
[8] See Thijssen, "Late-Medieval Natural Philosophy," pp. 177–185 for a survey of the most recent scholarly literature. The only book-length study of Buridan's metaphysics and natural philosophy, Ghisalberti's *Giovanni Buridano. Dalla metafisica alla fisica*, appeared in 1975. It conveniently assembled the material that was then known about Buridan's views (notably from Maier's studies), without, however, further exploring new aspects of Buridan's thought in these areas.

ceed to physics or natural philosophy (Chapters 4–6) and then to the particular questions of infinity and time (Chapters 7–8). From there, we consider psychology and ethics (Chapters 9–11), before finally closing with some reflections on the role played in Buridan's philosophy by doctrines from the "higher" faculty of theology (Chapters 12–14).

Nominalism is a doctrine for which Buridan was justly famous, though it appears in different guises throughout his writings. In "John Buridan's Solution to the Problem of Universals," Peter King considers the meaning of nominalism in Buridan's metaphysics. The doctrine has at least three interrelated aspects: (1) the ontological thesis that there are no non-individual entities in the world; (2) the psychological thesis that some concepts, though in themselves singular, are capable of representing a plurality of things; and (3) the semantic thesis that such common or universal concepts also function as common names in Mental Language. King suggests that it is the second thesis that does most of the work for Buridan, though as a solution to the problem of universals it raises further questions about how concepts can be said to represent one group of things more than another, and indeed, how concepts can be said to represent anything at all. King argues that Buridan has the resources to answer such questions by construing universal concepts as modes of thought, or else by treating resemblance as metaphysically primitive. Either way, King finds in the depth and complexity of Buridan's nominalism "a robust example of medieval philosophy at its finest."

Buridan's understanding of the nature of things is driven by his logic. In "Buridan's Theory of Definitions in his Scientific Practice," Gyula Klima examines some of the consequences of Buridan's *Summulae* treatment of nominal and quidditative definitions, suggesting that he was led to change his interpretation of the traditional definitions of the soul in Aristotle's *De anima* once he realized the implications of his account. In the first version of his lectures on this text, Buridan follows the older line of interpretation exemplified by Thomas Aquinas, according to which Aristotle offers first a nominal and then a real definition of the soul, such that the second can be proved from the first by means of a *quia*-type *a posteriori* demonstration. But a valid demonstration of this sort is simply not possible if we assume Buridan's "official theory" of definitions in the *Summulae*, which differs significantly from the older model found in Aquinas. Accordingly, the final version of Buridan's lectures on *De anima* treats Aristotle's first definition of the soul as a mere description and the second as a causal definition, thus preserving the consistency of his own interpretation of the Aristotelian definitions with his own theory of definitions, but at the expense of giving up the demonstrability of the quidditative definition of the soul from its nominal definition.

In "Buridan's Theory of Identity," Olaf Pluta surveys the different uses to which Buridan puts the concept of identity. Although the basic idea emerges from his logic, Buridan is forced to augment it considerably when he considers questions such as "Is Socrates the same today as he was yesterday?" in his commentary on Aristotle's *Physics*. After carefully distinguishing the temporal and numerical dimensions of the problem (he never uses the term "personal identity"), and noting that the latter can be understood as total, partial, or continuous, Buridan concludes that Socrates is the same in the second sense, but not the first (the third sense applies to continuous objects with diverse parts, such as the river Seine). Pluta proceeds to show that Buridan was also interested in the metaphysical conditions for identity when he asks whether the same thing could exist discontinuously in time, such as if God were to annihilate X now and then recreate the very same X later. Buridan's answer appeals to the principle that God's thinking it to be so – in this case, thinking that something is really identical across discontinuous stretches of time – is sufficient for making it so. The entire discussion testifies to Buridan's skill in dealing with theological questions which were strictly speaking beyond his mandate as an arts master.

In the late thirteenth century, theologians developed new patterns of argumentation in theological disputes that came to revolutionize the way we think about modality. Buridan was one of the first to take this new way of thinking out of the narrow confines of theology and apply it to the interpretation of Aristotle. In "Necessities in Buridan's Natural Philosophy," Simo Knuuttila examines Buridan's appropriation of three techniques: the concept of the *positio impossibilis*, the principle that anything follows from what is impossible, and the distinction between natural and unqualified or supernatural modalities (which he usually expresses in terms of the distinction between divine absolute and natural powers). In particular, Knuuttila argues that Buridan's insistence that the distinction between natural and supernatural modalities can be found in Aristotle was instrumental in his coming to view it as a systematic principle of natural philosophy itself. This went a long way towards normalizing the distinction in fourteenth-century natural philosophy and contributed, in later authors, to the introduction of new epistemological problems based on the logical contingency of the natural order.

In "The Natural Order in John Buridan," Joël Biard considers the different roles played in Buridan's natural philosophy by his concept of the natural order. This concept has two related aspects. The first is the idea – which can be traced to Albert the Great – of nature as an autonomous, law-governed structure whose principles can be discerned and validly applied using reason alone. But the second aspect reaches beyond the natural by contrasting it with the supernatural, explicitly recognizing the possibility of divine intervention in the course of nature as well as the thesis that created things are ordained to a single, supreme end "which is God himself." These aspects together give us a picture of scientific inquiry in which the regularities we observe in nature are necessary *ex suppositione*, i.e., on the assumption of the common course of nature. This conception is fundamental to Buridan's natural philosophy because without it, scientific knowledge would be impossible.

In "*Naturaliter principiis assentimus*:" Naturalism as the Foundation of Human Knowledge?," Gerhard Krieger looks at the epistemological basis of Buridan's conception of the natural order. How is it that, according to Buridan, we just "naturally" assent to principles which determine the way we think and reason about the world? Krieger argues that such acts of assent emerge from a fundamental desire for knowledge, which belongs not to the natural order but to the will. The appetitive origins of this natural inclination – which is natural in the sense that it is part of our creaturely endowment – explains why we can assent to principles without first needing them to be demonstrated, for "reason is agreeing with itself when it assents to principles." Krieger applies his hypothesis to Buridan's discussions of the principle of non-contradiction, showing how our immediate assent to the first principle can be traced back to this original and reflexive relation of reason.

Buridan was a major player in fourteenth-century controversies about infinity and continuity. In their jointly-authored paper, "John Buridan on Infinity," John Murdoch and Hans Thijssen illustrate the way in which Buridan developed and systematically defended his position on the composition of continuous magnitudes – a position Calvin Normore has nicely termed "pure divisibilism" – using analytical techniques borrowed from logic and semantics. Thus, he expresses the way in which continua are infinitely divisible by distinguishing between the categorematic and syncategorematic senses of the term "infinite," whereas the modal aspects of infinite divisibility are presented in terms of the distinction between the compound and divided senses of a proposition. The notion of God's absolute power figures prominently in Buridan's discussion of infinity, which, as in the related question of the eternity of the world, often seems to stray into territory already staked out by the theologians. But such quasi-theological topics arose quite naturally for Buridan in the context of teaching of Aristotle's *Physics*, where they were part of his distinctively philosophical effort to explain the natural world.

The application of these analytical techniques to the nature of time is the subject of Dirk-Jan Dekker's essay, "Buridan's Concept of Time: Time, Motion and the Soul in John Buridan's *Questions on Aristotle's Physics*." For Buridan, time is a successive thing (*res successiva*) which is identical to motion, such that the term "time" signifies the same as the term "motion," though connoting what it signifies as a measure. For most of us, time is measured via inferior and somewhat irregular sublunary motions, such as the length of shadows during the day and changes in crops during the seasons. Astronomers, however, conceive of time to refer to the motion of the outermost sphere of the heavens, the first and most proper sense of "time." But the existence of time does not depend on our actually measuring time or thinking of things in time. Buridan argues that even if God annihilated all intellective souls, there would still be the motion of the outermost sphere, and this is at least potentially applicable as a measure, i.e., if God were to create another intellective soul to do the measuring. Buridan uses the same sort of counterfactual thinking to refute Ockham's account of local motion in terms of mobile things acquiring one place after another: since it is possible for God to move the heavens without moving it to a different place by simply rotating it on its axis, motion must be a purely successive thing that is intrinsic to what is moved, not reducible to its external situation.

We move next to psychology, the extension of Aristotelian natural philosophy to things that are capable of moving themselves. In "On Certitude," Jack Zupko examines Buridan's psychological account of certitude (*certitudo*), a concept referring not to the objective warrant of a knowledge claim but to its qualitative character as stable or fixed. Certitude manifests itself on both sides of the knower/known relation. On the side of the object, propositions have "the certitude and firmness of truth" when they have a stable and fixed appearance by virtue of reflecting the order God has instilled in nature. The firm judgments we make on the basis of such appearances are thereby associated with the orderly movement of the cosmos. On the side of the subject, certitude refers to the quality of our assent to propositions we are said to know rather than merely believe or have an opinion about. In this case, the firmness of assent is conditioned by the proposition's appearance as evident, which is as much a product of our antecedent beliefs as it is of the proposition's objective warrant. The result is in keeping with Buridan's organic picture of how human knowledge is acquired, and explains as well his refusal to be swayed by skeptical arguments based on over-zealous application of the principle of non-contradiction.

Peter G. Sobol's essay, "Sensations, Intentions, Memories, and Dreams," nicely illustrates the sorts of practical questions that were addressed by late medieval psychologists. Medieval Aristotelians such as Buridan were faced with the difficult task of extending the theory of the soul to cover psychological phenomena where Aristotle's authoritative remarks are puzzling, minimal, or nonexistent. Buridan's skill at harmonizing the book of nature with the book of Aristotle made his own commentary on *De anima* into an authority for later philosophers. Insisting that sensation be conceived as the act of a living organism, Buridan explains the activity and passivity of the various sensory powers, the number of external senses, the nature and function of sensible species, and the origins of unsensed intentions such as our awareness of the passage of time. Rejecting the Galenic-Avicennian tradition of placing the common sense in the brain, he argues instead for the Aristotelian view that it is in the heart, which he believed must be proximately connected to the brain by a conduit of nerves. Throughout these discussions, we get a sense of the balancing act that was required of a philosopher who was determined to save the phenomena despite the limitations of the explanatory model he inherited from Aristotle.

Buridan's account of human free choice is the subject of Fabienne Pironet's essay, "The Notion of '*non velle*' in Buridan's Ethics." The nature of Buridanian volition has been a source of some controversy in the secondary literature because Buridan's use of Scotistic terminology to characterize certain acts of the will makes his otherwise strictly intellectualist account of free choice appear less deterministic. Through her careful analysis of the term "*non velle* (not willing)" and of the possible choices available to the will in each of its states of decision, Pironet shows that despite appearances, Buridan never uses the term in the Scotistic sense to signify a genuine act of self-determination on the part of the will. Rather, the will can choose to defer its act by not willing only if it remains in a passive state as regards willing (*velle*) something and nilling (*nolle*) it, or actively willing against it. Unlike the Scotistic picture, the Buridanian will can never act directly against reason because Buridan defines freedom of choice in terms of the will's capacity to suspend its act when faced with uncertain appearances in order to seek more information from the intellect. This freedom is sufficient to distinguish us from brute animals, which are driven to act unreflectively on the basis of whatever appears before them. Pironet closes by reflecting on how Buridan's views were shaped by his reading of the anti-intellectualist articles in the Condemnation of 1277.

Because he was a career arts master, Buridan's references to theologians and use of theological doctrines provide us with a unique perspective on the relation between philosophy and theology in the later Middle Ages. In "*Ideo quasi mendicare oportet intellectum humanum*: The Role of Theology in John Buridan's Natural Philosophy," Edith D. Sylla examines Buridan's treatment of such "theological" questions as the eternity or creation of the world and the possibility of a vacuum. Buridan's approach differs from that of philosophers such as Boethius of Dacia, who insisted that theology must be kept out of physics. Taking seriously his pledge upon incepting in the Arts Faculty to resolve questions touching on both the faith and physics on the side of the faith, and "to resolve the arguments for the other side accordingly as it seemed to him they should be resolved," Buridan replies to the naturalistic conclusions of Aristotle and Averroes using a variety of theological, metaphysical, and logical arguments. The result is not so much physics as speculation about possible worlds, since Buridan added certain hypotheses and conditions where it was necessary "so to speak to beg the intellect (*ideo quasi mendicare oportet intellectum humanum*)." Far from limiting rational inquiry about such questions, this strategy permitted the consideration of Aristotelian and Averroistic objections in terms that remained clearly philosophical.

Paul J. J. M. Bakker considers the difficult question of the metaphysics of the Eucharist in, "Aristotelian Metaphysics and Eucharistic Theology: John Buridan and Marsilius of Inghen on the Ontological Status of Accidental Being." Citing the same Arts Faculty pledge Sylla discusses in her essay, Bakker shows that the theological doctrine of transubstantiation forced Buridan the metaphysician to admit, *contra* Aristotle, the possibility of real accidental being. For if God could miraculously annihilate the substance of the bread while preserving its whiteness and other sensory qualities, then accidents could exist apart from substance, and whiteness would no longer be the same as some substance being white. Buridan thus alters his metaphysics to accommodate the separability of accidents, defining substance as a thing that exists on its own without inhering in any other thing, and accident as a thing that does not exist on its own in the common course of nature, although it could do so miraculously. The separability of accidents leads Buridan to another non-Aristotelian conclusion (which, ironically, is reached by Aristotelian arguments): there must be some added disposition by means of which accidents inhere in their subjects. Bakker proceeds to contrast Buridan's approach with that of Marsilius of Inghen, his younger associate at Paris who went on to teach at the University of Heidelberg. Marsilius makes no attempt to reconcile theology and metaphysics in his account of the ontological status of accidental being, and instead treats the separability of accidents in the Eucharist as completely miraculous, insisting that the term "being" signify substances and accidents equivocally. By contrast, Buridan introduces the distinction between the order of nature and the order of miracles at the most basic level of his metaphysics of substance and accident.

In "Philosophical Theology in John Buridan," Rolf Schönberger situates Buridan's concept of God within the broader tradition of natural theology in the later Middle Ages. Buridan follows Avicenna – and later, Aquinas and Scotus – in arguing that metaphysics, or the science of being *qua* being, is the most appropriate discipline for reasoning about the nature of God. The trouble with the Averroistic position that physics yields a more adequate conception of God is that even if God were shown to be the cause of all motion, it would not follow that God is the cause of immobile things, and hence, we would be unable to conceive of God as universal cause. Nevertheless, because he subscribed along with most late medieval nominalists to the idea that being *qua* being is empty, Buridan sought to establish a metaphysical concept of God as first cause using the physical concept of God as first unmoved mover. The latter concept did not always translate very well into metaphysics, however, especially when applied to psychological forms of motion, i.e., to thoughts and volitions. As a result, later authors tended to abandon Buridan's metaphysics of the first unmoved mover, though they eagerly embraced his general program in natural philosophy.

JOHN BURIDAN'S SOLUTION TO THE PROBLEM OF UNIVERSALS*

PETER KING

1. *The Failure of Realism*

> Thus it is pointless to hold that there are universals distinct from singulars if everything can be preserved without them – and indeed it can, as will be apparent ...[1]

Buridan issues this promissory note at the end of his critique of realist attempts to solve the problem of universals.[2] He began his negative case by attacking platonist theories, that is, theories identifying the universal as a separated form really distinct from the individuals it characterizes.[3] His next target was so-called moderate realist theories, which identify the universal as a form that is really distinct but not separate from the individuals it characterizes.[4] Finally, he turns to

---

* All translations are mine. See the Bibliography for abbreviations, editions, and references; when citing Latin texts I use classical orthography and occasionally alter the given punctuation and capitalization. I occasionally amend the text of citations in minor ways. For complete details on each of Buridan's works see Michael, *Johannes Buridan*.

[1] John Buridan, *In Metaphysicen* (Paris, 1518), Book VII, q. 16, fol. 51vb: "Et ideo frustra ponerentur talia uniuersalia distincta a singularibus si omnia sine illis possint saluari; et tamen possunt, quod apparebit ..." See also John Buridan, *Tractatus de differentia universalis ad individuum* [Szyller], q. 1, c. 2 (eighth argument), p. 152$^{16-19}$: "In natura non est ponenda pluralitas sine necessitate nec per consequens distinctio, cum distinctio non sit sine pluralitate; sed nulla necessitas est quod uniuersale sit praeter animam distinctum ab individuis praeter animam."

[2] See Ghisalberti, *Giovanni Buridano*, De Rijk, "John Buridan on Universals," and King, "Jean Buridan" for detailed analysis of Buridan's critique of realism as found in John Buridan, *In Metaphysicen* (Paris, 1518). There is as yet no discussion in the literature of John Buridan, *Tractatus de differentia universalis ad individuum* [Szyller] or the unpublished *Quaestiones super Isagogen* (See Michael, *Johannes Buridan*, II, pp. 457–463).

[3] John Buridan, *In Metaphysicen* (Paris, 1518), VII.15.

[4] This position is anonymous in John Buridan, *op.cit.*, VII.16 but attributed to Walter Burleigh in John Buridan, *Tractatus de differentia universalis ad individuum* [Szyller], q. 1, c. 2. Buridan explicitly refers to Burleigh at p. 138^{17} ("Gualterus in sua Expositione super primum Physicorum") as the author of eight of the twenty-five (!) arguments he recounts in favor of the claim that the universal exists outside the soul

Scotist theories, which identify the universal as a form that is only formally distinct from the individuals it characterizes, neither really distinct nor separable from them.[5] Buridan's discussion therefore follows a pattern similar to that found in William Ockham, where the arguments against one view are assumed in the critique of the next, cascading from more to less extreme versions of realism.[6] And indeed they reach the same negative conclusion: realism about universals, in any version, is bankrupt. What moral should be drawn? Buridan is explicit:

> First of all, we should note – as is sufficiently clear from what has been said – that whatever exists outside the soul does so in reality as an individual, that is, distinct from all else (whether belonging to its species or to others), such that it is nothing at all in reality apart from individual things and is not distinct from them.[7]

Everything is individual. More exactly, every being capable of existing *per se* is individual. There are no non-individual entities in the world, whether existing independently or as metaphysical constituents either of things or in things. (Individuals are individual all the way down.) Hence no real principle or cause of individuality, other than the individual itself, is required.

Individuality is a basic feature of the world. In its train comes distinctness: "Every thing exists as singular such that it is diverse from any other thing."[8] Even closely related individuals systematically differ from one another:

---

with a being distinct from that of individuals (p. 138⁴⁻⁵: "praeter animam secundum esse distinctum ab individuis:").

[5] John Buridan, *op.cit.*, p. 2, q. 2. Buridan doesn't explicitly refer to Scotus (unlike Burleigh), and he doesn't mention the formal distinction. He does talk about whether the unity of the universal is less than numerical unity, citing arguments from John Duns Scotus, *Ordinatio*, II d. 3, pars 1 q. 1 nn. 10–28. Most of Buridan's objections are devoted to showing that its unity is specific unity, and hence not a matter for metaphysical concern.

[6] William Ockham, *Scriptum in librum primum Sententiarum. Ordinatio* [Gál e.a.], d. 2, qq. 4–8.

[7] John Buridan, *Tractatus de differentia universalis ad individuum* [Szyller], q. 1, c.2, 1539⁻¹³: "Ad cuius euidentiam sciendum est primo quod, ut satis potest ex dictis apparere, quicquid praeter animam existit in re ipsum existit indiuidualiter, scilicet distinctum ab omnibus aliis tam suae speciei quam aliarum, ita quod ibi nihil est omnino praeter res quae indiuidualiter existunt nec est distinctum ab eis."

[8] John Buridan, *Quaestiones super octo Physicorum libros Aristotelis* (Paris, 1509), Book, q. 7, fol. 8ʳᵇ: "Immo omnis res singulariter existit ita ut sit diuersa ab unaquaque aliarum rerum." See King, "Jean Buridan" for an account of Buridan's theory of individuality.

> Individuals belonging to the same species, such as Socrates and Plato, differ substantially. That is, they differ by their substances, by their matter as well as by their forms, due to the fact that Socrates's form isn't Plato's form and Socrates's matter isn't Plato's matter.[9]

Buridan's world is therefore a world of individuals, each capable of existing *per se* and distinct from all else. They come in four kinds. First, there are substances: God and angels (who subsist *per se*) on the one hand, and less exalted traditional primary substances (such as Socrates and Plato, cats, and the like) composed of form and matter on the other. Second, at least some substantial forms – only one per composite; Buridan defends the unicity of substantial form[10] – can exist in separation from matter, namely human souls, and hence are themselves individual. Third, prime matter, which for Buridan is of itself a being, is capable of existing per se through divine power; as such it is an individual, though normally it exists in act only in combination with some form and as such is not individual.[11] Fourth, Buridan argues that real accidents may exist without inhering in any substance, at least by divine power, as in the case of the Eucharist; as such they are individuals.[12]

Whatever the merits or demerits of Buridan's list – I won't examine them here – the thesis that everything is individual only underlines Buridan's difficulty in making good on his promise that everything the realists did by postulating real universals in the world can be done without them. For if everything is individual, how does generality get into the world at all? The very convictions that led Buridan to argue against realism seem to undercut his attempt to work out a consistent non-realist alternative. But work one out he must if he is to have a solution to the problem of universals.

---

[9] John Buridan, *In Metaphysicen* (Paris, 1518), VII.17, fol. 52va: "Dicendum est quod indiuidua eiusdem speciei, ut Socrates et Plato, differunt substantialiter, scilicet per suas substantias tam per formas quam per materias ex eo quod nec forma Socratis est forma Platonis nec materia Socratis est materia Platonis."

[10] John Buridan, *In Metaphysicen* (Paris, 1518), VII.14.

[11] John Buridan, *Quaestiones super octo Physicorum libros Aristotelis* (Paris, 1509), Book I, q. 20.

[12] John Buridan, *In Metaphysicen* (Paris, 1518), V.8. Which accidents are real? Buridan countenances qualities such as whiteness; motions; perhaps magnitude or quantity; relations that are founded on real accidents; and the inseparable "added disposition" in virtue of which an accident informs a subject.

2. The Psychological Underpinnings of Nominalism

The challenge facing Buridan, then, is to show how "everything can be preserved" in a world of individuals without appealing to any non-individual entities. His strategy is to argue that generality, not found in the world, is present only in the mind. He therefore recasts the question in psychological terms:

> Since there are no universals outside the soul distinct from singulars, but, since every thing exists singularly, how does it come about that things are sometimes understood universally?[13]

Buridan's answer to the question formulated this way is that generality stems from the fact that the mind is fundamentally representational:

> Thus if we want to give a single reason (though not a sufficient one) why the intellect can understand universally even though the things understood neither exist universally nor are universals, I declare this to be the reason: Things are understood not because they are in the intellect but because likenesses that represent them are in the intellect.[14]

The plausibility of Buridan's strategy here is due, at least in part, to the fact that concepts are able to represent a plurality of things while remaining individual in themselves. This dual ontological aspect allows Buridan to appeal only to individuals in his account of generality, namely individual concepts, while nevertheless providing a foundation for generality in their representative features. Of course, this dual aspect is not unique to mental items (a statue in the park may be singular in itself while representing many people),

---

[13] John Buridan, *Quaestiones super octo Physicorum libros Aristotelis* (Paris, 1509), I.7, fol. 8va: "Ista quaestio continet dubitationes ualde difficiles. Una est cum non sint uniuersalia praeter animam distincta a singularibus, sed, quia omnis res existit singulariter, unde prouenit quod res aliquando intelliguntur uniuersaliter?"

[14] John Buridan, *Quaestiones in Aristotelis De anima, tertia lectura* [Zupko], Book III, q. 8, ll. 237–243: "Si ergo uolumus assignare unam causam, licet non sufficientem, quare intellectus potest intelligere uniuersaliter, quamuis res intellectae nec uniuersaliter existant nec uniuersales sint, ego dico quod haec est causa: quia res intelliguntur non propter hoc quod ipsae sint in intellectu, sed quia species earum, quae sunt similitudines repraesentiuae earum, sunt in intellectu." (Buridan notes that this isn't a sufficient reason because there can be concepts that represent only a single thing. Representationality is not in itself a guarantee of generality.) See also John Buridan, *Quaestiones super octo Physicorum libros Aristotelis* (Paris, 1509), I.7, fol. 8vb: "Dico ergo, sicut mihi uidetur, quod una causa est in hoc quod intellectus intelligit uniuersaliter, licet existat singulariter, et res intellecta singulariter, et intentio etiam singulariter. Et ratio huius est quia res intelliguntur non per hoc quod sunt apud intellectum, sed per suam similitudinem existentem apud intellectum."

nor do all mental items have it (complexive mental concepts are nonrepresentative). But concepts are also distinctive in another way: they are components of two systematic bodies of theory, a second "dual aspect" that makes them especially useful as an explanatory foundation for Buridan's solution to the problem of universals. On the one hand, concepts are psychological entities. They are literally the elements of thought: thinking of X just is having a concept of X, which manages to be "about" X in virtue of "naturally resembling" it.[15] Concepts are the primary building-blocks of the intellect. We acquire them from our interaction with the world, and an adequate psychological theory will detail the process of concept-acquisition, in light of the operation of other mental faculties (such as sense-perception). Since the basic conceptual apparatus of all humans is the same, psychology can be a universal natural science. On the other hand, concepts also have a semantic dimension. In particular, universal concepts in the intellect also function as common names in Mental Language (subject to certain qualifications); since Mental is an ideal language, concepts will be normatively governed and have semantic features that can be considered independently of their psychological properties.[16] Buridan is thus able to switch between the psychological and semantic features of concepts depending on the requirements of the case at hand. Whether we should call Buridan a "conceptualist" (since universals are representatively general concepts in the intellect) or a "nominalist" (since universals are common names in Mental Language) is moot: one and the same item, a representatively general concept, has a role in psychology as a mental item and in semantics as a common name.

On the semantic side, general concepts are plausibly identified as universals. A concept that is representative of many functions as a common name, and is thereby "predicable of many" – the many subjects it represents as a concept, that is. Hence it can appear in true Mental sentences as the predicate-term successively conjoined to different individual concepts acting as singular subject-terms; it can be used in such sentences to refer to (*supponere pro*) extra-mental

---

[15] John Buridan, *In Metaphysicen* (Paris, 1518), VI.12, fol. 41vb.

[16] See King, *Jean Buridan's Logic* for an account of Mental Language and its features as a "logically ideal" language. I now no longer think this picture of Mental can be sustained. Buridan did not think that concepts could simultaneously be elements in a descriptively adequate psychology and constituents of a normative ideal language, and he was right to think not. See the analysis given in King [forthcoming]. Nothing in the discussion here rides on the nature of Mental, though, so we can bypass the point for now.

items as well as signifying them via natural likeness. In short, the generality of language makes it reasonable to think that we could take concepts to be universals. If Buridan can credibly argue that there are general concepts filling the requisite semantic roles, much of his solution to the problem of universals will be in place.

To that end, Buridan proposes three psychological theses: (1) intellective cognition depends on sensitive cognition; (2) sensitive cognition is always singular; (3) intellective cognition can be singular and it can be universal. The payoff comes in (3), since universal intellective cognition is the key to the problem of universals, but Buridan's endorsement of (1) and (2) make (3) problematic. If intellective cognition depends on sensitive cognition, and the latter is always singular, where does generality enter the psychological realm? A closer look at each thesis is in order.

3. *Buridan's First Psychological Thesis*

The dependence in Buridan's claim that intellective cognition depends on sensitive cognition is causal: the intellect requires input from sense to function. (This is not to spell out how it functions, of course.) *Nihil in intellectu quod non prius fuerit in sensu*: there is nothing in the intellect not previously in the senses, as the Aristotelian maxim has it. Buridan treats the claim as sufficiently obvious to use as a minor premiss and not to need further proof.[17] This is remarkable in light of the fact that William Ockham denied it. In the first conclusion of his *Reportatio* II qq. 13–14, Ockham maintains that "... given a sufficient agent and patient in proximity, the effect can be postulated without anything else".[18] It is the nature of the sensitive and intellective souls that an object is both sensed and understood when it is present. For Ockham, sensitive cognition and intellective cognition are no more than independent distinct effects of the same

---

[17] John Buridan, *Quaestiones super octo Physicorum libros Aristotelis* (Paris, 1509), I.7, fol. 8va: "Et de hoc ponitur prima conclusio communiter concessa, scilicet quod necesse est hominem cognoscere prius esse singulariter quam uniuersaliter, quia necesse est hominem prius cognoscere aliquid cognitione sensitiua quam intellectiua; et tamen nos supponimus quod cognitione sensitiua nihil cognoscatur nisi singulariter; ergo etc." He also cites (1) at fol. 9vb: "Cum ergo dictum sit quod cognitio intellectiua dependet ex sensitiua ..."

[18] William Ockham, *Quaestiones in librum quartum Sententiarum (Reportatio)* [Wood e.a.], 5, p. 268 7–9: "Posito actiuo sufficienti et passiuo et ipsis approximatis, potest poni effectus sine omni alio."

cause, the former its proximate effect and the latter its remote effect; the intellect depends only on the proximity of the cause, not on the prior operation of the senses.[19] Buridan, however, endorses the general consensus that intellect depends on the senses. Its appeal isn't simply in its popularity, though. We can readily construct an argument for Buridan's first thesis, as follows. First, the analysis of the functioning of the sensitive soul applies equally to humans and the brute animals, who by definition lack intellective souls. Second, the intellective soul is immaterial (held on the grounds of faith if nothing else); that means it is not the form of any given sense-organ, or, to put the same point another way, the intellect has no means whereby to pick up information about the world. Hence any material processed by the intellect must already be in the soul, and the only way for it to get there is though the senses. Ockham is left postulating a causal claim (external objects have effects on the intellect) without having any mechanism for the cause to bring about the effect. Buridan's first thesis seems clearly preferable.

4. *Buridan's Second Psychological Thesis*

Given that intellective cognition depends on sensitive cognition, we need to determine, first, whether we can sense things universally (for sensitive cognition might be the means whereby we deal with individuals universally),[20] and second, what intellective cognition gets from sensitive cognition to work with. It turns out that both questions have a single answer, summed up in Buridan's second psychological thesis: sensitive cognition is always singular.[21] Therefore, we always sense things as singulars and never as universals. The deliverances of the senses to the intellect must therefore be singular, since this is the only kind of information sensitive cognition can provide.

---

[19] See King, "Scholasticism and the Philosophy of Mind" for an account of why Ockham was driven to this counterintuitive claim, as well as a discussion of the general problem of transduction in medieval philosophy of mind.

[20] Buridan calls this first question very difficult in John Buridan, *Quaestiones in Aristotelis De anima, tertia lectura* [Zupko], III.8, ll. 153–155: "Ista quaestio implicat in se plures maximas difficultates: scilicet utrum sensus possit sentire uniuersaliter uel solum singulariter …"

[21] See the end of John Buridan, *Quaestiones super octo Physicorum libros Aristotelis* (Paris, 1509), I.7, fol. 8va as cited in n. 17 above. Buridan thinks that Aristotle endorses (2) in *De anima*, III.7 431b1–20, which he summarizes in John Buridan, *Expositio et quaestiones in Aristotelis De anima* [Patar], Book I, q. 4, p. 196$^{82-83}$ as follows: "Sicut patet tertio huius: dicitur enim ibi quod sensus est singularium."

Now unlike (1), Buridan finds (2) in need of argument. This is surprising, since we might be inclined to grant (2) directly. After all, isn't it just a medieval version of the claim that we perceive only individuals? What realist, however committed, has thought that we perceive universals? Furthermore, even if we aren't inclined to grant (2) out of hand, we might think it follows directly from the fact that the sensitive soul is material and extended – a claim Buridan puts as follows:

> It seemed to some thinkers that sense doesn't have the nature for cognizing [its objects] universally, but rather singularly, in virtue of the fact that it has extension and a determinate location in a bodily organ.[22]

Yet Buridan rejects the inference from the sensitive soul's materiality to its singular cognition (as he indeed will reject the parallel inference from the intellect's immateriality to its universal cognition). His grounds for so doing also challenge our ready modern acquiescence to (2): (a) the indefiniteness of intentional activity; (b) problems with discernibility.

As regards (a): The sensitive appetite is just as material and extended as sensitive cognition. Yet sensitive appetite is not targeted at individuals. A thirsty horse wants some water, but no particular water more than any other. This holds generally: natural agents acting as causes seem not to single out individuals *qua* individuals. Fire heats up any wood in the range of its causal activity; it is not restricted to acting only on some particular piece of wood. The inference from materiality to singularity fails in these cases; why think it holds in the case of sensitive cognition?[23]

---

[22] John Buridan, *Quaestiones in Aristotelis De anima, tertia lectura* [Zupko], III.8, ll. 167–170: "Visum fuit aliquibus quod sensus, ex eo quod habet extensionem et situm determinatum in organo corporeo, non habet naturam cognoscendi uniuersaliter sed singulariter." See also John Buridan, *Quaestiones super octo Physicorum libros Aristotelis* (Paris, 1509), I.7, fol. 8va.

[23] John Buridan, *Quaestiones super octo Physicorum libros Aristotelis* (Paris, 1509), I.7, fol. 8^{va-vb}: "Tertio quia appetitus sensitiuus ita est extensus et materialis sicut sensus, et tamen equus et canis per famem et sitim appetunt modo uniuersali, non enim hanc aquam uel auenam magis quam illam sed quamlibet indifferenter; ideo quodcumque eis portetur, bibunt ipsum uel comedunt. Et est intentio posita uel appetitus ignis ad calefaciendum est modo uniuersali, non determinate ad hoc lignum sed ad quodlibet calefactibile indifferenter, licet actus calefaciendi determinetur ad certum singulare. Et ita potentia uisiua est modo uniuersali ad uidendum." Cf. John Buridan, *Quaestiones in Aristotelis De anima, tertia lectura* [Zupko], III.8, ll. 223–232: "Et iterum apparet quia uirtus materialis et extensa fertur bene in obiectum suum modo uniuersali, nam appetitus equi secundum famem aut situm non est singulariter ad hanc auenam uel ad hanc aquam, sed ad quamlibet indifferenter; unde

As regards (b): Our perceptual abilities do not seem to put us in touch with individuals. After all, Buridan notes, we cannot tell the difference between qualitatively indistinguishable substances unless we perceive them relative to one another; nor can we tell whether a given object is the same or different from one we saw previously, even for items that are merely similar rather than indistinguishable. Such failures of discernibility suggest that sensitive cognition does not in fact succeed at reaching to the individual rather to some qualitatively more general level.[24] Now (a) and (b) show that (2) needs argument. Yet Buridan cannot appeal to either the materiality of the sensitive soul in guaranteeing the singularity of sensitive cognition, or to the intrinsic singularity of sense-cognition. Deprived of the standard resources for defending (2), he offers instead an alternative original account of what it is to perceive something as singular:

> Let me therefore state that something is perceived singularly in virtue of the fact that it is perceived as existing within the prospect of the person cognizing it ...[25]

---

quamcumque primitus inueniret illam caperet. Et intentio naturalis uel appetitus ignis ad calefaciendum non se habet modo singulari ad hoc calefactibile uel ad illud, sed ad quodlibet indifferenter quod ipse posset calefacere; ideo quodcumque sibi praesentetur, calefaceret ipsum; ergo etc."

[24] See for example John Buridan, *Tractatus de differentia universalis ad individuum* [Szyller], q. 1, c. 2, p. 153[14–29]; John Buridan, *Quaestiones super octo Physicorum libros Aristotelis* (Paris, 1509), I.7, fol. 8[vb]; John Buridan, *Quaestiones in Aristotelis De anima, tertia lectura* [Zupko], III.8, ll. 263–274. There is a particularly clear instance at John Buridan, *In Metaphysicen* (Paris, 1518), VII.17, fol. 52[va–vb]: "Si essent duo lapides omnino similes in figura, in magnitudine, in colore, et sic de aliis, et successiue apportarentur in tua praesentia, tu nullam uiam haberes ad iudicandum utrum secundus apportatus esset ille idem qui primus apportatus fuit an alter. Et ita etiam de hominibus si omnino essent similes in figura magnitudine et colore et sic de aliis accidentibus; immo etiam hoc non solum ueritatem habet de substantiis immo etiam de accidentibus: si enim essent albedines consimiles in gradu et essent in subiectis consimilibus in figura magnitudine et caetera, tu non haberes uiam cognoscendi utrum esset eadem albedo an alia quae tibi prius et posterius praesentaretur."

[25] John Buridan, *Quaestiones super octo Physicorum libros Aristotelis* (Paris, 1509), I.7, fols. 8[vb]–9[ra]: "Dicam ergo, sicut magis uideri debet septimo Metaphysicae, quod ex eo aliquid percipitur singulariter quod percipitur per modum existentis in prospectu cognoscentis. (Ideo enim Deus omnia percipit distinctissime ac si perciperet ea singulariter: omnia clara sunt quia in prospectu eius.)" The same account is given in John Buridan, *Quaestiones in Aristotelis De anima, tertia lectura* [Zupko], III.8, ll. 298–303: "Ad soluendum illas dubitationes, debemus ex septimo Metaphysicae uidere modum percipiendi rem singulariter: scilicet quia oportet eam percipere per modum existentis in prospectu cognoscentis. (Ideo enim deus quasi per modum singularem cognoscit omnia distinctissime et determinate, scilicet quia omnia habet perfecte in prospectu suo per se.)"

We should be careful to avoid two misconceptions. First, Buridan is not merely saying that an object has to be present in the perceiver's sensory field to cause a perception. True enough, but this is only a necessary condition for singular cognition. Second, Buridan is not begging the question by assuming that sensitive cognition, triggered by the presence of the object in the perceiver's sensory field, must be singular. His point is somewhat more delicate: the singularity of perception is a function of the object's presence in the perceiver's sensory field. That is, the singularity of sensitive cognition does not stem from its inherent nature or from some characteristic feature of the object, but from the circumstances in which it occurs. Very roughly, singularity is due to the here-and-now conjunction of perceptible general features that make up an object. Buridan explains this carefully for the internal as well as the external senses:

> Therefore, because external sense cognizes what is sensible in the way that something exists within its prospect in a definite location, even if sometimes it does make a false judgment about its location (due to the reflection of appearances), it cognizes it singularly and distinctly, namely as this or as that.
>
> Although external sense cognizes Socrates or whiteness or a white item, then, this nevertheless occurs only in an appearance representing [the object] as fused together with the substance, the whiteness, the size, and the location according to which it appears within the prospect of the cognizer. Now sense cannot itself untangle that type of fusion, that is, it cannot abstract the appearance of substance and of whiteness and of size and of location from one another; hence it can only perceive the whiteness or the substance or the white item the way that something exists within its prospect, and so it can only cognize the aforementioned [objects] singularly.
>
> Again, although the [internal] common sense receives appearances from the external sense with this type of fusion and cannot untangle that fusion, it of necessity apprehends in a singular manner. Accordingly, in dreams we judge that something appears to us to be this or that, or to be here or there. Likewise, when an appearance fused together with location comes about from [external] sense in the [internal] power of memory, a "memorative cognition" occurs in us in a singular manner (though we judge with pastness that [its object] was this or that, here or there).[26]

---

[26] John Buridan, *Quaestiones in Aristotelis De anima, tertia lectura* [Zupko], III.8, ll. 304–326: "Sensus ergo exterior quia cognoscit sensibile per modum existentis in prospectu suo secundum certum situm, licet aliquando false iudicat de situ propter reflexiones specierum, ideo cognoscit ipsum singulariter uel consignate, scilicet quod hoc uel illud. Quamuis ergo sensus exterior cognoscat Socratem uel albedinem uel album, tamen hoc non est nisi secundum speciem confuse repraesen-

Sensitive cognition is above all the representation of a manifold: a buzzing and blooming confusion wherein the various deliverances of the senses are literally fused together (*confusa*): size, shape, color, and the like are all part of the appearance (*species*), indexed to a definite time and place – even if we happen to be wrong about the place, as Buridan notes. It is the mark of the senses to present us with a jumble of impressions fused together in the here-and-now: singular sensitive cognition. And as for the external senses, for instance vision, so too for the internal senses, for instance common sense (which unifies the deliverances of the external senses) or memory: their singular action derives from the singularity of external sensitive cognition. Like the external senses, the internal senses cannot untangle the fused sensory impressions that confront it. (As we'll see, only the intellect is capable of performing the necessary "abstraction.") Hence sense is necessarily singular; it lacks the requisite mechanism to transform its input into something appropriately general. The psychological legacy of sensitive cognition is the inexpressibly rich singular concept, intrinsically complex and the building-block of mental life.[27]

---

tatem cum substantia et albedine et magnitudine et situ secundum quem apparet in prospectu cognoscentis. Et ille sensus non potest distinguere illam confusionem: scilicet non potest abstrahere species substantiae et albedinis et magnitudinis et situs ab inuicem, ideo non potest percipere albedinem uel substantiam uel album nisi per modum existentis in prospectu eius. Ideo non potest cognoscere praedicta nisi singulariter. Item etsi sensus communis a sensu exteriori recipiet species cum tali confusione, et non potest distinguere confusionem, ipse de necessitate apprehendit modo singulari. Unde in somniis iudicamus quod apparet nobis esse hoc uel illud, et esse hic uel ibi, ita etiam etsi in uirtute memoratiua, species fiat a sensu cum tali confusione situs, cognitio memoratiua fiet in nobis per modum singularem, licet cum praeteritione iudicemus quod erat hoc uel illud, hic uel ibi." See also the parallel account in John Buridan, *Quaestiones super octo Physicorum libros Aristotelis* (Paris, 1509), I.7, fol. 9ra: "Sensus autem exterior obiectum suum apprehendit confuse, cum magnitudine et situ ad ipsum tamquam apparens in prospectu eius, aut longe aut prope, aut ad dexteram aut ad sinistram; ideo percipit obiectum suum singulariter tanquam demonstratum hic uel ibi. Sensus autem interior non potest speciem obiecti ut colorum uel soni ab huiusmodi confusione absoluere et abstrahere; ideo in somno per phantasiam et sensum communem apparet totum ita esse in prospectu sensus secundum determinatum situm sicut in uigilia; ideo etiam sensus interior non percipit nisi singulariter. Immo etiam in memorando, memoramur rem cum situ tamquam fuerit in prospecto nostro praesentata sensui secundum determinatum situm."

[27] Buridan explicitly says that such singular concepts deriving from sense are complex, John Buridan, *Expositio et quaestiones in Aristotelis De anima* [Patar], I.4, p. 195$^{64-66}$: "Dico quod talis conceptus quodammodo est complexus, quia est cum tali circumstantia quod non solum per ipsum concipitur res, sed etiam per ipsum concipitur rem esse talis figurae uel talis coloris." He even goes so far as to claim that individuals have an infinite number of properties and that we can therefore never grasp an individual perfectly: John Buridan, *op.cit.*, I.5, pp. 204^{90}–205^{19}.

Two features of Buridan's account of perception are worth mentioning briefly. First, we only possess singular concepts of those individuals we have directly encountered; we know all others only by description rather than by acquaintance.[28] Second, Buridan is a "descriptionalist" regarding mental acts. All cases of perception are intensional, since there is always an associated concept under which we perceive items; this may be more or less precise (*singulare uagum*), or it may be fully determinate; the relations among such concepts, especially in the account of the origins of cognition in sense, are highly complex.[29]

Buridan's response to (a) and (b) is straightforward. On the one hand, he agrees with (a) that the materiality of sense is not the ground of the singularity of sensitive cognition, and so bypasses its challenge.[30] On the other hand, he agrees with (b) that sensitive cognition involves the grasp of general qualitative features. But there is a

---

[28] This claim causes trouble for Buridan's semantics, since what appear to be logically singular terms cannot in fact correspond to singular concepts: "Aristotle" (if you have never met Aristotle), definite descriptions, and the like. The names of individuals with which one has never come into direct contact, Buridan holds, are not strictly discrete terms but rather disguised descriptions: "... to others who have not seen [Plato or Aristotle], those names are not singular, nor do they have singular concepts corresponding to them simply" (John Buridan, *Quaestiones in Aristotelis De anima, tertia lectura* [Zupko], III.8); we who have never come into direct contact with Aristotle "... do not conceive him as different from other men except by a given circumlocution, such as 'a great philosopher' and 'teacher of Alexander' and 'student of Plato', 'who wrote books of philosophy which we read', etc." (John Buridan, *Quaestiones super octo Physicorum libros Aristotelis* (Paris, 1509), I.7), which would equally signify and supposit for another individual if there were one having engaged in these activities. Put another way, the fact that "Aristotle" supposits only for Aristotle is not a matter of semantics but depends on the contingent historical fact that no other individual happens to fit the description, and so cannot be a discrete term. The same point may be made about descriptions generally, including definite descriptions: "... the expression 'the son of Sophroniscus' is not, strictly speaking, singular, since 'the son of Sophroniscus' is immediately apt to fit more than one if Sophroniscus produces another son" (John Buridan, *Quaestiones in Aristotelis De anima, tertia lectura* [Zupko], III.8). See further Perreiah, "Buridan" and King, "Jean Buridan" on the semantics of singular terms and descriptions.

[29] This is the topic of John Buridan, *Quaestiones super octo Physicorum libros Aristotelis* (Paris, 1509), I.7, John Buridan, *Quaestiones in Aristotelis De anima, tertia lectura* [Zupko], III.8 and John Buridan, *In Metaphysicen* (Paris, 1518), VII.20. See further Miller, "Buridan on Singular Concepts" and Van der Lecq, "Confused Individuals."

[30] There are still questions about how to analyze the particular cases mentioned under (a). Briefly, Buridan holds that desires are just as particular as perceptions, with the twist that the intentional nature of desire introduces a kind of opacity (intentionality produces intensionality): the horse wants some-water-or-other, which cannot be identified with any particular water, but is such that any particular water satisfies it. Natural causal agents can be analyzed in a similar fashion.

difference between perceiving individuals (the object of perception is individual) and being able to identify the individuals so perceived; failures of discernibility turn on the latter, not the former. Sensitive cognition can be thoroughly singular without guaranteeing that we can re-identify individuals previously sensed. Hence (2), the claim that sensitive cognition is always singular is secure.

5. Buridan's Third Psychological Thesis

Finally, Buridan holds that intellective cognition can be universal and it can be singular. Yet as noted above, if intellective cognition depends on sensitive cognition, and the latter is always singular, how is universal intellective cognition possible?

Begin with singular intellective cognition, which Buridan notes some thinkers call "intuitive."[31] He carefully discusses and argues against the view that intellective cognition must be universal precisely because the intellect is a separable and immaterial entity, and also against Aquinas's view that universal intellective cognition is primary and that singulars are only known indirectly through reflection on the phantasm.[32] His positive case is simple:

> The principal question can be settled by saying that the intellect cognizes things as singular before it does as universal, because sense, whether internal or external, only cognizes them as singular, namely as fused together with location and as existing within the prospect of the knower; therefore, etc. Sense thus represents a sensible object to the intellect with this sort of fusion. And just as sense primarily represents the object to the intellect, so too does the intellect primarily understand the thing. Therefore, the intellect is able to cognize the thing with this kind of fusion, and so as singular. (This is also apparent from what has been said, namely that by abstracting and so on the intellect understands as universal.) Furthermore, since the representation on the part of sense is in a singular manner, if the intellect were not to understand as singular on the basis of a representation of this sort, then we can't explain how it can understand as singular afterwards.[33]

---

[31] John Buridan, *In Metaphysicen* (Paris, 1518), VII.20, fol. 54va: "Et sic finaliter uidetur mihi esse dicendum quod nullus est conceptus singularis nisi sit conceptus rei per modum existentis in praesentia et in prospectu cognoscentis tamquam illa res appareat cognoscenti sicut demonstratione signata, et illum modum cognoscendi uocant aliqui intuitiuum."

[32] John Buridan, *Quaestiones super octo Physicorum libros Aristotelis* (Paris, 1509), I.7 and John Buridan, *Quaestiones in Aristotelis De anima, tertia lectura* [Zupko], III.8.

[33] John Buridan, *Quaestiones super octo Physicorum libros Aristotelis* (Paris, 1509),

The intellect has to begin with singular cognition, since that is the nature of the material passed along to it from sensitive cognition: "… we understand singularly before we do universally, since a representation fused together with size and location and other features occurs in the intellect before the intellect can untangle and abstract from that fused [representation]."[34] Singular intellective cognition is thus prior to all other forms of intellective cognition.

The process whereby singular intellective cognition is transformed into universal intellective cognition Buridan calls "abstraction" (perhaps involving other psychological mechanisms we need not explore here).[35] He describes the process as follows:

> I declare that when the intellect receives from the phantasm the appearance or understanding of Socrates as fused together with size and location, making the thing appear in the way something exists within the prospect of the cognizer, the intellect understands him in a singular manner. If the intellect can untangle that fusion and abstract the concept of substance or of whiteness from the concept of location, so

---

I.7, fol. 9[ra–rb]: "Et ex his apparet mihi quod determinari potest quaestio principalis dicendo quod prius intellectus cognoscit res singulariter quam uniuersaliter propter hoc quod sensus non cognoscit eas nisi singulariter, siue sit sensus exterior uel interior, scilicet cum illa confusione situs et per modum existentis in prospectu cognoscentis; ideo etc. Sic sensus cum huiusmodi confusione repraesentat intellectui obiectum sensibile. Et sicut obiectum primo repraesentat intellectui, sic intellectus primo intelligit rem. Ergo cum huiusmodi confusione intellectus potest cognoscere rem, et sic singulariter. Et hoc etiam apparet ex dictis, scilicet quod abstrahendo etc., intellectus intelligit uniuersaliter. Et iterum, cum repraesentatio ex parte sensus sit modo singulari, si intellectus ex huiusmodi repraesentatione non intelligat singulariter, non poterit postea dici quomodo possit intelligere singulariter."

[34] John Buridan, *Quaestiones in Aristotelis De anima, tertia lectura* [Zupko], III.8, ll. 411–415: "Dicendum est enim quod prius intelligimus singulariter quam uniuersaliter, quia prius fit in intellectu representatio confusa cum magnitudine et situ et aliis, quam intellectus posset distinguere et abstrahere illam confusionem." See likewise John Buridan, *Expositio et quaestiones in Aristotelis De anima* [Patar], I.4, p. 196[01–06]: "Dico quod conceptus talis causatur ex conceptibus primo modo dictis: unde prius concipitur homo cum talibus circumstantiis quam sine talibus circumstantiis. Et secundum hoc, si conceptus primo modo dictus dicatur singularis, et conceptus secundo modo dictus dicatur universalis, tunc necesse est, antequam intellectus habeat conceptum universalem, quod prius habuit conceptum singularem correspondentem illi conceptui universali."

[35] For example, Buridan declares in John Buridan, *Expositio et quaestiones in Aristotelis De anima* [Patar], I.4 that universal intellective cognition depends "causatively" on the phantasm (196[85–89]), and in John Buridan, *Quaestiones in Aristotelis De anima, tertia lectura* [Zupko], III.15 he argues, against Ockham among others, that the *species intelligibilis* is necessary for intellective cognition in general. An adequate account of Buridan's philosophy of mind should spell out exactly how these elements enter into intellective cognition.

that the thing is no longer perceived in the way something exists within the prospect of the cognizer, then it will be a common concept. Accordingly, once the concept of Socrates has been drawn out abstractly from the concepts of whiteness and of location and of other accidents or extraneous features, it will then no more represent Socrates than Plato: it will be a common concept, one from which the name "man" is derived.[36]

Abstraction is the process of isolating a feature from the others with which it is fused, in particular from its indexical features, such as location. Since these features are in themselves general, the feature that is isolated from the others and freed from its individualizing conditions will therefore be general. The intellect learns how to untangle the various features that are present in the singular intellective concept by recognizing that the accompanying features may vary: a stone may appear first here and then there; it may be at one time white and another black; and so on, until eventually the intellect is able to prescind from these accidental features, thereby producing a universal intellective cognition.[37]

The psychological process of abstraction sketched in these remarks, whatever we may think of its merits, is designed to explain how universal intellective cognition can occur; it supports (3) by showing how such a cognition can be generated within the mind. Yet whether it succeeds is a delicate question. The process as described surely produces one cognition from another: from the rich singular intellective cognition a single feature is drawn out and treated in

---

[36] John Buridan, *Quaestiones in Aristotelis De anima, tertia lectura* [Zupko], III.8, ll. 391–403: "Dico quod cum intellectus a phantasmate recipit speciem uel intellectionem Socratis cum tali confusione magnitudinis et situs, facientem apparere rem per modum existentis in prospectu cognoscentis, intellectus intelligit illum modo singulari. Si intellectus potest illam confusionem distinguere et abstrahere conceptum substantiae uel albedinis a conceptu situs, ut non amplius res percipiatur per modum existentis in prospectu cognoscentis, tunc erit conceptus communis. Unde cum elicitus fuerit conceptus Socratis abstracte a conceptibus albedinis et situs et aliorum accidentium uel extraneorum, ille iam non magis repraesentabit Socratem quam Platonem, et erit conceptus communis a quo sumitur hoc nomen 'homo'."

[37] See John Buridan, *Quaestiones super octo Physicorum libros Aristotelis* (Paris, 1509), I.7, fol. 9ra: "Sed iterum considerandum est quod intellectus – qui supra sensum est uirtus multo potentior et nobilior – potest distinguere huius confusionem, cum enim perceperimus quod iste lapis modo est hic, modo illic; modo albus, modo niger; sciemus quod hic lapis non determinat sibi quod sit hic uel illic, albus aut niger. Ideo intellectus poterit abstrahere speciem uel notitiam lapidis a specie uel notitia huius situs uel alterius: et sic intelligitur lapis: uel quantum ad hoc intelligendo de esse hic uel illic, et tunc indifferenter omnis lapis intelligitur conceptu communi non magis hic quam ille."

isolation, freed from its combination and fusion with other features. Whether the cognition thereby produced is non-singular, though, is another matter. The account of abstraction given here depends on the claim that a given feature is intrinsically general, or at least when set free from its individualizing conditions it is general. Now there is nothing question-begging in Buridan's claim that mental items are general, either intrinsically or under certain conditions. But there isn't much explanatory in it either. What does the generality of a mental item, already conceded to be an individual quality inhering in an individual intellective soul, amount to?

Buridan's answer is that mental items are general in virtue of being representational. The intellective cognition produced by abstraction is thus universal by representing many items, or, more accurately, by representing many distinct individuals indifferently:

> If an appearance of man in the imagination is stripped or divested of all extraneous features (or of all appearances of extraneous features), it will not determinately represent Socrates or Plato but instead indifferently represent either of them or other men. Thus the intellect doesn't understand this man determinately through the appearance but indifferently understands this man or that one or another: this is to understand man by a universal understanding.[38]

Representation can take at least two forms, namely determinate representation and indifferent representation; on the semantic side this corresponds to the distinction between proper names and other kinds of names (which may apply to more than one individual). Yet without an account of how representation takes place, this is no more than suggestive; what is it for a representation to be determinate or indifferent? (For that matter, what is it for a name to be proper or not?) Buridan adopts a traditional view of representation as a form of resemblance. Concepts represent things by resembling them:

> Hence it follows from the fact that representation occurs through likeness that what was representative of one item will be indifferently representative of them all (unless something happens alongside to prevent it, as will be discussed later). We ultimately conclude from this that whenever the appearance – the likeness – of Socrates was in the intellect and

---

[38] John Buridan, *Tractatus de differentia universalis ad individuum* [Szyller], q. 1, c.2, p. 155[29–35]: "Si species hominis fuerit in phantasia et denudetur seu praescindatur ab omnibus extraneis seu a specibus extraneorum, [quod] ipsa non repraesentabit determinate Socratem uel Platonem, sed indifferenter quemlibet ipsorum aut aliorum hominum; et ita intellectus non intelligeret per illam speciem hunc hominem determinate, sed indifferenter hunc uel illum uel alium. Et hoc est intelligere hominem uniuersali intellectione."

abstracted from the appearances of extraneous features, it will no more be a representation of Socrates than it is of Plato or of other men; nor does the intellect understand Socrates through it any more than it does other men. Instead, it thus understands all men indifferently through it by means of a single concept, namely the one from which we derive the name "man." And this is to understand universally.[39]

A concept produced by abstraction is equally a likeness of many items, and so indifferently represents them all. Of course, we have to grant that a mental item (a particular quality inhering in the intellect) can in some full-blooded way be said to resemble an external item, but that is as much a problem for singular as universal intellective cognition. If we swallow that camel, then what of the gnat: mental items simultaneously resembling many really distinct external objects? Why not?

One reason for hesitation is that the notion of "resemblance" has some theoretical baggage built into it that may not be warranted. Saying that one thing resembles or is a likeness of another is a success-verb or an achievement-verb: it cannot try to resemble but fail to do so. (It makes little sense to say that X only seems to resemble Y but in fact really doesn't.) How can we say whether a given mental item resembles Socrates and Plato but does not resemble a horse? Worse yet, resemblance seems to be a matter of degree: Socrates and Plato resemble one another more than Socrates resembles a horse. However, the boundaries of resemblance in any given case seem extremely context-dependent. Yet even if we put these worries aside, there is a deeper issue at stake, one having to do with the legitimacy of appealing to resemblance or likeness. Even if we grant that mental items can resemble non-mental items, how can non-individual items (likenesses) resemble individual items? In short, won't any universal intellective cognition misrepresent the way the world is, precisely in virtue of its generality?[40]

---

[39] John Buridan, *Quaestiones in Aristotelis De anima, tertia lectura* [Zupko], III.8, ll. 279–290: "Ideo consequitur ex quo repraesentatio fit per similitudinem quod illud quod erat repraesentatiuum unius erit indifferenter repraesentatiuum aliorum, nisi aliud concurrat quod obstet, sicut dicetur post. Ex hoc finaliter infertur quod cum species (et similitudo) Socratis fuerit apud intellectum et fuerit abstracta a speciebus extraneorum, illa non magis erit repraesentatio Socratis quam Platonis et aliorum hominum; nec intellectus per eam magis intelliget Socratem quam alios homines. Immo sic per eam omnes homines indifferenter intelliget uno conceptu, scilicet a quo sumitur hoc nomen 'homo'. Et hoc est intelligere uniuersaliter."

[40] There is no parallel issue about the legitimacy of sensitive cognition, since it is always linked to a particular external object (the one causally responsible for the sensitive cognition) the sensed features are taken to characterize.

Buridan's response is to explain how resemblance works to secure representative generality. It turns out that the legitimacy of universal intellective cognition rides on the real agreement of things:

> Now if it were the case that there are many items similar to one another, then anything similar to one of them, with respect to the feature in which they are similar, is similar to any one of them. Hence if all asses have in reality an agreement and likeness with one another, when the intelligible appearance represents some ass in the intellect by means of a likeness, it must simultaneously represent any given ass indifferently (unless something prevents it, as will be discussed later). An intention becomes universal in this way.[41]

Thus mental representation takes place through the presence of an item in the intellect that is a likeness of any member of a class of objectively similar items. Since thinking of X is just to have the concept of X in the mind, an intellective cognition that is a likeness of any one of $X_1,\ldots,X_n$ will thereby be a case of thinking of all of them. In short, the legitimacy of a universal concept is a matter of the real relations of agreement or likeness among things it is about: it will resemble any of them in virtue of resembling one of them, in accordance with the axiom Buridan enunciates at the beginning, since the objective agreement among things secures its resemblance to the rest.

With this last move Buridan has, I think, made a plausible case that the mind is capable of producing within itself items that are representatively general. He has sketched a psychological mechanism that produces such an item and explicated its generality through its resemblance to at least one singular (presumably the one from which it was derived) and the objective relations of agreement that item has to others. Whether it is an adequate account will depend on exactly how its details are spelled out, to be sure, but that should not detract from Buridan's success in offering his account.

---

[41] John Buridan, *Quaestiones super octo Physicorum libros Aristotelis* (Paris, 1509), I.7, fol. 8vb: "Modo si sit ita quod sint multa inuicem similia, omne illud quod est simile uni eorum, quantum ad hoc in quo sunt similia, est simile unicuique aliorum. Ideo si omnes asini ex natura rei habent adinuicem conuenientiam et similitudinem, oportet quod quando species intelligibilis in intellectu repraesentabit per modum similitudinis aliquem asinum, ipsa simul indifferenter repraesentabit quemlibet asinum, nisi aliud obstet, de quo postea dicetur. Ideo sic fit uniuersalis intentio."

6. Real Agreement

From psychology, we have to return to metaphysics. For universal intellective cognition is only legitimate to the extent it has some real basis in the world, as Buridan admits, since otherwise the universal concept would be fictitious.[42] But what is there in reality to ground the universal concept if there are no non-individual entities in the world? What, in the end, is the "agreement" among distinct individuals to which Buridan appeals?

It is clear that such agreement is an extra-mental feature of the world, namely a relation stemming from a thing's essence:

> External things have agreement and likeness among themselves in virtue of their nature and essence.[43]

According to Aristotle, there are three fundamental modes of unity: among substances, which is called "sameness;" among qualities "likeness;" among quantities "equality" (*Metaphysics*, V.15 1021a10–14). We might reasonably expect Buridan to have this doctrine in mind and be focusing on certain kinds of relations among objects: "agreement" as substantial sameness among individual substances, "likeness" as the sameness of quality among individual substances.[44] Naturally, such relations have their opposites: diversity and dissimi-

---

[42] See John Buridan, *Tractatus de differentia universalis ad individuum* [Szyller], q. 1, c.2, p. 152$^{22-25}$: "Deinde probo tertiam conclusionem quod uniuersale pro subiecto est praeter animam quantum ad aliquid sui, quia uniuersale pro subiecto est illud super quod fundatur intentio uniuersalitatis, sed ipsa uniuersalitatis fundatur super rem extra; aliter uidetur quod esset ficta." A similar claim is made in John Buridan, *In Metaphysicen* (Paris, 1518), VII.16, fol. 51rb: "Item oportet concedere quod conceptus uniuersales et singulares distinguuntur apud intellectum, et si non esset distinctio ex parte rei correspondens, illa distinctio apud intellectum esset falsa uel ficta, quod est inconueniens; ergo in re extra distinguuntur uniuersalia a singularibus."

[43] John Buridan, *Quaestiones super octo Physicorum libros Aristotelis* (Paris, 1509), I.7, fol. 8vb: "Res autem extra ex natura et ex essentia sua habent inter se conuenientiam et similitudinem." (This remark immediately precedes the passage cited in n. 41 above.)

[44] See John Buridan, *De dependentiis, diuersitatibus, et conuenientiis* [Thijssen], p. 245: "Sed conuenientia debet reduci ad illas relationes quas Aristoteles dicebat sumi secundum unum, cuiusmodi sunt: idem, aequale, et simile. Et e conuerso diuersitas debet reduci ad relationes sumptas secundum multa. Et hoc est ualde clarum." Note that there is a subtle point regarding likeness likely to be missed by modern readers. Strictly speaking, two qualities are alike, and we can speak of their likeness, but the same terminology was often used loosely (as here) to describe the relation between two substances each having the "same" quality: the likeness of Socrates and Plato in virtue of Socrates's whiteness and Plato's whiteness. This will be important when Buridan turns to the ontology underlying relational statements. There is no

larity, respectively. Buridan mentions the former while setting forth another aspect of agreement:

> I declare that two individuals of the same species have essential distinctness and they have essential agreement ... I say that Socrates and Plato wholly agree with one another of themselves by an essential agreement, and that they are wholly diverse from one another of themselves by an essential diversity.[45]

Which distinct individuals are related by agreement? The suggestion here is that all individuals of a given species are. (They aren't identical, though, so they are all diverse from one another as well.) Furthermore, since the agreement and diversity stem from the essences of the individuals, they must agree or be diverse necessarily – that is, in virtue of what they are. How do we know that there is such real agreement among things? Buridan has a surprising answer: the fact that we can distinguish individuals of the same species only through accidental differences points to their underlying substantial agreement:

> That there is [agreement among things] is clear, because, due to their agreement, you can have no way to perceive distinctness among things of the same species except through the diversity of extraneous factors, or because they are seen together externally and situated with regard to each other. For example, suppose there were two rocks completely similar in size and shape and other accidents and one were presented to you today and the other tomorrow; you couldn't know whether the later one is the same one presented to you before or the other one. And likewise for accidents too: if those stones were white in an equal degree of intensity and equally large and both spherical, and they were also similar with regard to all other accidents, you couldn't know whether it is the same whiteness or blackness, or shape, which is picked out as that which was previously shown to you.[46]

---

comparable usage for agreement, of course, though there is for equality (which Buridan does not mention in his discussions).

[45] John Buridan, *Tractatus de differentia universalis ad individuum* [Szyller], q. 1, c.2, p. 162[18–20] and 169[15–17]: "Dico quod duo indiuidua eiusdem speciei habent distinctionem essentialem et habent conuenientiam essentialem ... Dico quod Socrates et Plato seipsis totis conueniunt essentiali conuenientia et seipsis totis diuersi sunt essentialiter diuersitate."

[46] John Buridan, *Quaestiones super octo Physicorum libros Aristotelis* (Paris, 1509), I.7, fol. 8[vb]: "Quod ergo ita sit, patet, quia propter earum conuenientiam tu nullam uiam potes habere ad percipiendum distinctionem rerum eiusdem speciei, nisi propter diuersitatem extraneorum, aut quia simul uidentur extra inuicem situaliter. Verbi gratia, sint duo lapides omnino consimiles in magnitudine et figura et aliis accidentibus, et unus hodie tibi praesentetur et cras alius; tu non poteris scire de illo posteriori utrum sit ille idem quae tibi prius praesentabatur uel alius. Et similiter

The difficulties with discernibility noted earlier turn out to provide a positive basis for imputing real agreement to things. Our inability to correctly re-identify items is a logical ground for thinking them substantially in agreement.

Now in all these passages Buridan seems to be speaking of an arbitrary species, a point borne out by his remark that essential agreement and diversity are coordinated with genera and species.[47] Furthermore, agreement and diversity seem to admit of more and less, in crossing generic or specific boundaries:

> We hold that things of the same species or genus existing outside the soul singularly have of their nature a greater essential likeness or agreement than do those of diverse species or genera. For Socrates and Plato agree in reality more than Socrates and Brunellus do (even as regards their essences), and Socrates and Brunellus also agree more than do Socrates and this stone. Greater essential agreement of this sort comes from the fact that things belonging to the same species or genus come from the same or very similar causes more than do others, because in the order of beings they have the same rank, or a closer rank to one another than do others.[48]

From a contemporary point of view, then, essential agreement is an equivalence relation that partitions the class of individuals into their respective natural kinds, where members of different kinds are more or less in agreement depending on how "close" their respective natural kinds are. (Buridan's talk of things having distinct "ranks" is an oblique nod to the Great Chain of Being.) In short, things in

---

etiam est de accidentibus, quia si illi lapides sint albi secundum aequalem gradum intentionis et aequaliter magni et ambo sphaerici, et sint etiam similes quantum ad omnia alia accidentia, tu non poteris scire an sit eadem uel alia albedo aut nigredo, aut figura quae posterius ostenditur cum illa quae prius tibi ostensa fuit." The same point is made in John Buridan, *Quaestiones in Aristotelis De anima, tertia lectura* [Zupko], III.8 ll. 261–274: "Illa quae sunt eiusdem speciei specialissimae tantam habent essentialem conuenientiam quod tu non habes uiam ad percipiendum eorum distinctionem nisi per extraneam. Verbi gratia, sint duo lapides similes in magnitudine et figura et colore et aliis singularis accidentibus, et nunc uideas unum et quantum potes considerare ipsum. Demum, te recedente, auferatur ille et ponatur alius loco eius. Tunc tu rediens, iudicabis quod ille qui nunc est ibi sit idem quae ante uidebas. Et similiter, color quae in eo iudicabis sit idem ille color quae ante uidebas, et sic de magnitudine et figura. Nec tu habebis aliquam uiam ad sciendum an ille est idem lapis uel alter (et sic etiam de hominibus). Sed si uideas eos simul, tu iudicabis quod sunt alii per alietatem locorum uel situs."

[47] John Buridan, *De dependentiis, diuersitatibus, et conuenientiis* [Thijssen], p. 238: "... conuenientiae uel diuersitates essentiales seu quidditatiuae, cuiusmodi sunt conuenientiae uel diuersitates aliquorum secundum speciem aut secundum genus ..."

[48] John Buridan, *Quaestiones in Aristotelis De anima, tertia lectura* [Zupko], III.8,

the world seem to be sorted into natural kinds: they agree or differ depending on their natures.

This, of course, serves to point up the underlying metaphysical problem. How can things be objectively divided into natural kinds unless there is some extra-mental real commonness? Agreement seems to be not only a relation but a so-called "real" relation, that is, a relation that obtains independent of any mental states: Socrates and Plato agree with one another regardless of anyone's thinking so. Therefore, it should be something real too, and Buridan has to show that this approach does not introduce any real commonness or universality into the world.[49] What is more, the fact that distinct individuals may agree more or less with one another – so that an ass and a horse are more in agreement than an ass and a stone – seems to be an irreducible feature of the world, not easily explicable by appeal to anything less than real common features.[50] Yet if Buridan gives in to

---

ll. 246–261: "Tunc accipimus quod res extra animam singulariter existentes de eadem specie uel de eodem genere habent ex natura sui similitudinem seu conuenientiam essentialem maiorem quam illae quae sunt diuersarum specierum uel diuersorum generum. Plus enim conueniunt ex natura rei Socrates et Plato quam Socrates et Brunellus (etiam quantum ad suas essentias), et plus etiam conueniunt Socrates et Brunellus quam Socrates et ille lapis. Et huiusmodi maior essentialis conuenientia prouenit ex eo quod illa quae eiusdem speciei uel generis proueniunt ex eisdem causis, uel similibus magis, quam alia, propter quod in ordine entium sunt eiusdem gradus, uel propinquorum graduum ad inuicem, quam alia."

[49] Buridan takes this argument seriously: it is the first argument given in favor of the claim that there are relations outside the soul distinct from their foundations; John Buridan, *In Metaphysicen* (Paris, 1518), V.9, fol. 32rb (incorrectly paginated in the incunabulum as fol. 33): "Quaeritur nono utrum sit aliqua relatio praeter animam distincta a fundamento suo. Arguitur quod sic: Quia similitudo Socratis ad Platonem (si uterque est albus) est praeter animam, quia, quamuis nullus intelliget, adhuc Socrates esset similis Platoni et non sine similitudine; ideo similitudo esset licet nullus intelliget, et tamen illa similitudo est relatio et est res distincta a Socrate qua Socrates est similis, quia dictum est quod similitudo est una qualitas, scilicet illa albedo quae est distincta a Socrate; igitur."

[50] Buridan seems to have found this line of thought particularly compelling. It is the first argument he gives to show that agreements and diversities have an ontological standing independent of the things they apply to, in John Buridan, *op.cit.*, V.6, fols. 29vb–30ra: "Supposito enim quod equus et asinus habent adinuicem aliquam conuenientiam ex natura rei, propter hoc quod ad naturas eorum consequuntur accidentia magis similia quam consequantur ad naturam lapidis et asini. Oportet igitur concedere quod ex natura rei equus et asinus magis conueniunt quam asinus et lapis. Et cum hoc etiam certum est – quia ipsi ex natura rei sunt adinuicem diuersi magis quam essent duo asini adinuicem – omne modo igitur si conuenientia eorum adinuicem non sit res uel dispositio alia ab eis, tunc sequeretur quod idem erit conuenientia eorum adinuicem et diuersitas eorum abinuicem. Sed illud consequens est falsum, quia ex natura rei asinus et equus ratione suae conuenientiae ducunt ad

either of these lines of argument, his attempt to find a non-realist solution to the problem of universals will be a failure.

Buridan recognizes the challenge, and rises to meet in two works on the theory of relations.[51] His early polemical treatise *De dependentiis, diuersitatibus, et conuenientiis* [Thijssen] (dated in its colophon to 1332) examines whether agreements, diversities, and causalities have any independent ontological standing; the second of three theses he defends in it is that "... essential or quidditative agreements or diversities ... add nothing in the things that so agree or are diverse, apart from their essences."[52] He also devotes *In Metaphysicen* (Paris, 1518) V.6 to the question "... whether the agreements and diversities of things with one another are things or dispositions added to the things that agree or are diverse."[53] He declares that they are not:

> My first conclusion is that for any things said to agree or to be diverse of themselves, the agreements or disagreements in them are not things or dispositions added to them. This result is clearly established by the negative principal arguments.[54]

---

unum conceptum communem, scilicet animalis, et tamen ratione diuersitatis eorum non ducunt ad illum conceptum communem, immo ad diuersos conceptus specificos. Igitur huiusmodi conuenientia et huiusmodi diuersitas non sunt idem." In his resolution of the principal arguments, he defers his solution of this one to John Buridan, *op.cit.*, V.7, where, as far as I can tell, he never returns to it (fol. 30rb): "Illa prima ratio quae arguebat de illis conceptibus tangit magnas difficultates tractandas in septimo libro: ideo dimittuntur usque tunc."

[51] See the treatment of Buridan's theory of relations in Schönberger, *Relation als Vergleich*, which also deals with Buridan's polemical *Tractatus de relationibus* (see Michael, "Johannes Buridan," II, pp. 440–443).

[52] This is the complete version of the passage cited in n. 47 above: "Secunda: Quod conuenientiae uel diuersitates essentiales seu quidditatiuae, cuiusmodi sunt conuenientiae uel diuersitates aliquorum secundum speciem aut secundum genus, nihil addunt in rebus sic conuenientibus uel diuersis praeter suas essentias." The polemical John Buridan, *Tractatus de differentia universalis ad individuum* [Szyller], presents the same view (175$^{21-23}$: "Quarto dico quod huiusmodi conuenientiae uel diuersitates Socratis ad Platonem, et econuerso, nihil addunt in Socrate et Platone").

[53] John Buridan, *In Metaphysicen* (Paris, 1518), V.6, fol. 29vb: "Quaeritur sexto utrum conuenientiae et diuersitates rerum adinuicem sint res an dispositiones additae rebus conuenientibus uel diuersis." The date of John Buridan, *In Metaphysicen* (Paris, 1518) is disputed – estimates range from 1336 to 1350 – but it is generally conceded to be a work of Buridan's philosophical maturity, and so I shall concentrate on it. John Buridan, *op.cit.*, V.8 takes up the corresponding question about causality.

[54] John Buridan, *op.cit.*, fol. 30rb: "Prima conclusio est quod quaecumque dicuntur per seipsa conuenire uel esse diuersa in illis conuenientia uel disconuenientia non sunt res uel dispositiones ipsis additae. Et sic illa conclusio manifeste probatur per rationes quae iam factae sunt ad secundam partem quaestionis."

Agreements and diversities, because they stem from the essences of things, are not ordinary relations; Buridan discusses whether ordinary relations are real accidents, that is, whether they have any independent ontological standing, in John Buridan, *In Metaphysicen* (Paris, 1518), V.9. His arguments turn on special properties of such essential relations. First (putting aside an initial *ad hominem* theological argument), Buridan maintains that identity and diversity should be treated symmetrically, and that it is implausible to treat a thing's identity, which is just a case of self-agreement, as consisting in an added thing or disposition.[55] Next, if they were things, as soon as anything comes into existence the whole universe is remade anew, since in each existent being agreements and diversities are created in a kind of ontological "ripple effect." But this is implausible in its own right and seems to involve action at a distance.[56]

Buridan's final argument that agreements and diversities have no independent ontological standing is a cleverly concise version of Bradley's Regress:

> If Socrates is diverse from Plato by a diversity added to him, then that diversity is diverse from Socrates, and Socrates diverse from it. Then either (a) Socrates and that diversity are diverse of themselves from one another, or (b) they are diverse through another diversity. If (a), then by the same reasoning we should stop at the first stage. If (b), we proceed with regard to it as before, and so to infinity – which is unacceptable.[57]

---

[55] John Buridan, *op.cit.*, fol. 30ra: "Item: Si identitas non est res addita, ita nec diuersitas. Et tamen identitas non est aliquid additum rei quae est eadem, quoniam quacumque re accepta omnibus aliis circumscriptis, adhuc ipsa esset sibi eadem. Et etiam ex diuersitate: quaecumque enim duae res quarum haec non est illa, si concedantur esse et omnia alia circumscribantur, adhuc illae erunt diuersae abinuicem, quoniam ad aliqua esse diuersa sufficit hoc esse et illud esse et hoc non esse illud: igitur diuersitas non est res uel dispositio alia a rebus diuersis." A much briefer version of this argument is given in John Buridan, *Quaestiones in Praedicamenta* [Schneider], q. 10, p. 73$^{104-108}$.

[56] John Buridan, *op.cit.*, fol. 30ra: "Item: Ego pono quod Socrates est generetur de nouo. Constat quod quaecumque res alia de mundo efficitur de nouo diuersa a Socrate, quia antequam Socrates esset, nulla erat res diuersa ab eo; et non quaelibet alia est diuersa ab eo; igitur si ad esse diuersum ab aliquo requireretur dispositio addita, sequitur quod apud generationem Socratis generaretur in qualibet alia re quaedam dispositio sibi addita – quod est absurdum dicere, quia tunc oporteret Deum et Intelligentias mutari in recipiendo tales dispositiones."

[57] John Buridan, *op.cit.*, fol. 30^{ra-rb}: "Item: Si Socrates est diuersus a Platone per diuersitatem sibi additam, tunc illa diuersitas est diuersus a Socrate, et Socrates diuersus ab alia, et tunc: uel Socrates et illa diuersitas sunt abinuicem diuersi seipsis, uel per aliam diuersitatem: si seipsis, pari ratione standum erat in primus; et si hoc

(Buridan leaves the corresponding argument for the case of agreements as an exercise for the reader.) The conclusion Buridan draws from these three arguments, then, is that agreements and diversities aren't things or dispositions above and beyond the things that agree or are diverse. Hence they must be identified with those very things themselves. Buridan concludes that "… the diversity of Socrates from Plato is just Socrates, and conversely the diversity of Plato from Socrates is just Plato."[58] He surprisingly does not say, but presumably the agreement of Socrates with Plato is just Socrates and Plato, which is likewise the agreement of Plato with Socrates: then the agreement and the diversity of Socrates and Plato do differ but neither is in any sense "added" to already existent individuals. Hence agreements and diversities do not add anything to Buridan's ontology. Since Buridan hasn't countenanced any new entities, he *a fortiori* hasn't countenanced any new non-individual entities. His solution to the problem of universals, then, doesn't appeal to anything really common in the world. In the end, the real basis for universal concepts are the agreements and diversities that hold among individual items in the world, yet these are no more than those items themselves. He has finally made good on his promissory note that "everything can be preserved" without appealing to "universals distinct from singulars."

7. *Conclusion*

Or has he? I want to conclude by drawing some wider implications about Buridan's proposed solution to the problem of universals.

An obvious problem faces Buridan's account – one that he perhaps recognized and could not resolve.[59] It is this. Buridan has argued that diversities come in different grades: Socrates is less diverse from a horse than from a stone (say). But the diversity of Socrates from

---

sit per aliam diuersitatem, procederetur de illa ut prius, et sic in infinitum, quod est inconueniens."

[58] John Buridan, *op.cit.*, fol. 30rb ad 2: "Et breuiter ego credo quod diuersitas Socratis ad Platonem est Socrates; econuerso diuersitas Platonis ad Socratem est Plato." Buridan reiterates the point in John Buridan, *op.cit.*, V.9, fol. 32va: "Et aliquando ita est quod nullam aliam rem significat uel connotat praeter illas duas res, scilicet praeter illam pro qua supponit et illam ad quam est comparatio, sicut si ego dico 'Socrates est diuersus a Platone' uel si dico 'Socrates dependet a Deo' et tunc adhuc in isto casu credo quod eadem res est pro qua supponit terminus absolutus et pro qua terminus relatiuus etiam in abstracto sumptus: ita eadem est res quae est Socrates et quae est diuersitas Socratis ad Platonem, et dependentia Socratis ab ipso Deo."

[59] See n. 50 above.

a horse is just Socrates, and likewise the diversity of Socrates from a stone is just Socrates. Yet how is one diversity greater than the other? Both are just the same thing, namely Socrates.

A less obvious problem also faces Buridan's account. For Buridan has argued that there are no non-individual entities in the world, and that we should identify universals as mental items that represent many things in reality. Such a universal concept is not "fictitious" because it is grounded in reality by objective agreement among the substances it represents. This agreement, however, is not anything in the world above and beyond the individual substances themselves. But this line of argument threatens to lapse into triviality. Does it say anything more than that the universal concept applies to the individuals it does because it in fact does apply to them?

These are at bottom the same problem, I believe. They both address Buridan's identification of certain individual things as the real correlate answering to metaphysical truths. Turn it around: perhaps the kind of explanation Buridan is offering rejects the need to give some entity (*res*) in the first place: not merely an "added" thing, but any thing at all. We may be looking in the wrong place for Buridan's solution.

I can think of two ways to capitalize on this insight. Both have some support in the texts; neither is fully satisfactory; each manages to avoid the problems mentioned above.

We might, for instance, take Buridan's approach to agreement and diversity as being fundamentally modal.[60] The agreement between Socrates and Plato is not at bottom a matter of any thing they have or share, but rather a matter of the way they are. Socrates and Plato, as well as Socrates and Brunellus, are related in a certain fashion (*aliqualiter*) – or rather the former in one way and the latter in another – namely as being in agreement, or one pair being more in agreement than the other. But such ways or modes are not themselves part of the ontology: there are things and there are the ways things are, but there is no such thing as the way things are.

The modal approach has historical as well as philosophical merits. Buridan did recognize modes and speak of them; if less often than other contemporary philosophers, still often enough to make their deployment on the problem of universals a plausible move.

The disadvantage of the modal approach is that it seems a mere sleight of hand: modes by definition do not appear in the ontology,

---

[60] I have in mind modes taken roughly along the lines sketched in Klima, "Buridan's Logic and the Ontology of Modes."

but make a metaphysical difference to individuals that do appear in the ontology. How can the relation between Socrates and Plato be parsed in terms of some feature that does not exist, despite the fact that they have more of it than Socrates and Brunellus?

Alternatively, we might think of Buridan's approach as a roundabout way of getting at features we call nowadays "metaphysically primitive." It is a brute fact that Socrates and Plato agree with each other, and another brute fact that they agree more than do Socrates and Brunellus. (Not that there are facts, of course.) These are metaphysical truths, but truths that do not have any further explanation; they are primitive. Once we distinguish the truth from the truthmaker (whatever is responsible for making the truth true), we can easily see why Buridan should say that Socrates's essential agreement with Plato just is Socrates and Plato: the metaphysically primitive truth that they essentially agree requires them both to exist and follows from each being the very thing it is (in this case: human). Nor does countenancing metaphysical truths cause any ontological worries. There are truths, but truths are not things. Some truths are primitive, including those that describe how the world is ultimately sorted into natural classes: a fact that admits of no further explanation, or no further metaphysical explanation.

The disadvantage of this second approach is that the very facts that seem to prompt the problem of universals are in the end not explained but assumed: we are told that the facts hold rather than why the facts hold, a very different matter. In the end, Socrates and Plato agree because, well, there is no "because:" they just do. And that is not very helpful.

Buridan could well accept the disadvantages of either approach sketched here; he wouldn't be the first philosopher to defend a position known to have problems. (He could even adopt both views.) My suspicion – it is no more than that-is that Buridan is in the end a partisan of the second approach. The appeal to primitive metaphysical truths can be more or less rewarding, depending on how deep in the theory one has to go to find the appeal. In Buridan's case, the sheer wealth of close philosophical argument articulated in his proposed solution to the problem of universals suggests that we must go a long way indeed. And that, perhaps, is all we can ask of any philosopher. Buridan's nominalism is a robust example of medieval philosophy at its finest.

BURIDAN'S THEORY OF DEFINITIONS IN HIS SCIENTIFIC PRACTICE

Gyula Klima

1. *Introduction*

Just as the proof of the pudding is in the eating, so too, the proof of the viability of a theory of scientific method lies in whether it works in scientific practice. In this paper, I will address the question whether the "pudding" of Buridan's theory of definitions he cooked up in his *Summulae* proves "edible" in his scientific practice.[1]

In the next section, I will begin the discussion by presenting Buridan's theory as it is laid down in the *Summulae*. Then I will confront the theory with one of its particularly problematic applications in Buridan's scientific practice, namely, with Buridan's discussion of Aristotle's definition of the soul. Finally, I will conclude the discussion with a somewhat tentative evaluation of the historical significance of this application of Buridan's theory in his scientific practice.

2. *Buridan's Theory of Definitions in the* Summulae

In the eighth treatise of his *Summulae,* Buridan provides a systematic account of his theory of definitions. The theory presents us with four kinds of definitions, namely, nominal, quidditative, causal, and

---

[1] Of course, I mean "scientific practice" in the most relevant medieval sense, i.e., as referring to the application of a certain theoretical apparatus to a particular science – in this case, the application of the theory of definitions to psychology. In medieval science, there was no scientific practice in the modern sense, if by that we mean designing and conducting experiments using specialized instruments of various sorts, although even that should imply some thinking, when the scientist applies his or her general theoretical knowledge to a particular situation. (This is the process Buridan would refer to as the "appropriation" of the general theoretical principles of a more general science by a more specific field. Cf. for example, John Buridan, *Summulae. De demonstrationibus* [De Rijk e.a.], cap. 3, pars 6; cap. 8, pars 4.) Likewise, later medieval thinkers could be described as "empiricist" in their general approach to natural philosophy, but certainly not in the sense that later came to be associated with the British Empiricists, or the Logical Empiricists. For discussion of the latter point, see Zupko, "What Is the Science of the Soul?"

descriptive definitions. In addition to these, Buridan also touches on "composite definitions" (*definitiones complexae*), that is, definitions resulting from a combination of the aforementioned kinds. However, since the particular problem I am going to consider here will primarily arise concerning quidditative and nominal definitions, I will concentrate on these two in this account of Buridan's theory. (Although later on I will also refer to the other types, when they become relevant in the discussion of the problem.)

Buridan begins his discussion of nominal definitions by the following characterization:

> (1) A nominal definition is an expression convertibly explaining what thing or things the *definitum* signifies or connotes, and properly speaking it is called "interpretation." (2) It pertains to incomplex spoken terms to which there do not correspond simple concepts in the mind, but complex ones, whether these terms supposit for some thing or things or do not supposit for some thing or things.[2]

Thus, in accordance with this doctrine, a nominal definition provides precisely the meaning of a simple term that is subordinated to a complex concept.[3] Indeed, a correct nominal definition of a term is an expression that by its syntactical structure faithfully mirrors the semantic structure of the complex concept to which the term is subordinated. Therefore, the nominal definition and the term are

---

[2] John Buridan, *op.cit.*, 2.3: "(1) Definitio dicens 'quid nominis' est oratio exprimens convertibiliter quid vel quae definitum significat aut connotat; et nomine proprio vocatur 'interpretatio'. (2) Et convenit terminis incomplexis vocalibus quibus non correspondent in mente conceptus simplices, sed complexi, sive illi termini pro aliquo vel aliquibus sive nec pro aliquo nec pro aliquibus supponant." Passages quoted from this work are from my translation, forthcoming in *The Yale Library of Medieval Philosophy* series. For details on the text used for the translation, see the introductory essay of that volume. Subsequent references to passages from this work will be to the intrinsic divisions of the text.

[3] Cf. John Buridan, *Summulae. De syllogismis* [Hubien], cap. 1, pars 4: "Concerning the consequence from the exponents to what is expounded, I say that that is a formal consequence, but what is inferred is not different from the premises in intention, but only verbally. For to expound in this manner is just to explain the meaning of the name (*quid nominis*); but a definition that gives precisely the meaning of a name and the name thus defined have to have entirely the same intention corresponding to them in the soul. And the same goes for a proposition that requires some exposition on account of its syncategoremata, for the proposition and its exponents have to have entirely the same intention corresponding to them in the soul." Buridan here makes it entirely clear that as far his theory is concerned, synonymous expressions of spoken or written languages are not distinguished in mental language – a point which is still controversial in the secondary literature concerning Ockham's theory. Cf. Panaccio, "Connotative Terms" and Spade, "Synonymy and Equivocation."

strictly synonymous: the term is to be regarded as a mere shorthand for the complex phrase clearly explicating the structure of the corresponding concept.[4] Consequently, *only* such simple terms can have nominal definitions to which there correspond complex concepts in the mind. As we can read in the *Treatise on Suppositions*:

> ... to some incomplex utterances there correspond complex concepts, and to others incomplex concepts. Those to which there correspond complex concepts can, and should, be expounded as to their meaning (*quid nominis*) by complex expressions that are equivalent to them in signification. Those, however, to which there correspond incomplex concepts do not have definitions precisely expressing their meaning (*quid nominis*).[5]

In view of this doctrine, therefore, a simple term can have a nominal definition if and only if it is subordinated to a complex concept. Now, since in accordance with Buridan's account of the difference between absolute and connotative terms, all simple absolute terms are subordinated to simple concepts, this means that absolute terms cannot have nominal definitions.[6]

---

[4] Note that when we are talking about the structure of a complex concept this need not be imagined to be the same kind of syntactic structure as that of the corresponding expression: the complexity of a complex concept is mere *semantic* complexity as opposed to *syntactic* complexity. For this distinction, see my introduction to Buridan's *Summulae de dialectica*.

[5] John Buridan, *Summulae. De suppositionibus* [Van der Lecq], cap. 2, pars 4: "... vocum incomplexarum quibusdam correspondent conceptus complexi et quibusdam conceptus incomplexi. Et illae quibus correspondent conceptus complexi possunt et debent exponi quantum ad quid nominis per orationes illis aequivalentes in significando. Illae autem quibus correspondent conceptus incomplexi non habent diffinitiones praecise exponentes quid nominis." Note that strictly speaking, despite existing translational traditions to the contrary, *quid nominis* is not to be rendered as "nominal definition," for the latter is *definitio exprimens quid nominis*, i.e., "definition expressing the *quid nominis*;" therefore, since the phrase *quid nominis* alone is usually expounded as *quid significatur per nomen*, i.e., what is signified by the name, *quid nominis* may justifiably be rendered as 'the meaning of the name.' Accordingly, Buridan's point here is that significative utterances subordinated to simple concepts do not have nominal definitions, i.e., definitions precisely expressing their *quid nominis*, but of course they do have *quid nominis*, i.e., they do have meaning or signification, given that they are meaningful in virtue of being subordinated to a concept in the first place. Nevertheless, for the sake of simplicity, in its subsequent occurrences I will translate the whole phrase *definitio exprimens quid nominis* not as 'definition expressing the meaning of the name', but simply as "nominal definition." Furthermore, since even Buridan himself occasionally uses the shorter phrase *quid nominis* to refer to a nominal definition, on such occasions I will also translate *quid nominis* as "nominal definition."

[6] A connotative term is a term that signifies whatever it signifies in relation to some thing or things, its connotatum or connotata. Accordingly, such a term, when it occurs

To see this in more detail, suppose we have a simple absolute term A, which is not imposed to be a shorthand for a quidditative definition or some similar combination of other simple absolute terms.[7] This means that its signification and natural supposition are the same, taking these simply to be the set of its ultimate significata.[8] But suppose A has a nominal definition. Then this nominal definition has to be subordinated to a complex concept. However, since A was supposed to be absolute, the concepts making up this complex concept cannot be connotative, and cannot signify a connotation in respect of the significata of each other (as would be the case for instance in a genitive construction of two absolute terms, such as *hominis asinus*); therefore, the total signification of the complex concept would have to be the union of the sets of the significata of its components. But if this complex concept is made to be the subject of a mental proposition in which it has natural supposition, then it can supposit only for the intersection of the sets of the significata of its components. But then, since these components have to be different, they have to have different nonempty sets of significata. Thus, the intersection and the union of these sets will never be the same, which means that the total ultimate signification and the natural supposition of the same complex concept will never be the same. However, as was stated above, the total signification and the natural supposition of A have to be the same. Therefore, the simple term A cannot be subordinated to the same complex concept, and so A cannot have a nominal definition, contrary to our assumption.[9]

---

as the subject or the predicate of a proposition, supposits for what it signifies only by means of an oblique reference to what it connotes, i.e., using Buridan's terminology, by means of appellating its appellatum or appellata. By contrast, absolute terms are those that signify their significata not in relation to anything, and, accordingly, in a propositional context they simply supposit for their significata without appellating, i.e., obliquely referring to anything. For more on this distinction and its significance, see my introduction to Buridan's *Summulae de dialectica*.

[7] I have to provide this restriction to rule out the artificial cases when for example someone introduces the simple term B with the nominal definition "rational animal."

[8] In natural supposition a simple absolute term supposits for all its ultimate significata (past, present future, or maybe merely possible), so the signification and natural supposition of a simple absolute term can be adequately characterized in terms of the set of its significata. Note also that according to Buridan in the case of an absolute term this set is never empty, although the term may actually supposit for nothing. Cf. John Buridan, *Sophismata* [Scott], c. 1, 6th conclusion. (A new translation of the *Sophismata* will also be included in my translation of the *Summulae de dialectica*, as the ninth treatise of that work.)

[9] Cf. John Buridan, *Sophismata* [Scott], c. 1, 11th conclusion: "The reply to this is that this definition of 'man' is not a nominal definition, namely, one expressing what

Absolute terms, therefore, can have only other types of definitions, most importantly, quidditative definitions. Buridan describes quidditative definitions in the following manner:

> thing or things and in what ways the name 'man' signifies, but is rather a definition expressing what the thing is for which the name 'man' supposits, this being the same as that for which the expression 'rational mortal animal' supposits. But it is not necessary that those terms should precisely and adequately supposit for the things that they signify. So only a spoken term to which there does not correspond a simple, but a complex concept, is one which has a nominal definition in the strict sense, namely, [a definition] which signifies precisely what and how that term signifies. For the signification of such a spoken term is explicated by means of spoken terms corresponding to the simple concepts of which the complex concept corresponding to that term is composed. But when to some spoken term there corresponds a simple concept, as when to the term 'donkey' there corresponds the specific concept of donkey, which we assume to be simple, it is not possible for another spoken term to signify precisely and adequately the thing or things that that term signifies, unless it is entirely synonymous with it. Nor is it possible to posit a spoken expression consisting of terms of diverse significations, without there corresponding to them other concepts which do not correspond to that term. But where a spoken term to which there corresponds a simple concept is concerned, it is possible to provide a causal definition, or a description declaring what the causes or properties of the thing or things for which this term supposits are, or even a quidditative definition consisting of the genus and difference, to which there corresponds a complex concept, but which adequately supposits for the same things which are supposited for by the incomplex concept that corresponds to that spoken term. But these points had to be clarified in more detail in the treatise on definitions, divisions and demonstrations. And I am glad that I have understood these issues." – "Respondetur quod illa non est diffinitio 'hominis' exprimens quid nominis, scilicet quid vel quas res et quomodo hoc nomen 'homo' significat, sed est diffinitio exprimens quae res est pro qua supponit iste terminus 'homo', quia ipsa est eadem pro qua supponit ista oratio 'animal rationale mortale'. Sed non oportet quod illi termini praecise et adaequate supponant pro rebus quas significant. Unde solus terminus vocalis cui non correspondet conceptus simplex sed complexus habet proprie diffinitionem dicentem quid nominis, scilicet praecise significantem quid et quomodo ille terminus significat. Talis enim termini vocalis significatio explicatur per terminos vocales correspondentes conceptibus simplicibus ex quibus componitur conceptus complexus correspondens illi termino. Sed cum alicui termino vocali correspondeat conceptus simplex, ut huic voci 'asinus' conceptus specificus asini, posito quod sit simplex, non est possibile alterum terminum vocalem praecise et adaequate significare illud vel illa quae ille significat, nisi sit pure synonymus illi. Nec est possibile dare orationem vocalem ex terminis diversarum significationum constitutam quin eis correspondeant alii conceptus qui non correspondent illi termino. Sed talis termini simplicis cui correspondet conceptus simplex potest dari diffinitio causalis vel descriptio declarans quae sunt causae vel proprietates rei vel rerum pro qua vel pro quibus ille terminus supponit, vel etiam diffinitio quiditativa ex genere et differentia cui correspondet conceptus complexus, supponens tamen adaequate pro eisdem rebus pro quibus supponit ille conceptus incomplexus qui illi termino vocali correspondet. Et haec debebant magis declarari in tractatu de diffinitionibus, divisionibus et demonstrationibus, et gaudeo haec intellexisse."

(1) A quidditative definition is an expression indicating precisely what a thing is (*quid est esse rei*) by means of essential predicates. (2) These are the genus of the *definitum* and the essential difference, or differences, [which are added] until the whole definition is convertible with the *definitum*. (3) This definition responds precisely, most properly and truly to the question "What is it?" (4) And it presupposes the existence of the thing, if one has to reply to the question "What is it?"[10]

As Buridan explains, this description entails that no connotative terms can have quidditative definitions in this strict sense:

> ... let us assume that nothing is white, except a stone. Then a white [thing] is a stone, and it is nothing other than a stone, nor is it a whiteness, or an aggregate of a stone and a whiteness, if we express ourselves properly, but it is only that to which a whiteness belongs (*inest*), namely, the stone, just as a wealthy man is not his wealth, nor the aggregate of a man and wealth, but only a man to whom this wealth belongs (*adjacent divitiae*). But when I ask precisely "What is the white thing (*Quid album est*)?," I do not require that the reply should indicate what a whiteness is, or what an aggregate of a stone and a whiteness is, nor do I ask on account of what disposition a white thing is white; I only ask what the thing is which is white, and that is nothing but a stone. Therefore, in the case assumed above, I give a satisfactory reply to the question if I declare that the white thing is a stone; and if I add something else that signifies or connotes something other than the stone, then I provide more in my reply than what was asked for. Since, therefore, a purely quidditative definition should precisely indicate what a thing is, if the term "white [thing]" (*album*) has a quidditative definition, then it is necessary that it be the term "stone," or its quidditative definition, or an expression consisting exclusively of substantial terms.[11] But this is impossible, for

---

[10] John Buridan, *Summulae. De demonstrationibus* [De Rijk e.a.], 2.4: "(1) Definitio quidditativa est oratio indicans praecise quid est esse rei per praedicata essentialia. (2) Quae sunt genus definiti et differentia vel differentiae essentiales donec totalis definitio sit convertibilis cum definito. (3) Et haec definitio respondetur praecise, propriissime et vere ad quaestionem 'quid est?'. (4) Et praesupponit esse rei, si debeat responderi ad 'quid est?'."

[11] The critical text has here the following: *Cum ergo definitio pure quidditativa debeat indicare praecise quid est, necesse est, si ille terminus "album" habeat definitionem quidditativam, quod illa, si sit illius termini "lapis," vel ⟨sit⟩ eius quidditativa definitio,* [V 101va] *vel quod sit oratio constituta praecise ex terminis substantialibus* – "Since, therefore, a purely quidditative definition should precisely indicate what a thing is, it is necessary that if the term 'white [thing]' (*album*) has a quidditative definition, then that, if it is that of the term 'stone', should either be its quidditative definition, or that it should be an expression consisting exclusively of substantial terms." However, on the basis of the apparatus this can be amended as follows (providing the reading consonant with the Hubien-text, which I translated in the main text): *Cum ergo definitio pure quidditativa debeat indicare praecise quid est, necesse est, si ille terminus "album" habeat definitionem quidditativam, quod illa sit iste terminus "lapis," vel quod ⟨sit⟩ eius quidditativa*

after removing whiteness and retaining the substance subjected to it, the *definitum*, namely, "white [thing]," would not supposit for anything, for nothing would be a white thing, and the definition would still supposit for something, namely, for the thing that it supposited for before. Thus it would not be converted with the *definitum*, nor would it be truly predicated of it, which is impossible; therefore, it is impossible that "white [thing]" (*album*) should have a purely and properly quidditative definition.[12]

So, if a simple term is absolute, then it has to be subordinated to a simple concept, and hence it cannot have a nominal definition. Thus, any simple term that has a nominal definition has to be subordinated to a complex connotative concept, and so the term has to be connotative.[13] Nevertheless, an absolute term can have a quidditative definition in the strict sense, namely, a definition which convertibly *supposits* for the same things as its *definitum* does, but which always *signifies* more, for it can contain only other absolute terms, namely, the strictly quidditatively predicable genera and differences of the *definitum*.[14] However, since connotative terms cannot have quidditative definitions in this strict sense, it follows that a term can have a quidditative definition in the strict sense if and only if it is absolute.

---

definitio, [V101^va] *vel omnino* – John Buridan, *Quaestiones in Posteriorum Analyticorum libros* [Hubien], *quod sit oratio constituta praecise ex terminis substantialibus.*

[12] John Buridan, *Summulae. De demonstrationibus* [De Rijk e.a.], 2.4. Cf. John Buridan, *Quaestiones in Posteriorum Analyticorum libros* [Hubien], Book 2, q. 8. (unpublished edition of H. Hubien)

[13] Obviously, this much need not entail that all connotative terms have to have nominal definitions; indeed, since Buridan allows the possibility that some connotative concepts are simple, the terms subordinated to them cannot have nominal definitions in this strict sense. (Cf. John Buridan, *Summulae. De demonstrationibus* [De Rijk e.a.], 2.4, second doubt.)

[14] John Buridan, *op.cit.*, 2.4: "In respect of this property the quidditative definition also differs from the nominal definition, for the quidditative definition should signify much more, or something other, than the *definitum*. And this derives from the other difference between them, namely, that to the *definitum* of a nominal definition there should correspond a complex concept, whereas the species defined quidditatively has an incomplex concept; for it is not the species, but the definition of the species which is composed of genus and difference, whether in respect of utterance or concept." – "In hac etiam proprietate differt definitio quiditativa a definitione dicente 'quid nominis', quoniam illa vel multo plus vel aliud debet significare quam definitum. Et hoc provenit ex hac alia differentia quia definito definitione dicente 'quid nominis' oportet correspondere conceptum complexum, et species quiditative definita habet conceptum incomplexum; non enim species est, sed definitio speciei quae componitur ex genere et differentia, sive secundum vocem sive secundum conceptum."

But this conclusion imposes a rather severe limitation upon the scope of quidditatively definable terms. And since the middle term of the most powerful scientific demonstrations has to be a quidditative definition, this would also severely limit the scope of what is scientifically knowable. Therefore, Buridan immediately adds that in a less strict sense even a connotative term can have a quidditative definition:

> ... a connotative term can have a quidditative definition in a less proper sense, and one which is more broadly so-called. For it is by means of the subjects that their *per se* attributes are defined, as is said in bk. 1 of the *Posterior Analytics* and bk. 7 of the *Metaphysics*.[15] Therefore, a definition of a connotative term is called quidditative, because it indicates what it is (*quid est*) not only where the suppositum is concerned, but also the connotatum. For example, the definition of "pug" (*simum*) is "concave nose;" and by saying "nose" I say what it is and I likewise indicate what the term "pug" supposits for (since it is a concave nose, and a pug is nothing but a nose); however, by adding "concave" I indicate what that term appellates, for that is the very same thing, and none other, that the term "concave" signifies, namely, concavity. Similarly, if I define "pugness," then I say that it is the concavity of the nose; and when I say "concavity," then I indicate what it is that "pugness" supposits for, since it is concavity (because pugness is a concavity, and nothing else); but when I add "of the nose," I indicate what "pugness" appellates, for it is the nose, given that a concavity would not be a pugness, were it not in a nose.
>
> Therefore, in this connection we have to note that in the case of connotative terms, the genus is not predicated of its species *in quid* in the strictest sense, but broadly speaking it *is* predicated of it *in quid*. Thus, when I say: "A white [thing] is [a] colored [thing]" I do not say precisely what a white [thing] is, but I add what it is like, as I said earlier; therefore, this is not quidditative predication in the strictest sense. But broadly speaking it is admitted to be quidditative, because it indicates what it is for which "white [thing]" supposits, this being the very same thing as that for which "colored [thing]" supposits, and at the same time it indicates what "white [thing]" appellates, for this is the same as what "colored [thing]" appellates.[16]

---

[15] Aristotle, *Posterior Analytics*, I.3, 73a8 sqq.; *Metaphysics*, VII.5, 1030b29–35.

[16] John Buridan, *op.cit.*, 2.4: "... termini connotativi potest esse definitio quidditativa minus proprie et magis communiter dicta. Nam per subiecta definiuntur per se passiones eorum, sicut habetur primo Posteriorum et septimo Metaphysicae. Definitio igitur termini connotativi dicitur quidditativa quia indicat quid est non solum de supposito, sed etiam de connotato. Verbi gratia, definitio 'simi' sit 'nasus cavus'; per hoc quod dico 'nasus' ego dico quid est et similiter indico pro quo ille terminus 'simum' supponit, quia hoc est nasus simus; simum enim est nasus, et non est aliud; sed cum addo 'cavus', per hoc ego indico quid est quod ille terminus appellat, quia hoc

So, clearly, what allows Buridan to say that even simple connotative terms can have quidditative definitions is that it is possible to construct complex phrases which are convertibly predicable of such terms by means of quidditative or essential predication. That it is not only absolute terms that can be essentially predicated is clear from Buridan's general characterization of the distinction between essential and denominative predication in his treatise *On Predicables*:

> ... everything that is predicated of something is either predicated essentially, so that neither term adds some extraneous connotation to the signification of the other, or it is predicated denominatively, so that one term does add some extrinsic connotation to the signification of the other. This division is clearly exhaustive, for it is given in terms of opposites.[17]

Thus, whenever the subject of a predication is a connotative term, the predication is essential if and only if the predicate does not connote

---

est illud idem, et non aliud, quod iste terminus 'cavus' significat, scilicet cavitatem. Et similiter si definio 'simitatem', ego dico quod simitas est cavitas nasi; et cum dico 'cavitas', ego indico quid est pro quo 'simitas' supponit, quia est cavitas (simitas enim est cavitas, et nihil aliud). Sed cum addo 'nasi', ego indico quid est [E101^{vb}] quod 'simitas' appellat, quia est nasus; unde cavitas non esset simitas si non esset in naso.

Unde sciendum circa hoc quod in terminis connotativis genus non praedicatur in quid de sua specie, propriissime loquendo, sed communiter loquendo praedicatur de ea [K151^{rb}] in quid. Ut si dicam 'album est coloratum', ego non dico praecise quid album est, sed etiam quale est, sicut prius dixi; ideo non est propriissime praedicatio quiditativa, sed communius loquendo conceditur [V101^{vb}] quod sit quiditativa, quia indicat quid est pro quo 'album' supponit, quia est illud idem pro quo 'coloratum' supponit, et indicat cum hoc quid est quod 'album' appellat, quia est illud idem quod 'coloratum' appellat."

[17] John Buridan, *Summulae. De praedicabilibus* [De Rijk], cap. 1, pars 3: "Omne ergo quod praedicatur de aliquo vel praedicatur essentialiter, scilicet ita quod neuter terminus super significationem alterius addat extraneam connotationem, vel praedicatur denominative, scilicet ita quod unus terminus addat super significationem alterius aliquam connotationem; apparet enim quod haec divisio sit sufficiens, quia per opposita." Cf. John Buridan, *Summulae. In praedicamenta* [Bos], cap. 4, pars 1: "Now, that these predications are essential is obvious, for a relative concept is not only a concept of something, but also a concept of something [with respect] to something; therefore a relative term, in virtue of its proper signification and imposition connotes something [with respect] to something, whence the addition '[with respect] to something', construed with a relative term, amounts only to the explication of the connotation of that term, but it does not add some connotation extrinsic to that term, and so the predication is essential." – "Quod enim haec praedicationes sint essentiales apparet: quia conceptus relativus non solum est alicujus conceptus, sed alicujus ad aliquid conceptus; ideo terminus relativus, de propria significatione et impositione, connotat ad aliquid, et ideo ista additio 'ad aliquid', super terminum relativum cadens, non est nisi expressio connotationis illius termini, et non addit connotationem alienam super illud terminum, et ideo est praedicatio essentialis."

anything over and above the signification and connotation of the subject. The reason for this should also be clear. A connotative term is only contingently true of a thing it supposits for on account of the fact that it may become false of the thing in question simply because of a change in something else that it connotes. Thus, a connotative term is what in the modern parlance we would call a non-rigid designator of its supposita: it can cease to supposit for its supposita without their destruction, as a result of removing its connotata.[18] Take away a wealthy man's wealth, and the man, while he will still go on existing, will cease to be supposited for by the term "wealthy man." Nevertheless, it is still possible to form an essential predication in which the subject term is the connotative term "wealthy man," provided the predicate does not connote anything over and above the connotation of this term. Thus, the predication "A wealthy man is a man who possesses wealth" is an essential predication, in which whenever the subject supposits for something the predicate also has to supposit for the same thing, and thus which is always true, assuming the natural supposition of its terms.[19]

All in all, we can summarize Buridan's conclusions concerning what sorts of terms can have which of the two sorts of definitions discussed so far in the following table:

	Simple Spoken Absolute Term, Simple concept	*Simple Spoken Connotative Term, Simple concept*	*Simple Spoken Connotative Term, Complex concept*
Has nominal definition	No	No	Yes
Has quidditative definition	Yes	Strictly no, but improperly yes	Strictly no, but improperly yes

As can be seen from this table, only simple connotative terms subordinated to complex concepts can have both nominal and quidditative definitions, and they can only have quidditative definitions in the improper sense characterized above.

---

[18] For the significance of this point in the nominalist ontological program see Klima, "Buridan's Logic and the Ontology of Modes" and Klima, "Ockham's Semantics."

[19] For the example of "wealthy," see John Buridan, *Summulae. De suppositionibus* [Van der Lecq], 5.2. For natural supposition, *ibid.*, 3.4; 4.4.

3. *The Theory in Practice*

Now, in view of the conclusions of the foregoing discussion, one may find rather strange the things Buridan has to say about the Aristotelian definitions of the soul. The most striking formulation in this regard comes from Buridan's first redaction of his questions on the *De Anima*:

> ... the definition of the soul (*definitio animae*) is twofold: there is one definition of the soul expressing its *quid nominis*; but there is another, expressing not only its *quid nominis*, but also its *quid rei*. The definition of the soul expressing its *quid nominis* is that which Aristotle posits in Book II, Chapter 2 of this treatise, and it is this: "the soul is that by which we live, sense, are moved locally, and understand" (cf. 414a12). For this reason, this whole expression (*oratio*) and the term "soul" are equivalent in signification. But another definition, expressing the *quid rei*, is posited in Book II, Chapter 1 of this treatise, and it is the aforementioned definition [i.e., that "soul is the first, substantial, act of a physical, organic body potentially having life" (cf. 412a27–28)], and the present question is about this. Accordingly, this definition expresses not only the *quid nominis*, but also the *quid rei*, and not only the *quid rei*, but also the *propter quid*.[20]

According to this passage, we have two distinct definitions of the term "soul." The first is a purely nominal definition,[21] which, in accordance with the previous conclusions, establishes that this term is subordinated to a complex connotative concept, the structure of which is explicated by this definition. The second is also claimed to be a nominal definition, but also a quidditative definition, (and indeed, a causal definition as well). Again, if this is a nominal definition, then the term "soul" has to be subordinated to a complex

---

[20] John Buridan, *Quaestiones in Aristotelis De anima* [Patar], Book 2, q. 3, p. 242, ll. 71–82: "... duplex est definitio animae: quaedam est definitio animae exprimens quid nominis, quaedam autem est non solum exprimens quid nominis, verum etiam exprimens quid rei. Definitio animae exprimens quid nominis est quam ponit Aristoteles in II° capitulo tractatus Iⁱ huius, et est haec: 'anima est quo vivimus, sentimus et secundum locum movemur et intelligimus;' unde ista tota oratio et ista terminus anima aequivalent in significando. Alia autem definitio animae exprimens quid rei ponitur in I° capitulo huius II\ⁱ, et est definitio praedicta, et de illa est praesens quaestio. Unde ista definitio non solum exprimit quid nominis, sed etiam exprimit quid rei, et non solum exprimit quid rei sed etiam propter quid."

[21] To be sure, the definition I call here "the first" is the one that comes later in Aristotle's text, in the second chapter of the second book, whereas the other comes first, in the first chapter. However, the definition of Aristotle's second chapter comes first here in Buridan's text, and it is also prior insofar as it can be used to prove the other.

connotative concept explicated by this nominal definition. Indeed, since this nominal definition is obviously different from the first, the term "soul" as defined by the second definition has to be subordinated to a concept distinct from the one explicated by the first definition, so, apparently, this would establish that the term is equivocal.[22] Furthermore, the second definition is also claimed to be a quidditative definition. However, being the definition of a connotative term, it can only be a quidditative definition in the improper sense described in the previous section, namely, insofar as it supposits for and connotes the same things as the term it defines does.

But this is still not everything Buridan claims concerning these two definitions. Both in his running commentary on Aristotle's text and in the *Summulae* he claims further that the first definition can be used to prove the second. As he writes in the *Summulae*:

> … one definition of "soul" asserts that the soul is the principal intrinsic principle of living, sensing and understanding, as is clear from bk. 2 of *On the Soul*;[23] but another is that the soul is the first substantial act of an organic physical body, etc. And the second is demonstrated by means of the first thus: every principal intrinsic principle of living, sensing and understanding is the first substantial act of an organic physical body potentially having life; and every soul is such a principle; therefore, etc.[24]

However, together with the previously quoted passage, this passage gives us serious reasons to believe that this demonstration cannot produce a quidditative definition of the soul, not even in the im-

---

[22] Note that this equivocation is not the equivocation Buridan alludes to with reference to Averroes (John Buridan, *Quaestiones in Aristotelis De anima* [Patar] p. 43. n. 57), because that concerns the alleged equivocation between the concept of the intellective and that of the vegetative and sensitive souls.

[23] Aristotle, *De anima*, II.1, 412a27–28.

[24] John Buridan, *Summulae. De demonstrationibus* [De Rijk e.a.], 2.7: "Una ergo definitio 'animae' est quod anima est principium principale intrinsecum vivendi, sentiendi et intelligendi, ut patet secundo De anima; et alia est quod anima est actus primus substantialis corporis physici organici *etc*. Et demonstratur secunda per primam sic: omne principium principale intrinsecum vivendi *etc*. est actus primus substantialis corporis physici organici habentis vitam in potentia; et omnis anima est huiusmodi principium; ergo, *etc*." cf. John Buridan, *Quaestiones in Aristotelis De anima* [Patar], Book 2, c. 2, p. 55, ll. 96–103: "… illud quo animatum vivit, sentit, etc., est actus corporis physici organici vitam habentis in potentia; sed anima est qua animatum vivit, sentit, etc., sicut declaratum est in isto capitulo; ergo concluditur quod anima est actus corporis physici organici vitam habentis in potentia. Et notandum est quod ista demonstratio procedit ex notioribus nobis. Ista enim definitio animae *anima est qua vivimus, sentimus, etc*. est notior quoad nos, ex quo datur per operationes et effectus quos experimur in nobis, etc., qui effectus sunt notiores suis causis."

proper sense characterized earlier. To see this more clearly, let us lay out this demonstration marking the concepts subordinated to its terms by bracketing the terms in the following manner:

1. Every [principal intrinsic principle of living, sensing and understanding] is the {first substantial act of an organic physical body potentially having life}
2. Every [soul] is a [principal intrinsic principle of living, sensing and understanding]

Therefore,

3. Every [soul] is {the first substantial act of an organic physical body potentially having life }

Now, since the expression "principal intrinsic principle of living, sensing and understanding" is the first definition of the term "soul," with which it is strictly synonymous, the terms of this argument enclosed in square brackets are subordinated to the same complex concept, the structure of which is explicated by this expression. But then the third term, enclosed in curly brackets, being the other nominal, as well as quidditative, definition of the term "soul," has to be subordinated to a different, complex connotative concept. So, this demonstration demonstrates the predication of the correct quidditative definition of its *definitum* only if the first premise is a case of quidditative or essential predication, at least in the sense in which connotative terms can be essentially predicated. As we can recall, such a predication is essential if and only if neither term adds any extraneous connotation over the other. But a closer look at these two terms clearly reveals that the second connotes the subject of the soul not mentioned in the first, whereas the first connotes operations not mentioned in the second. Therefore, the predication clearly cannot be essential even in this sense.

To be sure, one might still say that in the conclusion the term "soul" occurs as subordinated to the concept which is expressed by the second definition, and according to that concept the predication of this definition is quidditative. As Buridan remarks elsewhere:

"... if the nominal definition is truly predicated of the *definitum*, then it is predicated of it quidditatively, and thus in the first mode, for this is a predication of the same about the same in intention, although not in utterance, for the *definitum* signifies or connotes nothing more nor less than does the definition ..."[25]

---

[25] John Buridan, *Summulae. De demonstrationibus* [De Rijk e.a.], 6.3: "Tamen ad

However, this defense would invalidate the demonstration. For then the demonstration would clearly be a case of the *fallacy of equivocation*, since then the term "soul" would appear in the conclusion according to a different concept than that according to which it occurred in the minor premise. Therefore, if the demonstration is valid, then its conclusion cannot essentially predicate the quidditative definition of the term "soul," taking this term in the subject according to the concept expressed by its purely nominal definition. On the other hand, if the conclusion is interpreted as essentially predicating a quidditative definition of its *definitum*, then the demonstration is not valid.

So, what went wrong here? Let us briefly survey the main points that have driven us into this predicament. These points can be summarized as follows.

1. The first definition of the soul is its nominal definition
2. The second definition is also its nominal and quidditative (and causal) definition
3. The second definition is provable by means of the first, using the first as the middle term
4. The conclusion of the proof has to be an essential predication of the quidditative definition of its *definitum*

The conclusion of the foregoing argument is that these points, extracted from the *prima lectura*, along with Buridan's theory of definitions discussed in the previous section, are inconsistent. Therefore, whoever wishes to maintain the theory has to reject at least one of these points. So, apparently, what went wrong was the combination of Buridan's theory of definitions in the *Summulae* and the strong claims of the *prima lectura* concerning the two definitions of the soul.

In fact, in the parallel passage of the *tertia lectura*, Buridan simply states the following:

> Some definition is expressing the *quid nominis*, another is purely quidditative, and yet another is causal, explicating not only what the thing is (*quid res est*), but also the reason why it is (*propter quid est*). And [a definition of] this kind is more perfect. And this definition [namely,

---

praesens volo dicere quod definitio dicens 'quid nominis', si vere praedicatur de definito, praedicatur de eo quiditative, et sic in primo modo, quia est sicut praedicatio eiusdem de eodem secundum intentionem, licet non secundum vocem, cum nihil plus aut minus significet aut connotet definitum quam definitio, prout alias dictum est."

"the first substantial act of an organic physical body potentially having life"] is of this kind.[26]

In this passage, the second definition is not claimed to be nominal, but quidditative and causal. The first definition, which in the *prima lectura* is claimed to be strictly nominal, is not even mentioned here.

Clearly, in this way it is easy to avoid the inconsistency derivable from the formulations of the *prima lectura*. But then the question naturally arises as to why Buridan provided that problematic formulation in the first place, when it so obviously gives rise to the inconsistency, assuming his theory of definitions of the *Summulae*. One might offer several hypotheses here, several of which may well be true, but not particularly intriguing. For example, one may assume that Buridan was simply inadvertent, or that he did not *really* mean the formulations of the *prima lectura* as strictly as they sound, etc. However, in conclusion I will risk a different hypothesis. Not because I think I can prove or disprove it here (so, it may well be false), but because it places the question of the viability of Buridan's theory in his scientific practice in an interesting historical perspective, deserving further exploration.

4. *The Viability of the Theory in Practice*

The hypothesis I would risk here is that in the problematic formulations of the *prima lectura* Buridan simply followed an older line of interpretation, which he later abandoned upon realizing its conflict with his "official theory" of definitions. According to what I call this "older line" of interpretation, the two Aristotelian definitions of the soul are related to each other as a nominal and a quidditative definition of the same *definitum*, the latter of which can be demonstrated in terms of the former, by means of a *quia*-type, *a posteriori* demonstration. In any case, this is how Aquinas interprets Aristotle's treatment of the two definitions of the soul provided in the *De Anima*.

In his commentary on the *De Anima*, Aquinas states that when the Philosopher proves the quidditative definition of the soul, he provides us with a proof from the better known effects to the lesser-

---

[26] John Buridan, *Quaestiones in Aristotelis De anima, tertia lectura* [Sobol], p. 35: "Item quedam est diffinitio dicens quid nominis, alia pure quidditativa, alia causalis, explicans non solum quid est res sed etiam propter quid est. Et talis est magis perfecta. Et huiusmodi est ista definitio."

known cause. As such, this is a *quia*-type demonstration.[27] However, we also know that according to Aquinas the middle term in such a demonstration has to be the nominal definition of the cause.[28]

Furthermore, Aquinas makes it clear that the definition thus demonstrated is a properly quidditative definition. Nevertheless, because of the incomplete nature of the thing defined, this definition has to contain an oblique reference to something extrinsic to the nature of the thing.[29] At the same time, however, he also insists that the correct quidditative definition should signify the same essence as the term defined.[30] Therefore, the well-established quidditative definition of the thing will always be essentially predicable of its *definitum*. Indeed, in a way it can also function as a nominal definition, by specifying what it is that anyone understanding both the term and the nature of the thing has to have in mind when using the term with understanding.[31]

---

[27] Aquinas, *In Libros De anima II et III*, Book 2, lc. 2: "… deinde cum dicit *dicamus igitur* incipit demonstrare definitionem animae superius positam, modo praedicto, scilicet per effectum. Et utitur tali demonstratione. Illud quod est primum principium vivendi est viventium corporum actus et forma: sed anima est primum principium vivendi his quae vivunt: ergo est corporis viventis actus et forma. Manifestum est autem, quod haec demonstratio est ex posteriori. Ex eo enim quod anima est forma corporis viventis, est principium operum vitae, et non e converso."

[28] Cf. Thomas Aquinas, *Summa Theologiae*, 1, q. 2, a. 2, ad 2um.

[29] Cf. Thomas Aquinas, *De ente et essentia*, c. 7: "Et quia, ut dictum est, essentia est id quod per diffinitionem significatur, oportet ut eo modo habeant essentiam quo habent diffinitionem. Diffinitionem autem habent incompletam, quia non possunt diffiniri, nisi ponatur subiectum in eorum diffinitione. Et hoc ideo est, quia non habent per se esse, absolutum a subiecto, sed sicut ex forma et materia relinquitur esse substantiale, quando componuntur, ita ex accidente et subiecto relinquitur esse accidentale, quando accidens subiecto advenit. Et ideo etiam nec forma substantialis completam essentiam habet nec materia, quia etiam in diffinitione formae substantialis oportet quod ponatur illud, cuius est forma; et ita diffinitio eius est per additionem alicuius, quod est extra genus eius, sicut et diffinitio formae accidentalis. Unde et in diffinitione animae ponitur corpus a naturali, qui considerat animam solum in quantum est forma physici corporis. Sed tamen inter formas substantiales et accidentales tantum interest, quia sicut forma substantialis non habet per se esse absolutum sine eo cui advenit, ita nec illud cui advenit, scilicet materia. Et ideo ex coniunctione utriusque relinquitur illud esse, in quo res per se subsistit, et ex eis efficitur unum per se; propter quod ex coniunctione eorum relinquitur essentia quaedam. Unde forma, quamvis in se considerata non habeat completam rationem essentiae, tamen est pars essentiae completae."

[30] See again in the quote above: "essentia est id quod per diffinitionem significatur."

[31] Cf. e.g. Thomas Aquinas, *Expositio libri Boetii de Hebdomadibus*, lc. 1; Thomas Aquinas, *In Libros Posteriorum Analyticorum*, Book 2, lc. 6.

But this interpretation will not cause the trouble for Aquinas we noticed in Buridan's *prima lectura*, because of the differences between their respective "background theories." In the first place, since for Aquinas a nominal definition is not the analysis of an underlying complex concept, providing the nominal definition will not commit him to assigning a complex concept to the term defined. On the contrary, in his conception we can obtain nominal definitions of terms subordinated to simple concepts by any sorts of indications somehow specifying what is meant by the corresponding term.[32] Therefore, in the argument laid out above, Aquinas is not committed to holding that the term "soul" and its nominal definition should be subordinated to the same concept. For him, the term is always subordinated to the simple essential concept by means of which we conceive of the simple nature of the soul. The same nature is indicated with reference to its effects by the nominal definition. The nominal definition is subordinated to a complex concept by which we have a rather inadequate grasp of the essence of the soul, not knowing *what it is*, but only *what it causes*. Finally, the same nature will be adequately grasped again by means of the complex concept expressed by the quidditative definition, which allows its essential classification in the system of the categories.

As we could see, Buridan seems to follow quite faithfully the pattern of Aquinas's interpretation in his *prima lectura*. He describes the first Aristotelian definition of the soul as a purely nominal definition, defining the soul in terms of its operations that are better known to us. He describes the second definition as being a nominal *and* quidditative definition, provable by means of the first in a *quia*-type demonstration. Finally, he states that this definition is also causal, insofar as it has to indicate the material cause, namely, the subject of the soul.

However, he soon must have realized that in view of *his* theory of definitions, he could not maintain the formulations of the *prima lectura*. Instead, he seems to have come to the conclusion that the best way to accommodate the two Aristotelian definitions in his "official theory" is to treat the first as a mere description,[33] and the second

---

[32] For the broader significance of this point in late-medieval metaphysics see Klima, "Buridan's Logic and the Ontology of Modes."

[33] John Buridan, *Summulae. De demonstrationibus* [De Rijk e.a.], 2.6: "(1) A description is usually defined thus: a description is an expression indicating what the thing is (*quid est esse rei*), the quiddity or essence of the thing] in terms of its accidents or effects that are posterior to it absolutely speaking (*posteriores simpliciter*). (2) Therefore, in a description the subject is defined in terms of its attribute or attributes and the

as a causal definition.[34] This solution does not place the strict logical requirement of synonymy between definition and *definitum* upon the first definition, or the requirement of quidditative predicability upon the second definition. Therefore, with this adjustment no inconsistency of the sort derived from the *prima lectura* can arise, even assuming Buridan's "official theory."

Now if my hypothesis is correct, then its broader significance can be summarized as follows. As I have argued elsewhere,[35] the new theory of signification and connotation utilized in the nominalist ontological program was intimately tied up with a subtle reinterpretation of the theory of definitions. From the point of view of this central programmatic concern, it was necessary to interpret nominal definitions as providing conceptual analyses, and quidditative definitions as providing the characterizations of their *definita* in terms of the latter's quidditative predicates. But then, this new conception inevitably had to lead to clashes with the usual classifications of well-established Aristotelian definitions originally interpreted within the semantic paradigm of the *via antiqua*. Assuming the correctness of my assumption about Buridan's motivations for revising the formulations of the *prima lectura*, Buridan's handling of the Aristotelian definitions of the soul would illustrate precisely this phenomenon.

To be sure, the foregoing considerations assumed the authenticity of the *prima lectura*. Thus, the inconsistency presented above may also be interpreted as providing evidence against the correctness of

---

cause in terms of its effect or effects. (3) However, a description incorporates items that are prior and better known to us (*ex prioribus et notioribus quoad nos*), but which are not so absolutely speaking (*simpliciter*). (4) Therefore, by means of descriptions one sometimes proves quidditative definitions or causal definitions to apply to their *definita*, not *propter quid* but only *quia*." Indeed, this squares very well with how the first redaction of the *Expositio* (Patar, Book 2, c. 2, pp. 47–48, ll. 85–99) treats the issue of the demonstration of the quidditative/causal definition of the soul.

[34] John Buridan, *op.cit.*, 2.5: "(1) A causal definition is an expression that convertibly indicates what the thing is [*quid est esse rei*, the quiddity or essence of the thing] and the reason why (*propter quid*) it is. (2) Such a definition is provided by means of terms which in the nominative case would supposit for the cause or causes of that thing or those things for which the term defined (*definitum*) supposits. (3) These terms are placed in those definitions in an oblique case." The last section and Buridan's subsequent comment on it clearly establish that this type of definition cannot be quidditatively predicated of its *definitum* even in the less proper sense, for it has to appellate the cause of the thing defined.

[35] In the papers referred to in n. 18, and in Klima, "Ontological Alternatives."

that assumption.[36] But the issue of authenticity should be the subject of a different study.

---

[36] The single manuscript on which the edition of Patar 1991 is based, Bruges 477, is anonymous, and scholars have variously attributed it to Blasius of Parma (Cf. Federici-Vescovini, *Les "Quaestiones de Anima"*), Buridan's student Dominic of Clavasio (Pattin, *Pour l'histoire du sens agent*), an anonymous compiler of Buridan's teachings (Marshall, "Parisian Psychology"), as well as to Buridan himself (Michalski, "La physique nouvelle;" John Buridan, *Quaestiones in Aristotelis De anima* [Patar]). An extensive summary of the controversy can be found in John Buridan, *Quaestiones in Aristotelis De anima* [Patar], 67*–98*.

BURIDAN'S THEORY OF IDENTITY[*]

Olaf Pluta

The question of identity seems to be a remote philosophical issue at first sight, but as soon as one speaks about national identity, for example, it is very easy to get into a heated political debate. Issues of male and female identity are also very controversial nowadays. Recently, I even came across a book entitled *Life on the Screen: Identity in the Age of the Internet*, whose author claims that the multiple virtual identities that we use in our virtual lives lead to the dissociation of our personal identity.[1] It seems that the question of identity is a burning issue in these fast-moving times. As this volume is devoted to Buridan's natural philosophy, I will not elaborate on these contemporary topics, but it should be kept in mind that the question of identity has far-reaching implications.

Among the different principles of natural philosophy, identity is arguably the most fundamental one. In traditional logic the three fundamental laws of thought are: the law of contradiction, the law of the excluded middle (or third), and the principle of identity. In three-valued and fuzzy logic the laws of contradiction and of the excluded middle both fail, and only the principle of identity remains intact. Thus, it is the topmost axiom in any scientific theory. But it is also a necessary prerequisite of scientific practice. If a given natural entity does not remain the same during a period of observation, the foundation of a scientific theory becomes all but impossible. In fact, one of the current problems of physics is that the exact status of an atom cannot be determined because the subatomic particles are changed in the course of observation – an effect known as the Heisenberg uncertainty principle (or indeterminacy principle).[2] Unless physicists find a new method of observation that is non-intrusive and does not alter the object of inquiry, the exact status of an atom will remain inaccessible to us.

---

[*] The writing of this article was made possible through financial support from the Netherlands Organization for Scientific Research (NWO), grant 200-22-295.

[1] Turkle, *Life on the Screen*, Chapter 10: Identity Crisis.

[2] The exact position and the exact velocity of a subatomic particle cannot both be measured exactly, at the same time. Any attempt to measure precisely the velocity of a subatomic particle, such as an electron, will alter its position, so that a simultaneous measurement of its position has no validity.

In Western philosophy the question of identity is connected with the name of Heraclitus, who formulated it in two sayings which have come down to us, the first and most famous being: "One cannot step twice into the same river."[3] It was already a famous saying at the time of Plato, who refers to it in his dialogue *Cratylus*.[4]

The second famous saying in this context is: "As we step into the same rivers, other and still other waters flow upon us."[5] Plausibly, this may have once served as the justification of the first fragment: One cannot step twice into the same river. For as one steps into (what is supposed to be) the same river, new waters are flowing on.[6]

It is worth noting that Heraclitus does not deny the continuing identity of rivers, which he takes for granted. However, the question is: what exactly makes up the identity of a given river? As the substance of the river, the water, is constantly changing, is it merely a structure or pattern that remains unchanged?

A similar observation can be made about the identity of individual human beings as Plato does in the *Symposium*, where he develops a Heraclitean insight:

> When a man is called the same from childhood to old age, he is called the same despite the fact that he does not have the same hair and flesh and bones and blood and all the body, but he loses them and is always becoming new. And similarly for the soul: his dispositions and habits, opinions, desires, pleasures, pains, fears, none of these remains the same, but some are coming-to-be, others are lost.[7]

The question of identity was kept alive in Athens over a long period of time by the ship of Theseus as Plutarch reports in his *Lives*:

> The ship on which Theseus sailed with the youths and returned in safety, the thirty-oared galley, was preserved by the Athenians down to the time of Demetrius Phalereus. They took away the old timbers from time to time, and put new and sound ones in their places, so that the vessel became a standing illustration for the philosophers in the mooted question of growth, some declaring that it remained the same, others that it was not the same vessel.[8]

---

[3] *Die Fragmente der Vorsokratiker* [Diels e.a.], fr. B 91 (cf. fr. B 49a).
[4] Plato, *Cratylus*, 402A.
[5] *Die Fragmente der Vorsokratiker* [Diels e.a.], fr. B 12.
[6] Cf. Kahn, *The Art and Thought of Heraclitus*, p. 169. Cf. also the following study on the river fragments: Vlastos, "On Heraclitus."
[7] Plato, *Symposium*, 207D.
[8] Cf. Plutarch, *Lives* [Perrin], vol. I, 49 (Life of Theseus, XXIII.1).

Over the years, the Athenians probably replaced each plank in the original ship of Theseus so as to keep it in good repair. Eventually, there was not a single plank left of the original ship. If a family of salvagers had collected all the old planks which the Athenians had replaced with new ones and later had reassembled the ship of Theseus from the original parts, who could have rightfully claimed to possess the ship of Theseus – the Athenians who had preserved its original appearance or the family of salvagers who had preserved its original components?

To give a more recent example which adds another twist to the problem of identity: in the popular Star Trek television series and movies the crew members of the starship Enterprise are "beamed" to distant locations, during which process they are briefly converted into energy, sent to another location, then reassembled into their original forms. (The transporter sports a Heisenberg compensator, which allows it to determine precisely both the motion and position of particles on a subatomic level.) In one episode, the captain's beam breaks up in transmission and he seems to be lost forever. Fortunately, the crew is able to recover him from a backup of his transporter pattern. If the writers of this episode are right in assuming that a human being is just a pattern, the question still remains of how faithful a copy can such a pattern possibly get? And how much slop can a copy have without making a difference?[9] At first glance, this example may appear remote in the context of this conference; however, some discussions of identity in Buridan are quite similar.

Following the outline of the problem of identity given above, the first part of my paper will briefly deal with Buridan's discussion of identity as a principle or law of logic, which he encounters in his questions on Aristotle's *Categories*. The second part will deal with identity as a principle of natural philosophy as it is presented in his questions on Aristotle's *Physics* and *Metaphysics*. The third part will focus on the problem of personal identity, which Buridan discusses in his questions on Aristotle's *De generatione* and *De anima*.

---

[9] Cf. Keith, "Deconstructing Star Trek," p. 104. Derek Parfit has made much use of the Star Trek fantasy of teletransportation. For example, see his *Reasons and Persons*, part 3.

1. *Buridan's Questions on Aristotle's* Categories

Logically speaking, identity can be defined as a relation. In fact, it is the most basic relation stating that a thing *a* is identical to itself ($a \equiv a$). In this sense, God is said to be identical to himself and different from anything else, and – as Buridan states in his Questions on Aristotle's *Categories* – given two things *a* and *b* without taking into account all other things, both would be said to be identical to themselves and each different from the other.[10]

Concerning the ontological status of relations in general, Buridan brings forth a striking argument, based on the principle of identity, against any theory that claims that relations are real in one way or another. If the relation of identity were something real which is added to a thing *a*, then *a* would be modified and hence *a* would not be identical to itself anymore.[11]

Consequently, being identical or being different cannot be separated from the thing itself. It is our way of looking at it that generates these different relations. We can look at the same thing absolutely (*absolute*), that is without connecting it to others, and denote it by an absolute concept (*conceptus absolutus*) like "man". Or we can look at the same thing relatively (*relative*), that is by connecting it to others, and denote it by a respective or relative concept (*conceptus respectivus sive relativus*) like "father" or "son".[12]

Logically speaking, we may conclude that identity is the only valid relation if we look at a thing without connecting it to others.

2. *Buridan's Questions on Aristotle's* Physics *and* Metaphysics

Now, natural philosophy deals with motion, and motion takes place in time, whether it is locomotion or change. Thus, in the context of natural philosophy the question of identity is connected with time.

---

[10] John Buridan, *Quaestiones in Praedicamenta* [Schneider], q. 10, p. 73$^{104-108}$: "Deinde etiam conclusio principalis patet de identitate et diversitate. Deus enim existens absolutissimus est idem sibi ipsi et diversus a quolibet alio, et non per identitatem vel diversitatem sibi additam et sibi inhaerentem; et quaecumque essent duae res A et B omnibus aliis circumscriptis, utraque esset sibi eadem et diversa a reliqua."

[11] Cf. John Buridan, *op.cit.*, p. 71. See also Schönberger, *Relation als Vergleich*, pp. 390 sq.

[12] Cf. John Buridan, *op.cit.*, pp. 73-74, 118-133.

In the final redaction of his questions on Aristotle's *Physics*, Buridan devotes an entire question to the problem of identity by asking, "Is Socrates the same today that he was yesterday?".[13]

The full title of the question in the final redaction reads: "Is Socrates the same today that he was yesterday assuming that (a) today something is added to him through nourishment which was transformed into his substance, or assuming that (b) today some part is removed from him, for example if a hand of his is cut off?".[14]

The title of the question makes clear that Buridan is referring to the numerical identity of Socrates as an individual *in a given continuum of time.*

It may be noted that an article concerning identity was condemned in 1277, namely "that man can become another person, both numerically and individually, through nourishment".[15] Thus, it is possible that this position concerning identity was actually upheld by some masters of arts at that time even though the sources have not been as yet identified.

Buridan starts his question with three arguments against identity, the first and the second referring to the two assumptions mentioned in the initial outline of the question, the third presenting an example similar to those used in fuzzy logic today.

---

[13] "Utrum Socrates est hodie idem, quod ipse fuit heri?" The Latin text closely follows the text of the MS København, Kongelige Bibliotek, Cod. Ny kgl. Saml. 1801 fol., which was written in Paris between 1377 and 1380 (based on the analysis of the seven watermarks of the paper used for this manuscript). For the arguments for using this manuscript as the base text see John Buridan, *Tractatus de infinito* [Thijssen]. We deviate from the text of this manuscript only where the manuscripts MS Frankfurt am Main, Stadt- und Universitätsbibliothek, Cod. Praed. 52, written in Vienna or Prague in 1368, and MS Città del Vaticano, Biblioteca Apostolica Vaticana, Cod. Vat. lat. 2163, written in Padua in 1377, both offer a different reading which improves the text. All spellings have been classicized.

It may be worthwhile to note that this question only slightly differs in content in the third redaction of John Buridan's Questions on the *Physics*. For the manuscripts of this redaction, see Michael, *Johannes Buridan*, vol. 2, pp. 574–577. Here the question reads "Utrum idem, quod heri erat Socrates, est nunc Socrates, et ponamus, quod per nutritionem aliquae partes de novo adveniunt Socrati, vel etiam ponamus, quod Socrates heri habuit digitum et abscisus est ei", and is numbered as question 8 of the first book, whereas in the final redaction it is numbered as question 10. Cf. MS Città del Vaticano, Biblioteca Apostolica Vaticana, Cod. Chigi E VI 199, fol. 6^{ra-vb}.

[14] Quaeritur decimo, utrum Socrates est hodie idem, quod ipse fuit heri, posito, quod hodie additum est sibi aliquid ex nutrimento et conversum in eius substantiam, vel posito, quod hodie est aliqua pars ab eo remota ut si amputata est sibi manus.

[15] "Quod homo per nutritionem potest fieri alius numeraliter et individualiter" (cf. Hissette, *Enquête*, p. 187).

First of all, it is argued that Socrates is not the same, because otherwise it would follow that the whole is identical to one of its parts. Now, if that which today has been newly added to Socrates is called *b*, and the rest is called *a*, then it is apparent that Socrates was that *a* yesterday, and if he is the same today, he still is that *a*, but nevertheless *a* now is a part of his different from *b*.

Second, if the hand which is removed today is called *b*, and the rest is called *a*, then it is apparent that Socrates was *a* and *b* yesterday, because the whole is identical to its parts, and he is not *a* and *b* today, because *b* has been cut off, consequently he is not the same that he was yesterday.[16]

Third, it would follow that a thing that has been completely destroyed remains the same as before. This is impossible, since it is said in the second book of Aristotle's *De generatione*, that something that has been completely destroyed cannot return as numerically the same. Now, let us assume that a vessel is filled with wine and that the wine is comprised of one thousand drops, then it would follow that if these one thousand drops had been destroyed, the wine would be completely destroyed, even if the wine remained the same. Buridan proves this conclusion by constructing a concatenation of events that finally leads to a complete replacement of the original wine. If in every hour one drop drains off at the bottom of the vessel and at the same time one drop of wine is added through the opening at the top of the vessel, then it is apparent that after the removal of the first drop and the addition of another one, it would be still the same wine as before – in the same manner as Socrates remains the same despite the fact that something has been added to him through nourishment and something else has been consumed by bodily heat. The same is true for every other event where one drop drains off at the bottom and another one is added at the top. In the end all the initial one thousand drops would be drained off within one thousand hours and the whole wine would be corrupted, while at the same time remaining the same.[17]

---

[16] A parallel argument is discussed in Zupko, "How Are Souls Related to Bodies?" (the discussion is on pp. 587–590).

[17] (1) Arguitur primo, quod non sit idem, quia sequeretur, quod totum esset idem cum sua parte et sic totum esset sua pars, cuius oppositum dictum est (fol. 17vb) in alia quaestione (cf. I.9: Utrum totum est suae partes). Consequentia probatur ponendo, quod illud, quod hodie additum est Socrati, vocetur *a*, et totum residuum vocetur *b*, (tunc) constat, quod Socrates heri erat illud *b*, et si hodie est idem, ipse adhuc est illud *b*, sed tamen *b* est pars eius distincta contra *a*.

(2) Item, si manus, quae hodie amputatur, vocetur *b* et residuum *a*, tunc Socrates

But Buridan presents twice as many arguments in favor of identity. They not only outnumber the arguments against identity, but they also refer to philosophical authorities that can hardly be neglected. It may be emphasized that Buridan directly refers to Heraclitus here, who is not very often quoted in medieval philosophical texts.

First of all, it is argued that Socrates is the same, because otherwise the saying of Heraclitus would make no sense, namely that it is not possible for the same man to step into the same river twice, because both Socrates and the river are constantly modified through nourishment.

Second, it would follow that the name "Socrates", which signifies an individual, is not a singular name anymore, because it would stand for several and diverse things earlier and later in time.

Third, that which is augmented remains the same according to Aristotle in the first book of his *De generatione*; now, Socrates is augmented through nourishment; hence, he must remain the same.

Fourth, it would follow that I have never before seen the person whom I see now, but that I have seen somebody else; consequently, any legal action would come to an end. That is to say, you would not be the same person who struck me yesterday, and I could therefore ask for no compensation from you; nor would you be the person who protected me yesterday from my enemies and to whom I should now be obliged to give a reward.

Fifth, by the same token it would follow that you, who is here now, have never been baptized, but somebody else; consequently, you would not be a Christian.

---

heri erat *a* et *b*, cum totum sit suae partes, et ipse hodie non est *a* et *b*, cum *b* sit ablatum, igitur non est idem, quod heri.

(3) Item, sequeretur, quod illud, quod totum esset corruptum, maneret adhuc idem, quod ante, quod est impossibile, cum dictum sit secundo *De generatione* (cf. Aristotle, *De generatione*, II.11, 338b16–17), quod corruptum non potest reverti idem in numero. Consequentia probatur ponendo casum, quod hoc dolium sit plenum vino et illud vinum ponatur continere centum vel mille guttas, tunc si illae mille guttae fuerunt corruptae, totum hoc vinum erit corruptum, et tamen remanebit hoc idem vinum. Probatio ponendo casum, quod qualibet hora una istarum guttarum defluat ad fundum et corrumpatur, et per os supra una gutta ad replendum apponatur, tunc constat, quod post remotionem primae guttae et appositionem alterius adhuc erit idem vinum, quod ante, pari ratione sicut Socrates est idem, licet sit aliquid appositum ante ex nutrimento et aliquid deperditum a calore consumente. Et pari ratione, si iterum auferatur una gutta et apponatur alia, adhuc erit idem vinum, et sic semper, tunc igitur per mille horas omnes illae mille guttae erunt corruptae, et sic illud totum vinum erit corruptum, tamen adhuc remanebit idem vinum.

Finally, it would follow that during one day many "Socrateses" are destroyed and many others are generated, because in any given hour Socrates would be different from the person he was in the preceding hour.[18]

At the beginning of the body of the question, Buridan states that he is not speaking of specific identity but of numerical identity, and that introducing a threefold distinction can easily solve the question of identity.[19]

Now, there are three common ways to say that something is numerically the same. First of all, something is numerically the same in its entirety (*totaliter*), if its integrity is completely preserved. In this sense, I am not the same that I was yesterday, because something was part of my integrity yesterday, which has already been used up, and something was not part of my integrity yesterday, which has meanwhile been added through nourishment. In this sense, Seneca says in one of his letters to Lucilius: no one is the same in youth and old age, not even the same yesterday and today, because our bodies are swept away as are the rivers. And in this sense, Heraclitus is also perfectly right in saying that we are continually changed in such a way that we cannot step twice into the same river as entirely the same each time. If one understands "numerically the same" in this way, the two arguments at the beginning of the question to the effect

---

[18] (1) Oppositum arguitur, quia reverteretur opinio Heracliti (cf. Heraclitus, fr. B 91, fr. B 49a), scilicet quod non contingeret eundem hominem intrare bis eundem fluvium, quia continue mutaretur per continuam nutritionem et fieret alius quam ante (cf. Heraclitus, fr. B 12).

(2) Item, sequeretur, quod hoc nomen "Socrates" non esset nomen discretum, quia supponeret pro pluribus et diversis, licet prius et posterius sicut hoc nomen "tempus".

(3) Item, quod augetur, manet idem, ut habetur primo (cf. Aristotle, *De generatione*, I.5, 321b11–15) et tamen augetur per appositionem aliquarum partium ex nutrimento.

(4) Item, sequeretur, quod ego numquam alias vidissem te, quem ego nunc video, sed vidissem unum alium et periret actio iniuriarum et retributio bonorum. Tu enim non es ille, qui heri me percussit vel qui heri me defendit ab inimicis, quare igitur peterem emendam (fol. 18ra) a te vel quare deberem tibi retribuere.

(5) Item, sequeretur, quod tu, qui es hic, non fuisti baptizatus, sed unus alius; igitur tu non es Christianus.

(6) Item, sequeretur, quod in eodem die corrumperentur multi Socrates et generarentur multi alii, quia in hac hora ille Socrates est et in hora praecedenti non erat, sed unus alius, qui modo non est, igitur ille hodie est genitus et ille idem est corruptus, cum generatio sit mutatio de non esse ad esse et corruptio e converso.

[19] Non quaerimus de identitate secundum speciem vel secundum genus, sed de identitate numerali secundum quam hoc esse idem illi significatur hoc esse illud. Et tunc illa quaestio faciliter solvitur per distinctionem.

that Socrates is not the same today that he was yesterday, are indeed conclusive.[20]

Second, something is numerically the same partially (*partialiter*), if a part and especially if the major or principal part is preserved.[21] Thus, Aristotle says in the ninth book of his *Ethics* that a human being is above all its intellect as in the state it is the souvereign that is held in the fullest sense to be the state. Accordingly, a human being remains the same throughout life, because the soul, its principal part, remains totally the same. And in this sense, it is true that you are the same who has been baptized nearly forty years ago, especially so as this act primarily befits the soul and not the body.[22] And in this sense, it is also possible for me to pursue you for some unjust treatment of my person or to feel beholden to reward you, because also the unjust or meritorious actions primarily stem from the soul and not from the body.[23]

---

[20] (1) Tripliciter enim consuevimus dicere aliquid alicui esse idem in numero. Primo modo totaliter, scilicet quod hoc est illud et nihil est de integritate huius, quod non sit de integritate illius, et e converso; et hoc est propriissime esse idem in numero. Et secundum illum modum dicendum est, quod ego non sum idem, quod ego eram heri, nam aliquid heri erat de integritate mea, quod iam resolutum est, et aliquid etiam heri non erat de integritate mea, quod post per nutritionem factum est de substantia mea. Et sic dicebat Seneca in epistula ad Lucilium, quae incipit: "Quanta verborum" (cf. Seneca, *Ad Lucilium* [Reynolds], 58 (VI.6), 22: "Nemo nostrum idem est in senectute qui fuit iuvenis; nemo nostrum est idem mane qui fuit pridie. Corpora nostra rapiuntur fluminum more."): nemo idem in iuventute et senectute, immo nec heri et hodie, corpora enim nostra rapiuntur fluviorum more. Et ad illum sensum locutus est bene Heraclitus, quod sic continue mutamur, quod non contingit hominem totaliter eundem bis intrare fluvium etiam totaliter eundem. Et ad hunc modum capiendi "idem in numero" procedunt bene rationes, quae fiebant in principio quaestionis ad probandum, quod Socrates non sit idem hodie, quod fuit heri.

[21] One may ask what happens when a human being loses a principal part of the body. To give a well-known example from the Middle Ages: when Abelard lost his genitals did he still belong to the masculine gender afterwards? And was he still the same person after he had become the victim of this violent amputation? In a letter to Abelard, his former teacher Roscelin mockingly expresses doubts as to his identity. He appears, says Roscelin, to be neither a cleric nor a lay person, and would he not even be lying if he dared to call himself by the Latin name of "Petrus"? For, as the letter concludes: "Having been deprived of that part which constitutes you a man, you are not to be called 'Peter', but rather 'incomplete Peter'." (cf. Henry, *Medieval Mereology*, p. 115).

[22] Looking at pictures of a person as an adult and as a child, one may doubt that it is indeed the same person, but this refers to the problem of identification, that is, of recognizing the identity of a given person through its changes in time.

[23] (2) Sed secundo modo aliquid dicitur alicui idem partialiter, scilicet quia hoc est pars illius, et maxime hoc dicitur, si sit maior pars vel principalior vel etiam, quia hoc et illud participant in aliquo, quod est pars maior vel principalior utriusque.

Third, something is numerically the same according to the continuous succession of diverse parts (*secundum continuationem partium diversarum in succedendo alteram alteri*). Thus, the same river is called "Seine" for nearly one thousand years, in spite of the fact that, strictly speaking, just now nothing whatsoever is part of the Seine that was part of it ten years ago. In this sense, both the sea and the sublunar sphere are called eternal. In this sense as well, the human body can be called the same throughout life. Thus, while strictly speaking it is not true that the river Seine, which I see now, is the same, which I saw ten years ago, this way of talking is conceded because the water which we see now is called "Seine", and the water which I saw ten years ago was also called "Seine", and all the water which made up this river in between was at every single moment called "Seine".[24] And it is because of this continuous succession that we can say that the name "Seine" is a discrete and singular name, although it is not discrete in the same strict sense as it would be if the river Seine remained totally the same all the time.[25]

Sic enim dicit Aristoteles nono *Ethicorum* (cf. Aristotle, *Nicomachean Ethics*, IX.8, 1168b31–33, 35), quod homo maxime est intellectus, sicut civitas et omnis congregatio maxime est principalissimum, prout allegatum est in quaestione praecedenti (cf. I.9: Utrum totum est suae partes), et exinde etiam proveniunt denominationes totorum a denominationibus partium. Et ita manet homo idem per totam vitam, quia manet anima totaliter eadem, quae est pars principalior. (fol. 18rb) Sic autem non manet equus idem immo nec corpus humanum. Et sic bene est verum, quod tu es ille idem, qui a quadraginta annis citra fuisti baptizatus, maxime cum hoc nobis conveniat principaliter ratione animae et non corporis. Et possum te prosequi super iniuriis vel teneor ad remunerandum tibi, quia etiam opera iniuriosa vel meritoria sunt principaliter ab anima, non a corpore. Et sic etiam non dicimus te generari hodie, quia non dicimus aliquid generari simpliciter nisi generetur secundum se totum vel secundum eius partem maiorem vel principaliorem.

[24] One may add that even if the name of the river Seine should have changed in the course of centuries, it still would be the very same river.

[25] (3) Sed adhuc tertio modo et minus proprie dicitur aliquid alicui idem numero secundum continuationem partium diversarum in succedendo alteram alteri, et sic Secana dicitur idem fluvius a mille annis citra, licet proprie loquendo nihil modo sit pars Secanae, quod a decem annis citra fuit pars Secanae. Sic enim mare dicitur perpetuum, et ille mundus inferior perpetuus, et equus idem per totam vitam, et similiter corpus humanum idem. Et iste modus identitatis sufficit ad hoc, quod nomen significativum dicatur discretum vel singulare secundum communem et consuetum modum loquendi, qui non est verus proprie. Non enim est verum proprie, quod Secana, quem ego video, est ille, quem ego vidi a decem annis citra. Sed propositio conceditur ad illum sensum, quod aqua, quam videmus, quae vocatur Secana, et aqua, quam tunc vidi, quae etiam vocabatur Secana, et aquae etiam, quae intermediis temporibus fuerunt, vocabantur quaelibet in tempore suo Secana et continuate fuerunt ad invicem in succedendo. Et ex identitate etiam dicta secundum huiusmodi continuationem dicimus hoc nomen "Secana" esse nomen discretum et singulare,

Buridan concludes the question with the remark that it is now evident what has to be said in response to the three arguments against identity at the beginning of the question, and he leaves this exercise to the reader.[26]

In his questions on the *Metaphysics*, there exists only a short passage that makes reference to the problem of identity. Nothing new is added here beyond what Buridan has already said in his questions on the *Physics*.[27]

Buridan's solution is pretty clear and straightforward. Compared with solutions of the thirteenth century, particularly in Thomas Aquinas[28] and Duns Scotus,[29] but also with solutions of the fourteenth century, particularly in William of Ockham,[30] Buridan's treatment is much more elaborate and presents a fresh view of the problem. But Buridan did not stop here. The last part of my paper will present some considerations which show that Buridan aimed at understanding precisely what constitutes identity in the first place.

3. *Buridan's Questions on Aristotle's* De generatione *and* De anima

In the preceding part of my paper, I dealt with the problem of the numerical identity of Socrates as an individual in a given continuum of time. A much more daring question would be to ask whether

---

quamvis non ita proprie sit discretum sicut esset, si maneret idem totaliter ante et post.

[26] Et per haec dicta apparet manifeste, quomodo sit dicendum ad rationes omnes, quae fiebant, et quomodo procedunt viis suis.

[27] John Buridan, *In Metaphysicen* (Paris, 1518), Book VII, q. 12, fol. 48va: "Similiter de identitate hominis a principio nativitatis usque ad articulum et instans mortis dico, quod non est totalis identitas, sed bene est identitas ratione principalissimae partis, scilicet animae intellectivae, et etiam in corpore est identitas, sicut fluvium Secanam diceremus esse eundem numero per multos annos, scilicet per successivam continuationem diversarum partium recedentium et advenientium. Et de hoc alibi dictum est.

Similiter de anima beati Petri: illam vocamus Petrum non quia sit idem totaliter, sicut cum Christo ambulabat, sed quia est pars principalissima incomparabiliter."

[28] Thomas Aquinas, *Summa Theologiae*, tertia pars, supplementum, q. 79, a. 1: Utrum idem corpus numero anima resumpta sit in resurrectione; a. 2: Utrum sit futurus idem numero homo qui resurget (pp. 178^a–180^a).

[29] John Duns Scotus, *Quaestiones super universalia Porphyrii*, q. 24: An haec sit vera, Socrates senex differt a seipso puero (pp. 298^b–300^a).

[30] William Ockham, *Quaestiones in librum quartum Sententiarum (Reportatio)* [Wood e.a.], q. 13: Utrum idem homo numero resurget qui prius vixit (pp. 257–277).

Socrates should be considered the same if he had been *discontinued in time*, that is if Socrates were annihilated and later on recreated. Is identity possible at all under such circumstances?

With respect to Buridan's example of the river Seine, which refers to the weak identity of the third mode, we could argue that there are rivers that dry out over the summer. So, as a matter of fact, for a certain period of time there is no river; only the traceable bed of the river, which the water once filled, remains. Later in the year, however, when there is sufficient water, the river starts to flow again.[31]

Buridan discusses this problem in one of his questions on Aristotle's *De generatione*, namely "Is it possible that something which has been completely destroyed returns as numerically the same?".[32] He refers to the possible total annihilation of the whole world and its subsequent restoration, hereby giving an example for the weak identity of the third mode.[33]

Buridan argues as follows: If everything were annihilated by God now, the resulting situation would be exactly the same as it was before the creation of the world. As neither God himself nor anything else would be different, nothing prevents us from assuming that – with respect to God's supernatural power – he could make all things he had made before, and not only similar things but the same things, because he has the power to do it now as he had the power to do it then, and the situation would be exactly the same as it had been before. In this way, the world would start to exist again like a river that had been dried out for a period of time.

Now, what about the partial identity of the second mode? Is it possible that Socrates remains the same given this understanding of identity, that is, if he is partially discontinued in time?

An example of this type of identity is discussed by Buridan in his questions on *De anima*. With respect to God's supernatural power, Buridan takes it for granted that it is possible to separate form and matter.[34] God is not only capable of separating an immaterial

---

[31] With respect to the example of the ship of Theseus, one could also argue that the Athenians completely disassembled the ship for restoration purposes, thereby only preserving the component parts. Therefore, there was no ship of Theseus for a certain period of time. After a special treatment of all the planks, which had formerly made up the ship, the Athenians reassembled it from the original parts.

[32] "Utrum illud, quod simpliciter est corruptum, possit reverti idem in numero." For the following cf. Braakhuis, "John Buridan," especially pp. 128–131 and the corresponding passage of Buridan's question on pp. 137–138.

[33] Surprisingly, this argument has not been analyzed by Braakhuis.

[34] Cf. John Buridan, *Quaestiones super octo Physicorum libros Aristotelis* (Paris, 1509),

form from its matter (that is a form not educed from the potency of matter), but is also capable of separating a material form, and conserve this form and put it into another matter. So, why, Buridan asks, should this not be possible for the human intellect?[35] Note that the problem which Buridan raises here only makes sense against the background of the theory, which he holds on natural grounds, that the human soul is a material form – a form educed from the potency of matter, and hence mortal.[36]

In this way, God can separate the form of a horse or a stone from its matter, and can conserve it separately. Thus, the stone or horse would continue to be, but it would not be a horse or stone anymore. Likewise, God can separate the form of a human being from its matter, and conserve it separately. But would this still be the same human being?

Buridan gives three possible answers to this question:[37]

Some concede that a human being, if its form is separated from its matter and conserved separately, is everlasting hereafter in such a way that it will always be. Nevertheless, it will not be a human being anymore, because the name "homo" connotes that body and soul are united in such a way that the soul inheres in the body.

Others, however, say that the substantial form is the principal part of a composite being. It is primarily by reason of the form that the name "homo" or "animal" is imposed on a composite being.

Book I, q. 9; Book I, q. 20.

[35] John Buridan, *Quaestiones in Aristotelis De anima, tertia lectura* [Zupko]: "Certum est, quod supernaturaliter Deus potest non solum formam (Zupko: formare) non eductam de potentia materiae, immo etiam eductam separare a sua materia et separatim conservare et ponere in aliam materiam. Quare igitur hoc non esset possibile de intellectu humano?" (Book III, q. 4, p. 37$^{203-207}$). Cf. *ibid.*, q. 6, p. 52$^{139-140}$.

[36] Cf. Maier, "Das Prinzip der doppelten Wahrheit," Pluta, "*Homo sequens rationem naturalem*"; Pluta, "Ewigkeit der Welt"; Pluta, "Einige Bemerkungen"; Pluta, "Der Alexandrismus." A similar position is upheld by Marsilius of Inghen who also holds that naturally speaking (*pure naturaliter loquendo*) the human soul is a material form, and hence mortal (cf. Pluta, "Die Diskussion der Unsterblichkeitsfrage."). Naturally speaking, one is not allowed to assume that there is an afterlife (*pure naturaliter loquendo non est ponenda vitam post hanc vitam*) (cf. Pluta, "Die Frage nach der *felicitas humana*."). A different interpretation is provided by Jack Zupko in his doctoral dissertation John Buridan, *Quaestiones in Aristotelis De anima, tertia lectura* [Zupko]. Cf. also his "How Are Souls Related to Bodies?", where Zupko claims in the very beginning that "In the case of human beings, he [i.e. Buridan] defends a version of immanent dualism: the thesis that the soul is an immaterial, everlasting, and created (as opposed to naturally generated) entity" (p. 575).

[37] Cf. John Buridan, *Quaestiones in Aristotelis De anima, tertia lectura* [Zupko], Book III, q. 6, pp. 52^{127}–53^{169}.

Hence, for the time during which that form is in that matter, the name is used to signify the composite of that matter and form, and for the time thereafter when that form is not in any matter, the name is used for the form alone, for example when we say "Saint Peter, pray for us!" even though Saint Peter is no longer composed of matter and form.

Still others say that while it is correct to state that the name "homo" is first and foremost used to signify a composite being, it is nevertheless applied to signify the form alone, on account of the form's pre-eminence over matter. Consequently, insofar as the name "homo" signifies a composite, a given human being will always be, but it will not be a human being due to the connotation. But insofar as the name "homo" signifies the form, a human being will always be, and it will always be a human being.

Buridan gives no final answer here, but simply states that the determination of this doubt pertains to the faculty of theology where questions like the following are raised: whether Christ was a human being during the three days (*Utrum Christus in triduo erat homo*), that is when his body was in the sepulchre without the soul, and his soul was among the dead, without the body.[38] Evidently, our author does not want to mix up natural philosophy with theology, because – as he states somewhere else – mixing theology with natural philosophy results in a very difficult argument (*miscendo theologiam cum naturali philosophia fit argumentum valde difficile*).[39]

In his questions on the *Physics*, however, he seems to opt for the second solution. In this case, the soul, the primary part of the composite being which makes up Socrates, guarantees identity while the matter from which it has been educed in the first place has been entirely replaced.

The most difficult question still remains, namely, of whether the strong or total identity of the first mode is thinkable if Socrates is discontinued in time? Such would be the case if Socrates were completely annihilated and later on recreated as exactly the same – and without a soul existing in the time in between which could serve as the vehicle of identity.

Let us assume that God not only has a blueprint of any individual human being before he creates it, but that he also has such a blueprint of every human being for every single instant in time. We may call these blueprints backups in the same sense as the trans-

---

[38] Cf. Thomas Aquinas, *Summa Theologiae*, p. III, q. 1, a. 4.
[39] Cf. Braakhuis, "John Buridan," p. 137.

porter in the Star Trek series makes a backup before a person is beamed to a distant location. Would such a backup of Socrates allow God to recreate him as exactly the same if he had been completely annihilated? If death means the complete annihilation of body and soul, would God still be able to resurrect Socrates as exactly the same on the day of the Last Judgment?

In the question mentioned above concerning whether something that has been completely destroyed can return as numerically the same, Buridan does not refer to the resurrection of the dead. However, he presents some general considerations that may give us a clue about how we could solve this problem.

As a kind of afterthought, he discusses the question of what is the origin of numerical identity and diversity (*unde proveniat originaliter identitas aut diversitas numeralis*).

Now, for the strong identity of the third mode, it generally holds that an interruption in being prevents identity (*interruptio essendi prohibet identitatem*). Previously, Buridan had discussed the example of water and heat, to which he now returns.

If water is heated by a fire, God can take away the heat and preserve it elsewhere. If the water is heated again by the fire and reaches the same temperature, the two heats are not the same in spite of the fact that they both have the same temperature, because one has a subject and the other not, and they occupy different locations in space. Even if the first heat would be destroyed and not preserved, the second heat would not be identical with the first one because it would have been generated at a different time.

Now, let us assume that God would destroy the first heat in an instant and would create a second heat of the same temperature at the same instant in such a way that there is no interruption. In this case, Buridan states, the heat must be called the same. At least, one may argue, it would be impossible for us to decide if it is the same heat as before or if God has created a second heat of the same temperature. One may further argue, that if God would completely annihilate and recreate the whole world in every single instance in a continuous fashion we would not be able to recognize it. As in a motion picture, the different worlds that God creates and annihilates would be seen by us as one continuous stream of pictures and would constitute a unified experience.

This brings us to the final thoughts of Buridan concerning the origin of numerical identity and diversity. Ultimately, Buridan states, it is God alone who is the source of identity and diversity (*nisi quod finaliter possemus recurrere ad primum principium omnium diversitatum et identitatum, quod est ipse Deus*). God knows all diverse things, past, present, and future; and it is impossible for a thing that God thinks to be diverse to become identical to any other thing. Likewise, God knows all identical things, past, present, and future; and it is impossible for a thing that God thinks to be identical to become diverse.

Thus, we may conclude, that even if Socrates had been *discontinued in time*, that is if Socrates were annihilated and later on recreated, he would be the same because in God's thinking he is identical.

Buridan does not further elaborate on this question, and he does not, in particular, elucidate on how God's thinking can span the gap in time between Socrates' annihilation and recreation, an interruption in being which, according to Buridan's own words, prevents numerical identity in the first place.[40] Nonetheless, it is remarkable how deeply Buridan speculates about the ultimate origin of identity in his writings on natural philosophy.

---

[40] It may be noted that some of Buridan's successors have completely denied the possibility that something can return as numerically the same; Marsilius of Inghen, for example, explicitly proves that this is an impossibility even at a supernatural level (cf. Braakhuis, "John Buridan," p. 127).

NECESSITIES IN BURIDAN'S NATURAL PHILOSOPHY

Simo Knuuttila

In his works on natural philosophy, John Buridan drew a systematic distinction between natural and supernatural or simple modalities which he believed to be of Aristotelian provenance. I shall analyse why Buridan thought in this way and how this distinction is related to Buridan's views of causality, the intelligibility of nature, and the tasks of natural philosophy.

1. *Indirect Proof in Natural Philosophy*

At the beginning of the seventh book of his *Physics*, Aristotle states that everything which moves is moved by something. This is said to be clear when a thing is moved by something else and also true of those things which are called self-movers and have no external mover. In anything moved, one can always differentiate between an active factor and a passive factor (241b34–242a49). Aristotle then puts forward a longer *reductio* argument purporting to prove that any sequence of causally dependent movers must terminate and that, consequently, there must be a first moved mover in any given sequence of movers and a first mover which is not a member of the finite dependent sequence of moved movers. The reductive premise is that if there is a finite movement, say A, during a finite time and an actually infinite hierarchy of simultaneous finite movers related to A, an infinite movement is performed in a finite time. This is impossible (242a49–242b53). Someone could object that assuming an infinite number of movements is not the same as assuming an infinite movement. Aristotle answers that a proximate mover is either in touch or continuous with what it moves; since the movers constitute a unity, the motion they execute is unitary, and since the movers are infinite, the movement is infinite (242b53–243a31).

Aristotle did not explain the details of the argument, which was considered problematic by ancient commentators and which has also caused difficulties for contemporary scholars.[1] In his questions

---

[1] For these discussions, see Wardy, *The Chain of Change*.

on Aristotle's *Physics* John Buridan takes several pages to discuss the argument.[2] In question three of the seventh book he deals with the modal part of the reduction. Aristotle's point was that since the movers and the things which are moved are continuous or contiguous, this disjunction must be embedded in the hypothesis of an infinite chain of causally dependent movers. Treating the hypothesis as a possibility yields an impossibility, and this shows that the hypothesis cannot be possible since nothing impossible results from postulating that a possibility is actual (242b72–243a31). Buridan states that someone might criticize Aristotle's argument as a reduction which is based on a *positio impossibilis*, i.e., that there could be a continuous infinite body formed of different kinds of bodies. If Aristotle's intention was to show that the opponent's position is false, i.e. that there is no need for a first mover and that the causal chains can be infinite, he should demonstrate that this view leads to a contradiction. But Aristotle introduces an impossible premise of his own. The whole argument then breaks down because from an impossibility anything follows. Before offering his own answer, Buridan refers to the views of Averroes and some others who tried to avoid problems of this kind by maintaining that Aristotle is speaking about the common principles pertaining to the movers and not about their applications. In Buridan's view this was a confused interpretive idea, implying that Aristotle's assumption was meant to be possible at a level which was irrelevant to the matter in hand (fol. 105^{ra-rb}).

Buridan's own answer on behalf of Aristotle runs as follows. When it is said that from an impossibility anything follows, one should remember that this holds true of the impossibilities which are impossible without qualification, but not of the impossibilities which, being possibilities in themselves, are merely natural impossibilities. Aristotle tried to demonstrate that the opponent's position was impossible, because it implied something obviously impossible. Aristotle's additional premise was that things which are moved can form a whole with their movers. This is naturally impossible, but not impossible as such. When this possibility is assumed to be actual together with the assumption that a chain of movers is infinite, an impossibility follows. Buridan says that because the additional premise is not simply impossible, it is not responsible for the impossible conclusion. It must be based on the position of the opponent.[3]

---

[2] John Buridan, *Quaestiones super octo Physicorum libros Aristotelis* (Paris, 1509), fols. 103va–105rb. (I have corrected the printing mistakes in the pagination of the edition.)

[3] John Buridan, *op.cit.*, fol. 105rb: "Videtur ergo michi dicendum quod ratio Aris-

This is how Buridan also reconstructs some other indirect proofs of Aristotle: that a proposition is impossible is shown by combining it with some other propositions which are said to be possible and this combination is claimed to imply a contradiction.[4]

The theoretical tools of Buridan's interpretation are the notion of the *positio impossibilis*, the principle that from an impossibility anything follows, and the distinction between unqualified simple possibilities and natural possibilities or, as Buridan more often says, between possibilities through the divine power and possibilities through the natural powers. None of these conceptual tools is mentioned by Aristotle in this context, but they were involved in a widely-known paradigm of argumentation which was developed in the medieval logic of obligations and in its theological applications in the thirteenth century. Let us have a look at this tradition, which seems to have influenced Buridan's terminology.[5]

Henry of Ghent argued that the Holy Spirit would not differ from the Son even if it did not proceed from the Son.[6] Aquinas's approach rendered this impossible. In fact, he considered this impossibility the main argument against the intelligibility of the Eastern position in the *filioque* controversy (*S. th.* I, q. 36, a. 2). It is understandable that Henry's position aroused interest – it was commented on by almost all Western medieval authors who wrote about the Trinity. Henry treated the question by using the terminology of *ars obligatoria* and, as Henry's arguments were often quoted, they also helped make the application of obligations logic to Trinitarian theology not uncommon in later medieval times.[7]

Henry asks us to think counterfactually that human beings are not able to laugh and that they differ from the beasts. He then asks whether this is acceptable as a *positio impossibilis* or whether it should be denied as a *positio incompossibilis*. Henry says that something is posited *per incompossibile* when accepting it would be accepting a

---

totelis pro tanto valet quia licet suppositio non sit possibilis per potentias naturales simpliciter est possibilis per potentiam divinam et tamen ex simpliciter possibili numquam debet sequi impossibile et tamen ex hec suppositione cum positione adversarii sequebatur impossibile. Ideo sequitur quod positio adversarii erat impossibilis."

[4] See, e.g., John Buridan, *Quaestiones super libris quattuor De caelo et mundo* [Moody], Book I, q. 15, pp. 69.35–70.27. Buridan does not explain why the additional possibility should be regarded as compossible with the position of the opponent.

[5] For the history of obligations logic, see Yrjönsuuri, *Obligationes*.

[6] Henry of Ghent, *Summae quaestionum ordinariarum* (Paris, 1520), a. 54, q. 9.

[7] For the historical details, see Knuuttila, "*Positio impossibilis*," Knuuttila e.a., "Innertrinitarische Theologie," Friedman, *In principio*.

contradiction, and something is posited merely *per impossibile* when it is impossible in a weaker sense of impossibility, for example by being a doctrinal impossibility. The distinction was important in the logic of obligations, which was meant to provide one with rules for maintaining consistency in a dialectical disputation. It is then stated that the Spirit does not belong to the concept of the Son as such, and therefore the assumption of the Holy Spirit's not proceeding from it is not contradictory as such. Accepting this proposition does not imply that one should deny that the Son differs from the Holy Spirit; they would still differ because of their different modes of proceeding from the Father. The conditional: "if the Holy Spirit does not proceed from the Son, the Holy Spirit does not differ from the Son" is not necessary in the sense that denying it would imply a contradiction. There were several Franciscan authors to repeat Henry's argument. Its early criticism was formulated by Godfrey of Fontaines, who stated that the position in Henry's argument is in fact incompossible and that it is futile to develop any argument which is based on a contradictory premise.[8]

In discussing the rules of a *positio impossibilis* disputation, William Ockham stresses that the starting point must not be logically impossible and that only those propositions must be granted as relevant with respect to the starting point which follow from it and/or the correctly granted propositions by natural and simple consequences which are evident and *per se notae*. The aim of employing this procedure in theology is to study the consequences of the denials of various doctrines and to analyse in this indirect manner the conceptual presuppositions embedded in them.[9] Even though Buridan made use of the obligational *positio impossibilis* terminology as it was applied in theological discussions, he did not apply this branch of logic more systematically in his interpretation of Aristotle's indirect proofs. This is understandable, for the obligational rules were not formulated from the point of view of proofs.[10] Buridan's main point was that a premise which is naturally impossible but simply possible is not as such the reason why a contradiction follows in a *reductio* argument.

---

[8] Godfrey of Fontaines, *Les Quodlibet cinq, six et sept* [De Wulf e.a.], pp. 294–295. See also Friedman, *In principio*, pp. 164–200.

[9] William Ockham, *Summa logicae* [Boehner e.a.], pp. 739–741.

[10] For the historical background of the early formulations of the *positio impossibilis* rules of obligations logic, see Martin, "Obligations and Liars," pp. 358–363.

2. *Natural and Supernatural Possibilities in Aristotle*

When Buridan refers to different kinds of impossibilities in question three of the seventh book of the *Physics*, he mentions that somebody might claim that Aristotle did not separate natural and supernatural modalities and that all impossibilities were of the same type in his view. Such an opponent could continue that even if Buridan's version of the indirect proof were correct, it was not what Aristotle had in mind.[11] Buridan argues that this claim is not true at all:

> I answer that Aristotle to a great extent agreed with our true faith and firmly believed that many things are impossible with respect to natural potencies and simply possible through supernatural potency (fol. 105rb).

Buridan substantiates his thesis by giving a list of several places in the *Physics* from which one can allegedly see that Aristotle consciously operated with the distinction between the notion of possibility which means that something can take place in the world through natural potencies and the notion of possibility which means that something is not contradictory and can take place in the world through a supernatural potency albeit not through a natural potency. He mentions that in the first book of the *Physics*, Aristotle states that in dealing with nature all authors have denied that anything can come into being out of nothing, "… as if he would like to say that it is possible supernaturally." Buridan continues with the following list of natural impossibilities in Aristotle: the heavens are made to stand still; the spheres are divided or made continuous; the velocity of the spheres is changing; a corruptible thing is subtler than fire; extensional things penetrate each other; there is an unmoved space outside the corporeal world or overlapping it. According to Buridan, these are examples of natural impossibilities which Aristotle himself maintained were simply possible, such that they could become actual through a supernatural power.[12] I think that Buridan was wrong and that there was

---

[11] Peter Aureoli was one of the authors who made use of a modal distinction similar to that of Buridan and who maintained that it was not found in Aristotle. According to Aureoli, there are natural impossibilities which cannot be actualized by any created power, and there are strong impossibilities which imply a contradiction and cannot be actualized by any power. In Aureoli's view, Aristotle believed that the eternal immutable structures of nature are simply necessary. The insight that this is not true is based on divine revelation and is delivered by Catholic faith. See Nielsen, "Dictates of Faith."

[12] John Buridan, *Quaestiones super octo Physicorum libros Aristotelis* (Paris, 1509), fol. 105rb: "Et hec omnia credidit non esse possibilia secundum potentias naturales sed

no real ground for ascribing this systematic distinction to Aristotle. However, it is remarkable that this was what Buridan believed about Aristotle, partly because he uncritically read some formulations in Aristotle as references to the distinction which he himself regarded as obvious, and partly because he thought that Aristotle's counterfactual assumptions in physical reduction arguments could not be simply impossible. Aristotle did not explicate them in this way, but Buridan thought that this is what he meant. Buridan concludes the question by saying that he feels great joy for having noticed that this is what Aristotle really taught.[13]

In his works on natural philosophy Buridan often refers to the doctrines of creation, the Eucharist, Jesus's resurrected body, or to the Parisian condemnations of 1277 as the reasons why Catholic believers regard some principles as merely naturally necessary. One might think that the purpose of these references was simply to admit that the theses of natural philosophy were inferior to Catholic faith and that alluding to miraculous events was not meant to be philosophically significant. Buridan tells himself that some theologians were not satisfied with his discussions of such philosophical questions, which were also theologically significant.[14] Perhaps he was only fortifying his position against possible attacks? But if this were true, why did Buridan make such a great issue of his discovery of the distinction between natural and supernatural modalities in Aristotle?

When Buridan believed he could show that the distinction just mentioned occurs in Aristotle, he could claim, pace Aureoli and some other theologians, that it is not based on divine revelation but is part of the cognitive structures of natural reason. Buridan's thesis could also be taken as a refutation of the condemned doctrine of double truth, but it seems that he primarily regarded it as an important systematic principle of natural philosophy itself. Buridan thought that this discipline formulates questions against the background of the distinction between what is possible and can be realized by natural powers and what is possible but cannot be realized by natural powers. Buridan was eager to give credit to Aristotle for having in-

---

alibi posuit hec omnia tanquam simpliciter possibilia ... hec omnia sunt simpliciter possibilia per potentiam divinam et non invenitur quod unquam posuerit aliquid ad arguendum tanquam possibile propter non repugnantiam secundum rationes communes aliquas nisi illud esset simpliciter possibile."

[13] John Buridan, *op.cit.*, fol. 105rb: "Et gaudeo gavisus sum quod illa michi apparuerunt quorum tamen subtilioribus et sapientioribus correctionem relinquo ..."

[14] John Buridan, *op.cit.*, Book 4, q. 8, fol. 73vb.

troduced this distinction, but he also believed that it was understood better by himself and his contemporaries than Aristotle did and that the same holds true of the logic of modalities in general.[15] Buridan's view of the history and development of modal logic was right. His own theory of the modal syllogistic was quite different from that of Aristotle and solved most of the problems involved in Aristotle's theory.[16]

3. *Natural Necessities and Universal Affirmative Statements*

In question 22 of the first book of the questions on Aristotle's *Physics*, Buridan analyses the terms "potency" and "possibility" as follows. One can take these terms to refer (1) to a proposition which expresses something that is possible, (2) to something which can be actualized by the interplay of an active potency and a passive potency which exist in nature, or (3) to something which can be realized by the supreme active potency which is God's omnipotence. He says that things which are possible in sense (2) are also possible in sense (3) and the same holds of (1) as far as the propositions are treated as expressing something which is realizable (fol. 26^{rb-va}). Even though Divine omnipotence is the ultimate executive power, the possibilities which can be realized are possible by themselves. As for the unrealized possible beings (*possibilia*), Buridan states that they have no kind of existence and are not founded on anything. They are said to be something in the sense that they can be actual (1.15, fol. 19^{ra-rb}, 1.20, fol. 24va, 1.21, fol. 25vb). This kind of modal metaphysics was sketched by John Duns Scotus and was widely known in the fourteenth century.[17]

Buridan mostly characterizes different modalities by referring to natural powers and to a supernatural power. This terminology was associated with his view that there are often several possible and understandable answers to the questions of natural philosophy and that by distinguishing and determining them, we can tell what can or cannot take place in nature by natural power. Natural necessities are static or dynamic invariances which, if included in the actual

---

[15] Buridan thought that the Catholic doctrine of miracles invited attention to some possibilities which did not come within Aristotle's purview. See, e.g., John Buridan, *In Metaphysicen* (Paris, 1518), Book VII, q. 2, fols. 42vb–43ra.

[16] For Buridan's modal logic, see Hughes, "The Modal Logic," Knuuttila, *Modalities*, pp. 167–175, Lagerlund, *Modal Syllogistics*, pp. 141–179.

[17] See Knuuttila, "Duns Scotus."

world, cannot be changed by natural powers, though they are not intrinsically necessary. They are unchangeable on the supposition that nature follows its normal course (*communis cursus naturae*).[18] In his questions on the second book of the *Physics*, Buridan develops a general picture of how the active and passive potencies work in the universe. The highest active potencies are the powers of God and the heavenly bodies on the one hand and the primary passive power is the prime matter. Between these limits there is a fine yet complicated structure of natural active and passive principles and powers which determine the transformations and mixtures of the basic elements, the structures and functions of the plants and the animals, and the activities of the human soul (2.4–5, fols. 31rb–33vb, 2.13, fols. 39ra–40vb). Buridan did not find anything wrong in Aristotle's view that natural powers are not in vain and that the generic potencies prove their mettle through actualization. In describing the behaviour of created things, the notion of unrealized alternative possibilities is relevant only with respect to agents which have a free will.[19] The crucial question associated with Buridan's conception of natural necessity is whether there is in nature a power of realizing something or preventing something. A generic natural possibility is thought to be realized at some time, and therefore a true universal affirmative statement with a natural supposition is necessarily true by a natural necessity. By "natural supposition" Buridan means that the terms stand for past, present, and future things.[20] When Buridan sometimes says that the necessarily true propositions of natural philosophy cannot be falsified by any case, he means that nothing in nature can falsify them.[21]

4. *Substance and Power*

In describing the functions of nature, Buridan treats the substantial forms as the most central active principles. They are active by

---

[18] John Buridan, *In Metaphysicen* (Paris, 1518), fols. 8vb–9ra. For this kind of view, see also John Duns Scotus, *Ordinatio*, I, d. 8, p. 2, q. un., n. 306; William Ockham, *Summa logicae* [Boehner e.a.], p. 83$^{49–53}$.

[19] See John Buridan, *Quaestiones super libris quattuor De caelo et mundo* [Moody], I, 23–6, pp. 112–128.

[20] John Buridan, *Quaestiones super decem libros Ethicorum Aristotelis* (Paris, 1513), Book VI, q. 3, fols. 122vb–123ra.

[21] See the formulation in John Buridan, *Quaestiones super libris quattuor De caelo et mundo* [Moody], I.1, p. 6$^{22–24}$.

determining and maintaining the structures of beings and by controlling the behaviour of the wholes and of their components. These are the same tasks with which Aristotle provided the substances in the seventh book of the *Metaphysics* and in the eighth book of the *Physics*. The substantial forms of the basic elements are responsible for their remaining in their natural places or conditions when there is no external influence, and for their returning to their natural conditions after having been deflected from them by a prevailing external influence. As for the animals, their instinctual behaviour is in a sense programmed in their substantial form by divine art (*Quaest. super Phys.* 2.5, fol. 33^{ra-rb}, 2.13, fol. 40ra). The substantial forms are also called the natures of things in their capacity as the principles or causes of movements and alterations. They are principles because they are ontologically prior to the things which they determine, and they are causes because they influence the behaviour of the things which are dependent on them (2.4, fol. 31va). The immediate causes of the movements and alterations of actual substances are the various potencies which are associated with the properties kept together by the substantial forms (2.5, fol. 33^{rb-va}).

In his paper on the causal structure of the physical, Hilary Putnam says that there is a medieval notion of causation according to which what is normal, what is an explanation, what is a bringer-about is all in the essence of things themselves and not contributed by our knowledge. The purpose of this remark is to refer to a deep difference between the discussions of causality before David Hume and after Hume's criticism of the traditional approaches to causality.[22] In their book on causal powers, Rom Harré and E. H. Madden describe Hume's stance as follows. In the Humean analysis of causality, the causal relations are exhausted by actual or hypothetical regularities between independent entities. It is assumed that when the notion of necessity is employed in this connection, it is a psychological phenomenon. There is no causal natural necessity in the world. In order to make his point clear, Hume asks us to consider some sequence of events which we believe to be causally connected. Could we imagine performing the initiating procedure and the expected outcome simply not occurring? He thought that the answer in many cases is "Yes, quite easily." This shows that at least the alleged natural necessity is not a logical necessity expressed in the principle of non-contradiction. It is more like a habit of thought.[23]

---

[22] Putnam, *Realism*, p. 88.
[23] Harré e.a., *Causal Powers*, pp. 27–43; cf. Harré, *Laws of Nature*, pp. 80–83.

According to Harré and Madden, the crucial element which makes an action causal is a powerful particular. Natural necessity characterizes the relation that holds between the nature of a particular and the occasion for the exercise of any of its powers, on the one hand, and the manifestation of that power in observable effects, on the other.[24] The authors are not impressed by the Humean argument. They write that "... if something which is apparently identical in kind with whatever usually produces a certain effect is introduced into a suitable environment, and the usual effect does not follow, provided that we can find no extrinsic circumstance to explain the deviant reaction, we can say that there was a change in the nature of the thing, and that it is no longer what it was." There is an inconsistency between the assumption that a scientifically sanctioned effect does not follow in the proper circumstances and the assumption that the natures of things involved are unchanged and all constraints on their action removed. This inconsistency is a formal contradiction in reasoning. The nature of things is discovered *a posteriori*, but, once discovered, we can see that things of a certain nature must have the dispositions they do have and must produce the effect they do produce. The causal powers belong to the primary constituents of the natures of things. The authors exemplify their thesis by a discussion of a man in a fire not being burnt to death which they found in al-Ghazali and Averroes. They say that in truth, a fire that has no heat is no fire at all, but simply another particular with a superficial resemblance.[25]

If questioning the nature of the necessity which is associated with causal relations is typical of the Humean approach, it does not differ from Buridan's view in this respect. The distinction between logical necessities which we cannot imagine to be otherwise and natural regularities which we can imagine to be otherwise (though there are not natural powers which could bring about the deviations) was a standard late medieval assumption and one of the most frequently used conceptual devices in Buridan's natural philosophy. In this particular sense, he was closer to Hume than to what Putnam calls the medieval conception of causation.

Buridan thought that the link between a composite substance and its permanent properties, other than its components, is naturally but not logically necessary, and in his view the same holds true of

---

[24] Harré e.a., *Causal Powers*, pp. 5–6.
[25] Harré e.a., *Causal Powers*, pp. 44–47. For contemporary discussions of causality, see also Sosa e.a. (eds.), *Causation*.

the interactions between the active and passive natural potencies.[26] Natural philosophy is interested in the invariant features of nature, but they only form a subclass of the imaginable invariant factors of reality. Because of this discrepancy between the possible and actual features of the world, the task of natural philosophy could be characterized as an attempt to identify the real world among the possible ones. One should learn to know the causes which belong to the causal structures of the actual world. In this respect, Buridan thought like Harré and Madden, though his views of the metaphysics of beings and the achievements of natural philosophy were different. Buridan believed that the general features of the world are known to natural philosophers, but the more detailed causal connections are to a great extent unknown. Buridan regarded acquiring new scientific knowledge of this kind as a cumbersome task. It should take place through extensive empirical and inductive studies, and he could not imagine that there would be millions of people doing basic research of this kind. He was inclined to think that human knowledge of the fine structures of nature would always remain at a pretty modest level.[27] Buridan's attitude was based on his habit of comparing the actual order of things with the possible modes of existence; the number of the established truths was small in comparison with the possible truths. Aristotle's view of natural philosophy was much more optimistic in this respect. He did not operate with the distinction between natural and logical possibilities and correspondingly he was inclined to think that the majority of the knowable truths are known in most areas of knowledge.

Buridan's awareness of the practical problems of acquiring new demonstrative knowledge is reflected in his epistemological views. While discussing the question of whether we can have certain knowledge about nature, Buridan says that such a certainty demands firmness of truth and firmness of assent. The firmness of truth is divided into two types: there are propositions whose truth is firm in the sense that they cannot be false at all, and there are propositions whose truth is firm on the supposition of the common course of nature. An example of the latter type of firmness is the statement that fire is hot. It cannot become false on the supposition that nature functions in its habitual manner. The firmness of assent to a proposition

---

[26] For the separability of the predicates outside the category of substance, see John Buridan, *In Metaphysicen* (Paris, 1518), VII.1–2, fols. 42ra–43ra.

[27] John Buridan, *Quaestiones super octo Physicorum libros Aristotelis* (Paris, 1509), I.5, fol. 7rb.

representing the conditional firmness of truth is conditional in the same manner. Therefore, when the proposition "Every fire is hot" is accepted as a natural necessity, the firmness of assent to it is qualified by the addition that it is not necessary in the sense that the proposition "Some existent fire is not hot" would be contradictory. The basic difficulty in making scientific discoveries is associated with the ascertainment of the conditional certainty of the new results.[28]

Before John Buridan, John Duns Scotus and William Ockham put forward the idea that natural necessities are those invariant aspects of the created order which cannot be changed by natural agents, although thinking that they were different does not imply a contradiction. Scotus and Ockham mention as an example that fire is hot or burning.[29] Buridan also thought that fire could be cold and remain fire. Late medieval authors would not have agreed with Harré and Madden that this is not possible. In their view, substances are what they are by substantial forms, but they determine their physical behaviour by a merely natural necessity which is not reducible to an *a posteriori* known conceptually necessary relationship. The assumption of the logical contingency of central structural and causal connections brought new kinds of epistemological problems into natural philosophy. At the same time the explanatory role of substantial forms tended to become enigmatic.

---

[28] John Buridan, *In Metaphysicen* (Paris, 1518), 2.1, fols. 8vb–9ra.
[29] John Duns Scotus, *Ordinatio*, I, d. 8, p. 2, q. un., n. 306; William Ockham, *Summa logicae* [Boehner e.a.], 1.24, p. 80$^{60-61}$.

THE NATURAL ORDER IN JOHN BURIDAN[*]

Joël Biard

The concept of a "natural order" or "order of nature" has two correlative aspects, both of which I aim to evaluate in the context of Buridan's thought and its crucial role in the development of fourteenth-century natural philosophy. First is the idea (to put it somewhat anachronistically) of the law-governed structure of nature, i.e., of the status of physical laws. Second is the question of the consistency of nature to the extent that it is distinguished from other orders, which can be said to be supernatural. On the side of the supernatural, there is the question of divine intervention, which in the time of Malebranche or Pascal came to be called "the order of grace," but which must be considered here in its whole amplitude, i.e., not only via the question of the salvation (which is not a Buridanian question), but also through every form of divine omnipotence, as well as the order of human liberty, which was becoming increasingly important in Buridan's time.[1]

1. *Consistency and Autonomy*

Questioning the consistency of nature would of course be closest to the medieval way of thinking about this problem. With respect to the Creator, nature stands in a contradictory relationship of dependence and relative autonomy, whose formulation changes throughout the Middle Ages. Already in the twelfth century, we have the concept of an autonomous law-governed structure delegated to creation by the Creator. If, on the one hand, the symbolic approach of the world runs the risk of reducing it to a mere vestige, a trace bearing no agreement to that of which it is the sign, the sanctity of the created world is, on the other hand, an invitation to study nature in all of its details. And finally, the creative abundance manifests itself in a dynamic nature. Nature is not only, as it is in Aristotle, the internal principle

---

[*] Translated by Jack Zupko.
[1] In his *In Metaphysicen* (Paris, 1518), Book II, q. 2, John Buridan distinguishes several senses of *naturale* besides the two mentioned here. But the senses I am concerned with can be found in John Buridan, *In Metaphysicen* (Paris, 1518), II.2, fols. 9vb–10ra.

of the movement of a substance that is always individual,[2] but the creative energy of a great living being considered as a whole, in the way suggested by Plato's *Timaeus*, often the subject of commentaries from the period.

Aristotelian physics as such does not force us to consider the problem of the consistency of nature in its autonomy relative to a creator God – in fact, the prime mover has a completely different function in Aristotle's physics and metaphysics. But the question arises immediately in the confrontation with Neoplatonism and monotheist theologies, first in the Muslim-Arabic world, and then in the Christian Middle Ages.

Returning to Buridan, we must remember that the Picard master – his origins are now well known – is very much a part of a tradition of Parisian thought which had been evolving since the time of Albert the Great. This tradition is marked by the tendency to develop a natural (or naturalist) discourse having its own proper field of rationality and legitimacy, without recourse to theology, or without even assuming itself to be in conversation with theology, but claiming that its own modes of reasoning are specific to it. This claim, which is broader than the field of the natural philosophy alone but which concerns the whole of philosophy, including ethics and, to a large extent, metaphysics, is at the heart of the debate during the years 1260–1270, in which Bonaventure, Thomas Aquinas, and Siger of Brabant were among the chief protagonists. By asserting the autonomy of philosophical reason in questions such as the eternity of the world, the idea was promoted of reason as a natural faculty, exploring its object with its own proper means. It is thus hardly surprising that we should find this idea conveyed in Latin by the use of adverbs such as *philosophice* (philosophically), as well as the expressions *loqui naturaliter* (to speak naturally) or *naturaliter loquendo* (naturally speaking).[3] The origin of these expressions can be found in Albert the Great. In this sense, the concept of reason as a natural faculty implies the relativization of physical discourse and even its subordination to superior sciences, though, at the same time, it defines a field in which the philosopher or natural scientist can move with relative independence. This attitude and the expressions associated with it seem to be more easily accepted in the fourteenth century

---

[2] Aristotle, *Physics*, II.1.192b21–23: "... nature is a principle or cause of being moved and of being at rest in that to which it belongs primarily, in virtue of itself and not accidentally."

[3] See Bianchi e.a., *Le verità dissonanti*, Chapter 2: "Loquens ut naturalis."

than they were some decades earlier, e.g., in the crisis of the 1270s. In any case, it must be emphasized that Buridan never ceases to assert the natural character of philosophical discourse in order to define its field of validity and autonomy: "… in natural philosophy we must understand actions and attributes as if they always occurred in a natural way; thus, God is no less the cause of this world and the order found in it than he would be if the world were eternal."[4]

But Buridan most often contrasts the word *naturaliter* with *supernaturaliter*. Thus, on the question of whether the heavens are generable and corruptible, he remarks that "we must hold on the basis of faith that the heavens are supernaturally (*supernaturaliter*) created, and also that they are capable of being annihilated. But it must also be said that the heavens are not naturally (*naturaliter*) generable or corruptible."[5] "Supernatural" refers here, first, to divine intervention in the course of nature. But if we compare this with the theological distinction between absolute and ordained power, we can note two modifications. First, if Buridan's concept of the supernatural involves this distinction – and in the notion of "ordained power" there is already the idea of created order – it is being used not so much in its properly theological formulation as in a way that would make sense from the perspective of Masters in the Faculty of Arts, who used it to establish domains of competence. More than deepening the distinction between the two powers properly speaking, Buridan regularly appeals to "omnipotence." This is either (1) to introduce imaginary hypotheses contradicting Aristotelian principles – e.g., the appeal to "… a certain Parisian article ⟨from the Condemnation of 1277⟩ (*quidam articulus Parisiensis*)" in arguing that God could move the world even if there were no other body relative to which it could be

---

[4] John Buridan, *Expositio et quaestiones in Aristotelis De caelo* [Patar], Book II, q. 9, pp. 423–424. (also John Buridan, *Quaestiones super libris quattuor De caelo et mundo* [Moody], p. 164): "Modo in naturali philosophia nos debemus actiones et dependentias accipere ac si semper procederent modo naturali, unde non minus Deus est causa huius mundi et ordinationis eius quam si iste mundus fuisset aeternus."

[5] John Buridan, *Expositio et quaestiones in Aristotelis De caelo* [Patar], I.10, p. 278 (John Buridan, *Quaestiones super libris quattuor De caelo et mundo* [Moody], p. 42): "Primo ergo de generatione vel de corruptione, tenendum est secundum fidem quod caelum est *supernaturaliter* creatum, et etiam est annihilabile. Sed dicendum est etiam quod caelum nec est generabile *naturaliter* nec corruptibile." See also John Buridan, *Expositio et quaestiones in Aristotelis De caelo* [Patar], I.24, p. 280 (John Buridan, *Quaestiones super libris quattuor De caelo et mundo* [Moody], p. 118), where Buridan contrasts "speaking from the perspective of divine power" and "speaking in accordance with natural powers."

situated[6] – or, most frequently, (2) to leave room for affirmations of faith apparently contradicted by rational arguments. The systematic appeal to divine omnipotence was considered a theologian's way of thinking.[7] As such, it established a domain the philosopher must avoid entering, in keeping with the *modus vivendi* of the statute of April 1, 1272: "We state and ordain that no Master or Bachelor of our Faculty ... should presume to determine, or even to dispute, any purely theological question where doing so would transgress the boundaries assigned to him."[8]

But what does the methodological autonomy of physical discourse imply from the perspective of the things themselves, i.e., from what Buridan calls, in contrast to "supernatural cases (*casus supernaturalis*),"[9] the "common course of nature (*communis cursus naturae*)?"

2. *Order and Necessity*

The question of the natural order emerges primarily in developments surrounding the concept of necessity. There is nothing sur-

---

[6] See John Buridan, *Quaestiones super octo Physicorum libros Aristotelis* (Paris, 1509), Book III, q. 7. Buridan wants to show that local motion is distinct from place and the moved body: "Ad secundum dico quod non plus reputarem inconveniens quod esset motus et nihil moveretur, quam quod esset albedo et nihil esset albus. Neutrum est possibile naturaliter, et utrumque est possibile supernaturaliter" (John Buridan, *op.cit.*, III.7, fol. 50vb).

[7] See, e.g., Buridan's discussion in the *Quaestiones super decem libros Ethicorum Aristotelis* (Paris, 1513) of how two distinct things (in particular, intellection and enjoyment) are separable by divine power: "Quia tunc, ipso manente aliis actibus ablatis, sicut auferri possunt per potentiam dei absolutam (ut aliqui theologi dicunt), adhuc homo esset felix sicut lapis esset albus manente in eo albedine, omnibus accidentibus remotis ... Et confirmatur fortius, quia dicunt alii theologi quod in anima Sortis cum clara Dei visione posset Deus formare tristiciam sine delectatione, et odium Dei sine amore" (John Buridan, *op. cit.*, X.4, fol. 209rb).

[8] Denifle e.a. (eds.), *Chartularium Universitatis Parisiensis*, I, n. 441, p. 499: "Statuimus et ordinamus quod nullus magister vel bachellarius nostre facultatis aliquam questionem pure theologicam ... determinare seu etiam disputare presumat, tamquam sibi determinatos limites transgrediens."

[9] See, e.g., John Buridan, *Quaestiones super octo Physicorum libros Aristotelis* (Paris, 1509), III.12, fol. 54vb: "Utrum omnis motus est subiective in mobili vel in movente vel in utroque." See also John Buridan, *op.cit.*, II.13, fol. 39ra, where the supernatural does not suggest anything other than the direct and determinate action of God: "Licet deus per suam infinitam potentiam et voluntatem liberam posset sine aliis causis concurrentibus producere et creare diversos effectus contrarios in eodem tempore, sive in diversis, et hoc modo supernaturali et miraculoso, tamen modo naturali non esset possibile."

prising here. Aristotelian or graeco-arabic necessitarianism had been a target of conservative theologians since the 1270s. These theologians relied more on a concept derived from Alfarabi and Avicenna of the emanation of the universe from God and the first intelligence, than on Aristotle himself, for whom the chance has a place in sublunary world. The picture of a world in the grip of natural necessity is nevertheless opposed to the Christian idea of free creation by divine decree, extending not only to the actual contingency of the created order – for authors such as William of Ockham show that regularity is certainly consistent with the global contingency of the created order – but also to the possibility that God would intervene in this order. This possibility is in no way denied by Buridan. On the contrary, it is precisely what is indicated by the hypothesis of the *casus supernaturalis*, which is assimilated to a miracle.[10] But the question of what sort of necessity is at issue here becomes a problem to be considered once the supernatural is left aside.

The question of necessity does not immediately take the form of knowing how to redeem laws or rules.[11] On the one hand, it seems to be admitted a priori, at least in a certain field whose limits can be defined. Thus, in Book II, Question 13 of his *Questions on the Physics*, Buridan asks about the origin of necessity in natural operations as if this were a matter of course: "Does the necessity that occurs in natural operations have a final or a material cause? (*Utrum in operationibus naturalibus necessitas provenit ex fine vel ex materia?*)." On the other hand, this leads to the question of the order of the universe in the sense of the organization (whether hierarchical or not) of the substances that compose it – a question that should be resolved in the interplay of efficient and final causality.

Book II, Question 13 is initially concerned with the relation between what acts by nature and what acts by will. We see already that nature refers to a field (it would be better to say "an order," though not in the sense of an ordered disposition, but in the way Pascal speaks of three orders),[12] defined by the difference that is for him constitutive, in the first instance, not of the order of the miraculous, but of human free will. Nevertheless, if "acting by will" leads us to posit a sort of liberty of indifference (I will return to

---

[10] See, e.g., John Buridan, *op.cit.*, II.5, where Buridan sets aside "miraculous or supernatural" acts from his discussion of the role of substantial forms or dispositions in actions or operations.
[11] I will return to this below.
[12] Blaise Pascal, *Pensées* [Lafuma], fragment 308.

this), "acting by nature" invites us to ask about what it is that governs natural operations if we do not admit any ontological necessity in the relation between natures and if we emphasize the singularity of first substances.

Buridan begins to entertain such an order by presenting "materialist" arguments, which he will later refute.[13] These are mechanist arguments, which see in the dispositions of things the only foundation for their operations and movements.[14] In this way, rain would follow necessarily from changes in elevation, insofar as it would be made lighter through rarefaction, then fall as it became heavier through condensation. His reasoning opposes final and material causes, and refers to Aristotle, though it presupposes a total reevaluation of the significance and place of final causation. Our author remarks elsewhere that "this question is difficult enough," precisely because of its relation to the question of final causation. He gives a long list of details in the form of theses or "conclusions (*conclusiones*)," which I will consider in order, giving particular emphasis to those that concern the question presently under discussion.[15]

First, the order of natural things comes "primarily and principally (*primo et principaliter*)" from one sort of supreme end, which is God himself:

> Every order that is good and right in the operations and dispositions of natural beings arises primarily, principally, and originally from that best end for the sake of which everything else exists and acts or is acted upon in its first intention, viz., from God himself.[16]

This supreme end for the sake of which things are what they are, and are ordered as they are ordered, is the first and principal cause of everything that is good in the universe. In Book II, Question 5, Buridan also insists that God is the first and principal agent of everything that happens, a role to which Avicenna gave the name *dator formarum*:

---

[13] John Buridan, *op.cit.*, II.13, fol. 38vb: "Arguitur primo quod ex materia, rationibus Antiquorum …"

[14] The proposed model is that of rain, which cannot but remind us of the teaching of Lucretius. See Lucretius, *De rerum natura*, II.v. 221–222.

[15] I here summarize material from my article already published under the title "L'idée de nature."

[16] John Buridan, *op.cit.*, II.13, fol. 39rb: "Omnis ordo bonus et conveniens in operationibus et dispositionibus entium naturalium provenit primo et principaliter a primo ab illo fine optimo gratia cuius omnia alia sunt et agunt vel patiuntur prima intentione, scilicet ab ipso deo."

One [end] is the giver of forms (*dator formarum*), as Avicenna states, who acts in relation to everything that happens as the common and first and altogether most principal agent, and this is supreme God.[17]

But second, this supreme end or first cause is not sufficient to explain the order of the universe, since it would be impossible to explain the diversity of effects and events by referring only to a first cause that is by definition absolutely simple. This leads Buridan to emphasize that, if we leave aside cases of the supernatural or miraculous, other agents must contribute. The argument here reasons from divine unity and perfection to the insufficiency of these attributes in the explanatory order. The problem is clearly that of moving from the one to many, a remote echo of the Neoplatonic side of medieval Aristotelianism,[18] as well as of the thesis, deriving from Alfarabi and Avicenna, that God can immediately produce only a first intelligence.

This question of the move to the many is linked, in a curious way, with that of omnipotence. The latter principle, which is theological in origin and nature, is brought out here in the context of physics, in terms of the possibility that God could, "by virtue of his infinite power and free will," create or produce different contrary effects at the same time "without other contributing causes."[19] In others words, God can bring about directly everything that normally occurs by means of secondary causes. Paradoxically, we find here, forming an integral part of natural philosophy, Duns Scotus's expression of the properly theological definition of omnipotence, which is opposed to the more philosophical conception of an infinite power.[20] At the same time, it is here, to begin with, that we move away from a purely "logical" model for the sake of an "operative" model, adopting the useful distinction proposed by Randi.[21] The logical model, of the sort

---

[17] John Buridan, *op.cit.*, II.5, fol. 32va: "Unus est dator formarum, sicut dicit Avicenna, qui ad omne quod fit agit tanquam agens commune et primum et omnino principalissimum, et ille est deus supremus."

[18] Nevertheless, it is to Aristotle that Buridan refers when he assumes that it would be impossible to go from the one to the many, a move reminiscent of the second book of *On Generation and Corruption*. See John Buridan, *op.cit.*, II.13, fol. 39rb; cf. Aristotle, *On Generation and Corruption*, II.1.

[19] John Buridan, *op.cit.*, II.5, fol. 32va.

[20] Duns Scotus opposed the philosophical definition of omnipotence as infinite power, defined by the force and amplitude of its causal power and manifested in the extent of its effects, to the properly theological (and indemonstrable) conception, according to which God can produce directly and without intermediary everything that occurs naturally through secondary causes. See John Duns Scotus, *Ordinatio*, I, d. 42, "Ad questionem," vol. VI, pp. 362–363.

[21] See Randi, *Il sovrano e l'orologiaio*.

developed in its most unadulterated form by William of Ockham, considers omnipotence as a purely methodological instrument to test the metaphysical structure of reality and to imagine the radical contingency of the created world. But in Buridan we find a sort of naturalist version of the operational or "juridical-political" model of omnipotence, characterized by God's ability to intervene *de facto* in the *de jure* order of nature, a naturalist version of omnipotence built around the opposition between the "supernatural and miraculous" and "natural" modes. This supernatural mode continues to have a methodological function entirely analogous to omnipotence. Far from being an established and common fact that would destroy all regularity in the world, it permits us to define the metaphysical status of this regularity.[22]

Taken by itself, the "natural mode" requires the presence of secondary causes for the reason indicated, viz., the need for multiplicity at the level of causes: "... nevertheless, in the natural mode it would not be possible for the same simple and invariable thing to produce different and contrary effects, now these effects and tomorrow others, unless there were other, diverse, causes contributing to them."[23] To sum up, "in the natural mode (*modo naturali*)," God alone is not sufficient to explain the diversity of effects.

The third thesis emphasizes that it would also be insufficient to introduce a second principle, which would be the matter. Since matter is everywhere "of the same nature (*eiusdem rationis*)," we cannot use it to explain why we have air here and water there, a horse here and a goat there: "... it cannot give an adequate account of the cause of such a diversity of transformations and natural effects."[24] Consequently, we must assume the existence of other causes that are diverse in themselves, some of which are determined (and determination should be understood here in terms of precision) to generate this particular effect, others that.[25]

---

[22] See Reina, "L'ipotesi," pp. 683–690. The author treats especially of Buridan's use of omnipotence in the problem of the certitude of knowledge.

[23] John Buridan, *op.cit.*, II.13, fol. 39rb: "... tamen modo naturali non esset possibile quod ab eodem simplici et invariabili provenirent effectus diversi et contrarii, et nunc tales et cras alii, nisi essent alie cause concurrentes diverse."

[24] *Ibid.*: "... non potest sufficienter reddi causa talis diversitatis transmutationum et effectuum naturalium."

[25] As G. Federici Vescovini has emphasized ("La concezione," pp. 616–624), Buridan confers upon matter a dignity equal to that of form through its dispositions to receive form. Nevertheless, matter is not a sufficient cause unto itself.

The fourth thesis introduces celestial bodies. This is essential in order to situate Buridan in the evolution of Parisian thought, beginning with graeco-arabic Aristotelianism and the synthesis of Albert the Great. The *Questions on the Metaphysics* returns to this point, which I will consider here only briefly. Celestial bodies are causes: "It is obvious that celestial bodies are ordaining causes with respect to the diversity of the many transformations and effects occurring in this world."[26] What we have here is actually a multiplicity of causes and effects. But are they causes of a superior order? Of course, Buridan often refers to bodies in the sublunary world as "inferior" beings, and analogously, to God and celestial bodies as "superior agents." But do celestial bodies have a degree of being or constitute a species of intermediary cause between the absolute simplicity of God and the multiplicity of material beings? The examples immediately following this passage do not permit such an interpretation. They concern natural causes of meteorological sort: the length and shortness of days and differences of climate. Celestial bodies really do exert an influence. It is in this connection that Buridan alludes to the causality of the stars, naturalized through interference between terrestrial and celestial causes.

Nevertheless, the apparatus of God, matter, and celestial bodies is not complete. We must add to it (and this is the fifth thesis) the network of particular terrestrial agent causes: "… other diverse particular agents ordaining this [to occur]."[27] It is necessary to diversify further because neither prime matter nor the heavens can account for the fact that in the same corner of the earth is generated such and such an animal, or such and such a tree. We are not dealing here with a process of emanation, i.e., with a continuous outgrowth of complexity from the primary cause. We are faced with a different, more horizontal order of causality, where the aforementioned influences intervene in the production of the effect, but are not of themselves sufficient for it. The "particular causes" determining a certain portion of matter to be one thing or another are, e.g., the "seeds" of animals and plants, or the "diverse dispositions of matter" by virtue of which flies or frogs are born: "… it is apparent that diverse and dissimilar effects often arise from diverse material dispositions

---

[26] John Buridan, *op.cit.*, II.13, fol. 39va: "Manifestum est quod corpora celestia sunt cause ordinative diversitatis plurimarum transmutationum et plurimorum effectuum in hoc mundo evenientium."

[27] *Ibid.*: "Alia agentia particularia diversa hoc ordinantia."

by means of the same agents."[28] This idea of a material disposition can without a doubt be traced to Albert the Great, and through him ultimately to Avicenna. Albert is elsewhere cited in relation to the concept of the *dator formarum*, reinterpreted here in terms of Plato's ideas, in order to express why we need to ascribe a role to forms. We must remember, however, that Albert rejects the theory of the *dator formarum* in its Platonist interpretation in favor of a complex theory of the drawing out of forms (*eductio formarum*), in which, through a process internal to the matter, the form is "educed (*educta*)" or, let us say, "aroused," by a mover univocally with respect to what is generated.[29]

In view of the importance assigned to material dispositions, it is also apparent that the difference with the aforementioned materialist conception, though real, is very subtle. The replies given to the initial arguments, developed on behalf of a materialist conception of necessity, were based on the following idea: since prime matter is of itself indifferent, material causes must be accompanied by a diversity of agents. Therefore, order and diversity are not explained by the interplay of material causes alone. The role played by agents thus has a certain priority, even though this is in the dispositions of the matter exhibiting it, transcribing the interplay of agent causes. This complex interplay of forms and qualities or dispositions of corporal substances is treated at length in Book II, Question 5, of Buridan's *Questions on the Physics*.

The question remains, however, of the knowledge of how and why agents act. Here efficient and final causality are conjoined, but it is final causality that is decisive for establishing the existence of an order of nature: "… all of the order and diversity in natural changes and effects comes from the order and diversity of ends." Following Averroes, Buridan distinguishes between two senses of "end:" that according to first intention and that according to second intention.[30]

The end of first intention (*prima intentione*) is that which is first in the order of being, goodness, and perfection. It is that for which, or for the sake of which (*gratia cuius*), something or someone acts. For example, it can be the man for whom a house is constructed. If we consider the whole universe, it is God who is in this sense the end

---

[28] John Buridan, *op.cit.*, I.5, fol. 32vb: "Apparet quod ex diversis dispositionibus materie proveniunt sepe diversi et dissimiles effectus ab eisdem agentibus."
[29] See De Libera, *Albert le Grand*, p. 127.
[30] See John Buridan, *op.cit.*, II.7; also Biard, "Le système des causes," pp. 491–504.

of everything. It is God who has brought order and necessity to the system of the world:

> For he necessitates and orders the heavens in their motions, and consequently, those inferior beings are more principally necessitated and ordered by the heavens and the motion of the heavens than by particular agents.[31]

Two points should be emphasized here. The first concerns the relation between "to order (*ordinare*)" and "to necessitate (*necessitare*):" although the necessity is not merely that of efficient causality, as in classical science, order is very much a factor in necessity here. Second, this order is realized through particular agents, where we again find ends of second intention (*secunda intentione*) deliberately sought by agents. But the cornerstone of this building is God himself. Therefore, God (here a purely metaphysical and philosophical God) does not appear so much the omnipotent and even arbitrary God of the theologians as the originator and guarantor of the order of the universe. That is why the order of created nature itself will realize (give reality to) this divine order, just as, at another level, the end of first intention is resolved in the nature of the agent.[32]

3. *Freedom and Natural Rules*

This conception of the natural order presents us with two questions. The first concerns freedom; the second concerns the procedures we should use in order to establish the rules.

If Buridan admits that it is possible for supernatural cases to arise and disturb the ordinary course of nature, then, in a general way, the order of the nature is not opposed to an order of grace but to

---

[31] John Buridan, *op.cit.*, II.7, fol. 35rb: "Necessitat enim et ordinat celum in motu suo, et consequenter per celum et motum celi necessitantur et ordinantur ista inferiora principalius quam per agentia particularia."

[32] This reduction of ends to agents is expressed in certain passages that appear enigmatic at first glance. Thus, when proposed in a passage speaking of God as the supreme end, it seems immediately generalized: "For in this way the Commentator says in *De substantia orbis* that the end signifies the agent by a necessary signification (Sic enim dicit Commentator in *De substantia orbis* quod finis significat agens significatione necessaria)" (John Buridan, *op.cit.*, II.13, fol. 39rb). As a matter of fact, in order to demonstrate that order and diversity come from ends, Buridan explains: "For since they come from agents, as has been stated, and agents are final causes of their actions of first intention and not conversely ... the proposed thesis follows (Cum enim proveniant ex agentibus, ut dictum est, et agentia sunt causae finales suatrum actionum prima intentione et non econverso ... sequitur propositum)." (*Ibid.*)

the order of human freedom. The question inevitably arises because the premises seem to orient us toward a necessitarian vision of the universe, as opposed to the radical contingency defended, e.g., by Franciscan theologians.

Only the advent of free agents – it being essential to refute the accusation of "necessitarianism" leveled against philosophers by the University Censors in 1277 – breaks the chain of necessity. Only free agents act in accordance with ends they set for themselves. On the other hand, there is no place for such goal-directed intentions among other natural beings. This point is largely developed in the same Question 13 of the Book II of Buridan's *Questions on the Physics*, beginning with the remark that the appearances are here against Aristotle and the Commentator. In the case of artificial beings, the end is reduced to the intention of the agent, which is absent in the case of beings that act by nature:

> But as far as natural things are concerned, I believe that a swallow mating, nesting, and laying eggs does not cognize any more when it produces chicks than a tree does when it produces branches and flowers. Nor do the mating, nesting, and egg-laying activities of the swallow depend for their being and order on those chicks. Rather, the converse is true. And those chicks do not determine the swallow to act in this way, but the form and nature of the swallow, celestial bodies at certain times of the year, and supreme God in his infinite wisdom, together determine the swallow to mate, from which the production of eggs consequently follows. And again, when the swallow is so disposed, the nature of the swallow together with the celestial bodies and God determine it to nest and lay eggs, to hatch and produce chicks, and finally, to nourish them, etc. All of this comes about by divine artifice, celestial bodies, and particular agents, both extrinsic and intrinsic [to the subject of the action], which are the substantial forms of these same natural agents.[33]

---

[33] John Buridan, *op.cit.*, I.13, fol. 40rb: "Sed de naturalibus ego credo, quod hirundo coiens, nidificans et ovificans nihil plus cognoscit pullos generandos quam arbor frondes et flores producens cognoscit fructum generandum. Nec hirundinis coitum, nidificatio et ovificatio dependent in esse et ordine eorum ab illis pullis sed e contra. Nec illi pulli determinant hirundinem ad sic operandum, sed forma et natura hirundinis et corpora caelestia determinatis temporibus et Deus supremus per suam sapientiam infinitam determinant hirundinem ad coitum, ex quo consequenter sequitur generatio ovorum, et iterum hirundine sic disposita, natura hirundinis cum corporibus caelestibus et Deo determinant illam ad nidificandum et tandem ad ponendum ova, covandum et generandum pullos et ulterius ad nutriendum etc. Haec ergo omnia proveniunt ab arte divina et corporibus caelestibus et agentibus particularibus tam extrinsecis quam intrinsecis, quae sunt formae substantiales ipsorum naturalium."

Several Questions from the *Questions on the Metaphysics* develop the same idea.[34] The operations of beings acting by nature are governed by necessity, and the latter is what grounds the ordering of natural phenomena:

> necessarily as regards the actions of non-free agents, from those things that exist and proceed along certain lines there follows the being and order of everything that happens later.[35]

But this necessity on which the order is based must be conjoined with the action of free agents. This explains the apparently restrictive formulae – which actually express a penchant for precision – that are endlessly put forward by Buridan:

> Whenever sufficient causes have been posited in such a way as to be sufficient, viz., such that there is no impediment as regards the fact that something is produced, it must be produced.[36]

What sort of impediment is at issue here? Except for the miraculous intervention of God, which is usually – at least since the thirteenth century – opposed to the common course of nature, but whose possibility is circumscribed, it is in the power of a free agent to act or not act. In other words, a free agent is not necessitated to act, as is confirmed by well-known passages in the *Questions on the Nicomachean Ethics*. The essential point, however, is that this antinomy of freedom and necessity does not destroy the order of causes:

> His second proposition [i.e., the proposition at n. 36 above, which Buridan ascribes to Plato] would not be denied as regards a free agent, but would be conceded as regards a non-free agent, viz., the proposition that when sufficient causes have been posited in the way that they are sufficient to produce something, the production would follow of necessity, such that nothing would impede it.[37]

---

[34] See especially John Buridan, *In Metaphysicen* (Paris, 1518), VI.5, fols. 35va–37rb: "Utrum omne futurum de necessitate eveniet." See also John Buridan, *op.cit.*, IX.4 and John Buridan, *op.cit.*, XII.1, fol. 65rb: "It should be noted that some things, such as God, intelligences, and the intellective soul, are free agents. Others are natural and non–free agents (Notandum est quod quedam sunt agentia libera ut Deus et intelligentie et anima intellectiva. Alia sunt agentia naturalia et non libera)."

[35] John Buridan, *Quaestiones super octo Physicorum libros Aristotelis* (Paris, 1509), II.13, fol. 39vb: "... de necessitate quantum ad actiones non liberorum agentium, ex eis que sunt et procedunt sequitur esse et ordo omnium que posterius eveniunt."

[36] John Buridan, *In Metaphysicen* (Paris, 1518), VI.5, fol. 35vb: "Quandocumque posite sunt cause sufficientes, eo modo quo sunt sufficientes, scilicet quod non sit impedimentum ad hoc quod aliquid fiat, oportet quod illud fiat."

[37] John Buridan, *op.cit.*, VI.5, fol. 36vb: "Sua secunda propositio negaretur de agente libero, sed concederetur de non libero, scilicet illa propositio quod posi-

So despite the crucial question of human liberty, what we have here is in fact the assumption of a natural order, or better, of a necessary natural order – necessary under certain conditions.

I will not develop at length the question of the status of natural necessity in Buridan, which has already been examined by Simo Knuuttila.[38] I will only emphasize the sense in which it is a conditional necessity. It is conditioned in the first place because it rests upon the divine will which has created such an order, which is not necessitated internally, meaning that it would be impossible to deduce it metaphysically. This is, then, actual necessity. It is conditioned in the second place because it is realized only under certain epistemic conditions, i.e., in abstraction not only from supernatural or miraculous cases,[39] but also from the unquestionable presence of free agents who insert themselves, from another perspective, into this network of efficient or final causes. These conditions are not so much restrictive as constitutive, belonging to the order of nature as against the divine order and the order of freedom.

It is nothing other than this conditional order that makes possible a certain scientific evidentness "on the assumption of the common course of nature." In the first question of Book II of his *Questions on the Metaphysics*, Buridan asks "whether the comprehension of the truth of things is possible for us."[40] He distinguishes several different kinds of necessity. After mentioning the necessity of propositions

---

tis causis sufficientibus, eo modo quo sunt sufficientes ad aliquid producendum, sequitur necessario productio ita quod nihil sit impediens."

[38] Knuuttila, "Natural Necessity," pp. 155–176.

[39] Buridan explains this clearly in his *In Metaphysicen* (Paris, 1518), when he has occasion to discuss a logico-semantic problem arising from the supposition of the terms, John Buridan, *op.cit.*, V.2, fol. 27^{va-vb}; emphasis mine: "There is another necessity, however, which is called 'natural' and which is not ⟨the same as⟩ absolute necessity. This is necessity *leaving the supernatural cases aside*, such as when we say 'A donkey is an animal,' because *leaving the supernatural cases aside*, it is always necessary that there be some donkey, and so the proposition, 'A donkey is an animal' is ⟨true⟩ per se, because the subject determines the predicate to itself (Alia autem necessitas est que vocatur naturalis, que non est simpliciter necessitas. Sed est necessitas *circumscripto casu supernaturali*, sicut dicendo 'asinus est animal,' quia *circumscripto casu supernaturali* necesse est semper esse aliquem asinum et sic illa propositio est per se 'asinus est animal,' quia subiectum determinat sibi predicatum).' In his *Quaestiones super octo Physicorum libros Aristotelis* (Paris, 1509), Buridan emphasizes that the natural philosopher must not appeal to God: "… unless having recourse to the universal source of generation, i.e., to God – which in the matter at hand is not to have recourse to nature (… nisi recurrens ad universale generans, scilicet a Deum, quod in proposito non est naturalis recursus)." (John Buridan, *op.cit.*, III.5, fol. 46vb)

[40] John Buridan, *In Metaphysicen* (Paris, 1518), II.1, fol. 8ra: "Utrum de rebus sit nobis possibilis comprehensio veritati."

expressing first principles, which the intellect cannot deny, he continues:

> But in another way, "evidentness (*evidentia*)" is understood to be "relative (*secundum quid*)" or "on the assumption (*ex suppositione*)," such as when it was said earlier that it would be observed in beings in the common course of nature. And in this way it would be evident to us that every fire is hot and that the heavens are moved, even though the contrary is possible by God's power. And evidentness of this sort suffices for the principles and conclusions of natural science.[41]

The appeal to divine omnipotence reveals the actual character of this order: there is no essential necessity here. As for the examples – especially the second – they clearly have a propositional nature which leads to the formulation of rules or laws, at least by the attribution of properties to subjects. The affirmation is unambiguous: necessity *ex suppositione* is the norm of natural science. Is it not a hypothetico-deductive necessity, but a necessity that is operative when we assume the absence of miraculous intervention and set aside the order of free choice.

This leads to a second question, which I shall only be able to answer in outline: how can we find and formulate these regularities? What is the scientific process allowing us to establish natural laws? This question concerns induction. I will not review the Buridanian theory of induction – Hans Thijssen has done it elsewhere[42] – but only recount how this question is articulated together with that of the natural order.

Buridan is wholly aware of the difficulties of induction, if we consider it not only to be the intuition of a universal nature via singulars – even via just one singular – but also the inference from singular propositions to a universal proposition. In Book II, Question 1 of his *Questions on the Metaphysics*, on the possibility of comprehending truth, one of the negative arguments presents the difficulties of induction:

> Again, it is argued as far as principles are concerned that principles are known by experience and that these experiences are fallible, as is obvi-

---

[41] John Buridan, *op.cit.*, II.1, fols. 8vb–9ra: "Sed alio modo accipitur evidentia secundum quid sive ex suppositione, ut prius dicebatur, quod observaretur in entibus communis cursus naturae, et sic esset nobis evidentia quod omnis ignis est calidus et quod celum movetur, licet contrarium sit possibile per potentiam Dei. Et huiusmodi evidentia sufficit ad principia et conclusiones scientie naturalis."

[42] Thijssen, "John Buridan and Nicholas of Autrecourt." See also Zupko, "Buridan and Skepticism."

ous via Hippocrates. Second, it is proved that they are fallible because experiences do not have the power to lead us to a universal principle except by way of induction over many, and a universal proposition never follows from induction unless it has been reached on the basis of every singular of that universal, which is impossible.[43]

In his *Questions on the Prior Analytics,* he develops this problem of sufficient enumeration:

> Again, an induction is not a good consequence unless all of the singulars are enumerated in it. But we cannot enumerate all of them because they are infinitely many.[44]

After establishing that such a consequence is never valid *simpliciter* but only *ut nunc,* Buridan replies to the problem of enumeration on two fronts. In many cases, we have a finite number of objects. But in other cases, in which it is impossible to enumerate the relevant singulars, we add the clause "and so on for the others (*et sic de aliis*)" as long as there is no reason to think that it is different in cases other than the ones enumerated. However, one cannot stop here without begging the question.

In his *Questions on the Prior Analytics* II.19, as well as in a corresponding passage of his *Questions on the Metaphysics* II.2, Buridan discusses the gradual process by which some non-demonstrative propositions are established by induction so that they can be used as principles in the domains of practical art (which concerns singulars) and natural science. By sensation and then by memory, we come to have an experience, which serves as mediator between the singular and the universal. In fact, experience permits us first to formulate a singular judgment from other singular judgments; thus, I say that this fire warms before I have approached it. The leap that brings us to a universal judgment is, as Thijssen has shown, based on this curious "natural inclination to the truth (*inclinatio naturalis ad veritatem*)," which Buridan attributes to our intellect. I quote here the passage from the *Questions on the Prior Analytics* in its entirety:

---

[43] John Buridan, *In Metaphysicen* (Paris, 1518), II.1, fol. 8va: "Item arguitur quantum ad principia, quia principia sunt nota per experientias et ipse experientie sunt fallaces ut patet per Hippocratem. Secundo probatur quod sunt fallaces quia experientie ad concludendum universale principium non habent vim nisi per modum inductionis in multis, et nunquam ex inductione sequitur universalis propositio nisi sit inductum in omnibus singularibus huius universalis, quod est impossibile."

[44] John Buridan, *Quaestiones in Priorum Analyticorum libros* [Hubien], Book II, q. 19: ("Utrum inductio sit bona consequentia"): "Item, inductio non est bona consequentia nisi ibi enumerentur omnia singularia; sed illa non possumus enumerare, quia infinita sunt."

Other principles need induction to become evident, and these principles are universals, such as that every fire is hot and that all rhubarb purges bile. For these principles are known to us through an induction based on sense, memory, and experience. For when you have often seen rhubarb purge bile and have memories of this, and have never found a counterexample in the many different circumstances you have considered, then the intellect, not as a necessary consequence, but only from its natural inclination to the truth, assents to the universal principle and understands it as if it were an evident principle based on an induction such as "this rhubarb purged bile, and that ...," and so on for many others, which have been sensed and held in memory. Then the intellect supplies the little clause, "and so on for the ⟨other⟩ singulars," because it has never witnessed a counterexample (even though it has considered many circumstances) nor is any reason or dissimilarity apparent why there should be a counterexample. And then it reaches the universal principle as a conclusion.[45]

Is this inclination a merely verbal reply to the conceptual difficulty of what some call the "problem of the induction," given that it remains to be proved that the inductively inferred universal proposition can be legitimately be regarded as true? In fact, if we place this idea in its context, particularly in Book II of the *Questions on the Metaphysics*, we can ask whether an inclination of this sort does not itself have an objective basis.

The first question of Book II relates different varieties of "firmness of assent (*firmitas assensus*)" and "firmness of truth (*firmitas veritatis*)." With regard to the latter, the *firmitas veritatis*, Buridan distinguishes between an absolute mode, which concerns propositions that cannot be falsified in any case, such as "God exists," and another

---

[45] John Buridan, *op.cit.*, II.20: "Alia principia indigent inductione ad hoc quod fiant evidentia, et illa principia sunt universalia, ut quod omnis ignis est calidus, et quod omne rheubarbarum est purgativum cholerae. Illa enim principia sunt nobis nota per inductionem supponentem sensum, memoriam et experientiam. Cum enim saepe tu vidisti rheubarbarum purgare choleram et de hoc memoriam habuisti, et quia in multis circumstantiis diversis ⟨hoc⟩ considerasti, numquam tamen invenisti instantiam, tunc intellectus, non propter necessariam consequentiam, sed solum ex naturali ejus inclinatione ad verum, assentit universali principio et capit ipsum tamquam evidens principium per talem inductionem 'hoc rheubarbarum purgabat choleram, et illud,' et sic de multis aliis, quae sensata fuerunt et de quibus memoria habetur; tunc intellectus supplet istam clausulam 'et sic de singulis,' eo quod numquam vidit instantiam, licet consideravit in multis circumstantiis, nec apparet sibi ratio nec dissimilitudo quare debeat esse instantia, et tunc concludit universale principium." See also John Buridan, *In Metaphysicen* (Paris, 1518), II.2, fol. 9rb, where Buridan emphasizes that this movement is not "on account of the form (*gratia forme*)," and John Buridan, *Quaestiones super octo Physicorum libros Aristotelis* (Paris, 1509), I.15, fol. 19rb (quoted by Thijssen, *op.cit.*, p. 247).

mode, which concerns propositions that could be falsified by divine omnipotence:

> But again, there is firmness of truth (*firmitas veritatis*) assuming the common course of nature, and in this way it would be a firm truth that the heavens are moved and that fire is hot, and so on for other propositions and commonalities of natural science, notwithstanding the fact that God could make a cold fire, and so falsify this proposition: "Every fire is hot."[46]

Here again we have the "common course of nature (*communis cursus naturae*)." Moreover, just after these remarks we find the passage quoted above, which made this common course foundational to the science of nature in order to distinguish the different forms of assent (*assensus*). This confirms the fact that it is not a possible irregularity in natural phenomena which leads to the relativization of universality, but only the hypothesis of a supernatural or miraculous mode of divine intervention.

But if this is true, then we do not simply have the assumption of a pre-established harmony between what I am inclined to think is true and what is the case in the real world. More precisely, we have the assumption or supposition that there is in fact a natural order, limited and defined by this very limitation, on the one side by the supernatural order, and on the other side by human freedom, as I said before. And this order legitimates – grounds, if you wish – the leap which represents the passage from singular to universal proposition in the inductive process. If there is no contrary reason in the cases enumerated and considered, I may infer – not by a formal inference, but by a process based on the assumption of the common course of the nature, and thus of a regular order – the universal proposition, which I will be able to use as a principle of reasoning in the physical sciences.

Perhaps it will be thought that from a post-Kantian perspective – and it is from this perspective that the problem of induction was formulated at the end of the nineteenth and beginning of the twentieth centuries – there is a pre-critical confidence in the harmony of being and thought. But the historical and intellectual interest of the Buridanian worldview is different. In articulating the question of

---

[46] John Buridan, *In Metaphysicen* (Paris, 1518), II.1, fol. 8vb: "Sed etiam est firmitas veritatis ex suppositione communis cursus nature, et sic esset firma veritas quod celum movetur, quod ignis est calidus, et sic de aliis propositionibus et communibus scientie naturalis, non obstante quod Deus posset sic facere ignum frigidum, et sic falsificaretur ista 'omnis ignis est calidus'."

induction (rigorously formulated in terms of the inference of propositions) and the question of natural regularity, it is concerned with affirming the idea of the natural order as a presupposition of all scientific investigation.

NATURALITER PRINCIPIIS ASSENTIMUS:
NATURALISM AS THE FOUNDATION OF HUMAN KNOWLEDGE?[*]

Gerhard Krieger

The following reflections present a contribution to the question of skepticism in John Buridan. This problem has already received wide attention in Buridan scholarship. The present essay will not address previous work in detail, however. That is because my concern here is with the foundation of the principle of non-contradiction and the absolutely first principles of knowledge and science to which it is related, a question that has only been indirectly treated in the aforementioned research.[1] Likewise, the historical aspect of the problem will not be my main focus, though I will provide a starting point for such an investigation. This means that a precise account of the concept of skepticism will also be left aside.

The word "naturalism" in the subtitle makes reference to the concept of nature at work in Buridan's remark that "we assent to principles naturally (*naturaliter principiis assentimus*)." Buridan is here referring to a claim from Aristotle's *Metaphysics* (980a21). The word "naturalism" suggests a particular content or object independent of and prior to human understanding and knowing. Nature is what determines us and not what results from our interaction with this object or content. The contrary idea is that of a determination subject to our power alone. This capacity for self-determination leads one to think of the will. If the question arises whether Buridan answers the problem of skepticism in terms of naturalism, a further question is raised, i.e., whether Buridan bases our knowledge of the absolutely first principles of knowledge and science on the claim that

---

[*] Translated by William Edelglass and Oliver Baum.

[1] Cf. King, "Jean Buridan's Philosophy of Science," especially p. 122; Thijssen, "John Buridan and Nicholas of Autrecourt," especially p. 250 f; Zupko, "Buridan and Skepticism," especially pp. 205, 216. There is a greater attention to this particular question in Schönberger, "Evidenz und Erkenntnis," especially pp. 14–19. I agree with the interpretation of A. Maier in "Das Problem der Evidenz" that for Buridan "evident appearance always remains bound to truth (*Evidenz immer an Wahrheit gebunden bleibt*)." In contrast to this claim, however, I will show here that Buridan reverses the relationship of truth and evident appearance in favor of the priority of the latter. A detailed discussion of the literature can be found in my inquiry *Subjekt und Metaphysik*, which is currently being prepared for publication.

we naturally desire knowledge, or whether it is exclusively through self-determination that we arrive at such knowledge.

The general nature of this question points to the fact that Buridan bases teleology in human practice. Hence, we begin with reflections on Buridan's general determination of the end of human action in ethics. In the second part, we will investigate his corresponding observations about the human desire for knowledge in metaphysics. The third and fourth parts will consider the foundation of the absolutely first principles of knowledge and science. With this background, we will discuss the problem of evident appearance in the cognition of principles in the fifth part.

1. *The Human Will as Original and Rational Self-Reflection*

Buridan begins his discussion of teleology in human action by distinguishing between the determination of the absolutely universal and fundamental end as such, and its content.[2] (1) With regard to the determination of the end in general, reason judges by the power of its own nature immediately, i.e., not in any discursive fashion that would require cognition. (2) This cognition is due to a "rectitude of appetite (*rectitudo appetitus*)," meaning that in a moral sense, one is striving with rectitude toward an end. (3) From this it follows that what we must do or leave undone in order to realize this end is determined discursively. (4) Thus, the application of (1)–(3) in a proof determines the rectitude of moral judgment in relation to what needs to be done or left undone in order to realize the absolutely universal end. (5) Accordingly, from this we have the moral determination of our concrete desire. (6) From this, concrete action follows.[3]

---

[2] John Buridan, *Quaestiones super decem libros Ethicorum Aristotelis* (Paris, 1513), Book VI, q. 5, fol. 121va: "... finis principalis et ultimatus, puta felicitas humana, potest dupliciter apprehendi, uno modo secundum eius communem rationem, scilicet secundum quod est quid optimum et delectabilissimum, alio modo secundum eius specialem rationem, scilicet considerando quae sit illa res, quae dicitur felicitas, scilicet quae est optima et delectabilissima."

[3] John Buridan, *op.cit.*, VI.5, fol. 121vb: "... sciendum est quod intellectus recte iudicat de fine, saltem quoad communem rationem per eius naturam absque ratiocinatione. Secundo hanc iudicii rectitudinem sequitur naturaliter rectitudo appetitus respectu illius finis. Tertio ex eo quod ad finem tendit appetitus, intellectus movetur ad ratiocinandum, ut inquirat ea, quae ad illum finem valent. Per appetitum igitur finis derminatur intellectus ad ratiocinandum. Quarto per huius ratiocinationem determinatur ad rectum iudicium de ordinatis ad finem desideratum. Quinto hanc

Two points should be raised in this connection. First, as regards the teleology of human desire, Buridan speaks of a judgment of reason arising immediately, and therefore not discursively. Second, the absolutely universal end consists in the application of the discursive process to determining what is to be done. We will begin by making a number of observations with regard to the first point (a), and then explain the second point (b).

(a) Given Buridan's emphasis that the judgment of reason does not arise discursively here, one might suppose that he is thinking instead of an original and simultaneously reflexive cognition of the will. For this interpretation, it is crucial that Buridan is speaking of cognition in relation to human desire. Moreover, this interpretation assumes that the determination of the absolutely universal end results from the judgment in question: to the extent that we speak of desire, it can be assumed that one is focusing one's attention on the human will. To the extent that we are considering the determination of the absolutely universal end, the orientation of the will as such, and hence willing as such, becomes significant. As these are traced back to a judgment, it appears that the human will determines itself in an original and reflexive manner.

That Buridan is referring here to a deliberate self-reflection is made clear in another account of this mode of cognition. Specifically, he states that for us, teleological determination itself is naturally governed insofar as (1) one cannot be deceived in this judgment, and (2) the will is incapable of determining itself in such a way as to contradict it. Accordingly, if it must will, the will must will the teleological determination of the ultimate end.[4]

That Buridan is speaking of a deliberate self-reflection can first of all be shown by the fact that he is speaking of a judgment (*secundum intellectum*). He does not consider any rationale independent of reason. The fact that error is excluded from such judgments can only mean that it involves a cognition presupposed in, though not temporally prior to, the action it determines. Otherwise, error could not be ruled out. So, whenever there is an act of willing, it occurs

---

iudicii rectitudinem sequitur rectitudo appetitus circa ordinata in illum finem. Sexto per huiusmodi appetitum incipimus operari."

[4] John Buridan, *op.cit.*, VI.19, 135vb: "... ultimus finis agibilium determinatus est nobis per naturam secundum eius communem rationem in tantum, quod nullus sic mentitur de eo secundum intellectum, quin sit amandus et prosequendus ... Et voluntas non potest ferri in oppositum secundum oppositum rationem ... Ideo necesse est, si voluntas debeat transire in actum circa vitam optimam sub ratione optimitatis quod velit eam."

on the assumption of the cognition in question. That willful action follows this judgment of necessity can only mean that this judgment concerns what is present in every willing insofar as it is an act of willing. In this way, the judgment can only concern willing as such. That is why Buridan speaks in the passage quoted (in n. 4 above) of the will's knowledge of the determination of its own action, through which and in which the will determines itself in an original and exclusively rational way.

(b) The second point is that the content of the absolutely universal end is limited according to its method of demonstration. Thus, one can assume that this is how Buridan arrives at the unrestricted obligation of the universal and necessary form of knowledge and science by means of the entirely rational self-determination of human volition. This assumption can be established by noting that in the text we are considering, the method of proof alone is applied to determine the concrete action.

A second argument for this interpretation comes from another text, in which Buridan ascertains that one can acquire from each singular judgment about what is to be done or left undone the universal requirement that everyone must at all times act in a similar fashion. This cognition, Buridan says, arises by virtue of the middle term and a similarly discursive process (*ex omni iudicio singulari potest habere iudicium universale virtute medii et ratiocinationis consimilis*).[5] Now, when Buridan says that the generalization in question arises "by virtue[6] of the middle term," he is not thinking of it as a predicate that characterizes a particular action, because if he was, the generalization could only be about this one deed, and not, as he claims, every concrete prescription. The middle term refers instead, classically speaking, to the cognition of the essence of the good as the reason for the joining of the subject and predicate affirmed in the conclusion.[7] In this way, the middle term refers to the relational structure of the subject and predicate in a prescription. The structure of strict commitment or normativity is in keeping with the uniqueness of this cognition as a cognition of the essence of the good. The middle term can therefore

---

[5] John Buridan, *op.cit.*, VI.17, fol. 131rb: "... videtur quod ex omni iudicio singulari potest statim habere iudicium universale virtute medii et ratiocinationis consimilis. Si enim per prudentiam tu iudicas me nunc debere hoc facere, constat quod ita iudicandum est quemcumque quocumque tempore debere simile opus facere similibus circumstantiis occurentibus."

[6] On the interpretation of "virtue" as "power" see Thomas Aquinas, *Summa contra gentiles* [Albert e.a.], Volume 4, 1: XVII.

[7] Cf. Aristotle, *Posterior Analytics*, II.2.

only be asserted as a prescription of conduct regarding the mode of discovery relative to strictly obligatory prescriptions of action. For this reason, the cognition of the essence of the good, which concerns the foundation of the prescription of action, is asserted with the help of the expression, "middle term," i.e., with the help of a technical term of syllogistic logic. Similarly, this identification of the middle term is followed by a proof, which enables us to determine the prescription of conduct in question. In other words: the principle for discovering moral content is connected to the application of the method of proof.

Finally, it can be said that this middle term, or the cognition of the essence of the good, consists in nothing other than Buridan's demand that one should act as everyone at all times ought to act, i.e., one should perform an act that strictly conforms to the rule.[8] Nothing else guarantees the proof. When Buridan says here that one can derive the requirement of strict regularity of action from any concrete prescription, he means that one who demands that something be done or left undone does so justifiably because in doing so, he determines himself by virtue of his demand for the necessary and universal form of knowledge and science.

The previously analyzed texts suggest that Buridan bases human willing on an original and rational self-reflection of the will. Accordingly, the teleology of human action results from the fact that it is willed exclusively through this rational determination, as we saw above. Hence, acting according to teleological determination means acting in a strictly law-like fashion, or according to the criterion of the necessary and universal form of knowledge and science.

Before moving on to the foundation of the desire for knowledge in Buridan's metaphysics, let us first sketch the significance of this basis for the will for the foundation of the absolutely first principles of knowledge and science. Then we can investigate a text from Buridan's ethics suggesting that this is what he must have had in mind.

With its original and reflexive self-cognition, the will determines itself both in its origin and in its foundation: as regards its origin, to the extent that it alone can be a source for the will; as regards its foundation, because the will determines itself in its volition exclusively through reason. Insofar as the latter is the case, there follows simultaneously the original cognition of the form of knowledge and

---

[8] J. J. Walsh has the same interpretation in "Buridan on the Connection," e.g.: "... portentous for the thought of Kant," p. 478.

science. With respect to the object, the absolutely first principles of knowledge and science are comprehended in an act of deliberate self-reflection. Insofar as the will in its willing determines itself in its origins, the a priori foundation of the cognition of reason or knowledge will follow.

The text from Buridan's ethics that supports this claim – viz., that Buridan, in the course of his defense of voluntary activity, conceives of the notion of the a priori nature of reason and knowledge – is one in which he articulates the distinction between theoretical and practical knowledge. Buridan traces the difference in its principle and original sense (*magis principaliter et originaliter*) back to the appropriate object of knowledge. As far as the foundation of knowledge is concerned, this means that it is not in our power to produce the theoretical or practical character of determinate knowledge. However, given one particular determinate piece of knowledge, we are able either to orient it to, or else to refrain from, a particular application. Buridan illustrates this with the example of moral philosophy, which is pursued only theoretically, or for the sake of knowledge. He supports this with a reference to Aristotle's remark that those who are interested in moral philosophy from a purely theoretical perspective often do not act appropriately.[9]

One could try, Buridan argues further, to embrace the teleological determination of knowledge as the basis of the distinction between science (*scientia speculativa*) and concrete knowledge (*istud scire*). Thus, scientific knowledge has a theoretical character, whereas concrete knowledge is capable of being applied and hence has a practical nature. Buridan takes exception to the objection that this is the only way we can distinguish between abstract and concrete knowledge. This effort, then, depends on a distinction of knowledge through itself, not on a distinction of knowledge as regards the end.[10]

From what we have seen so far, Buridan has also shown that the context in which the knowledge is applied is independent of the

---

[9] John Buridan, *op.cit.*, I.2, fol. 3rb: "Credo quod magis principaliter et originaliter sit haec distincio per obiecta, quia in potestate nostra non est scientiam unam facere speculativam aut practicam. Nos autem possumus eandem scientiam ordinare ad opus aut non ordinare, sed stare in scire, ut istam scientiam ethicorum. Propter quod Aristoteles secundo huius reprehendit eos, qui statum huius scientiae ponunt in scire, dicens quod multi quidem non operantur, sed ad rationem fugientes existimant philosophari et sic forte studiosi."

[10] *Ibid.*: "Item, scientia [speculativa] ab isto scire non distinguitur nisi sicut abstractum a concreto, et ita scientiam speculativam distingui per finem a practica est ipsam per seipsam distingui a practica."

knowledge itself. Knowledge seems essentially theoretical. Nevertheless, the context of its application does not explain the fact that the knowable is also potentially effective and hence of a practical nature. To illustrate this, one need only consider that it is not simply the desire for a house that leads to the capacity to build. Rather, the causal power emerges from the potential to be known. Therefore, knowledge is bound up with the practical determination of ends. But because knowledge is not by itself restricted to a theoretical or practical nature, the explanation of this difference must be located in the object of the knowledge in question. If the subject (*subiectum*) cannot exert any causal power through human beings, then they lack the power to relate it practically to themselves (*ex eo quod subjectum non est a nobis operabile, non stamus in scire*).[11] Thus, the crucial question is: why, or with what kinds of assumption, can human beings recognize what is causally efficacious as knowable?

This question can be answered by looking at the original self-determination of the will. More specifically, the effective object reveals itself as knowable through the assumption that what is or ought to be done is done in keeping with the structure of knowledge, or in a strictly universal manner.

This answer lies at the heart of the claim we are considering. It becomes clear when we note that Buridan distinguishes here between the sensory determination of knowledge and knowledge itself. Whether a piece of knowledge is related (or not) to practical or theoretical interests has nothing to do with the knowledge itself. Thus, some prior ground or criterion is needed to determine what can be known with a practical or theoretical intention. This is given in the aforementioned rational determination of the will, since that is how it is possible to make the very first judgment *that* something should be known at all.

But this criterion also permits us to pass judgment on *what* should be known. Even though Buridan ties knowing to objectivity, it still occurs in the context of the knowledge itself (*ex eo quod scibile est a nobis operabile, nos scientiam eius ordinamus ad opus*). In this way, knowledge of knowledge is necessary in order to be able to grasp the knowable as such. One may assume that this knowledge of knowledge is the a priori form of reason, or knowledge.

---

[11] John Buridan, *op.cit.*, I.2, fol. 3[rb-va]: "Item, scientiam ordinari ad opus vel non ordinari non est causa eius quod est scibile esse a nobis operabile vel non operabile, sed econverso ex eo quod scibile est a nobis operabile, nos scientiam eius ordinamus ad opus, et ex eo quod subiectum non est a nobis operabile, nos stamus in scire."

2. *The Practical* Locus *of the Foundation of Human Knowledge*

The preceding reflections support the claim that Buridan sees the human will as grounded in its original and rational self-reflection. Moreover, they suggest that when he considers the foundation of knowledge and science, Buridan has the aprioricity of reason and knowledge in mind.

With this background, we turn to the foundation of the desire for knowledge in Buridan's *Questions on the Metaphysics* (Book I, q. 5). First, we need to ask how Buridan derives the foundation of human action from his idea of its practical application. Second, we need to clarify whether Buridan bases knowledge and science in the aprioricity of reason. The first question will be answered in the second part of this study; the second question will be discussed in the third and fourth parts.

Our point of departure in the second part is the distinction between "natural appetite (*appetitus naturalis*)" and "animal appetite (*appetitus animalis*)" in Book I, q. 5. According to Buridan, animal appetite arises through cognition, and thus strives toward the good. As this mediation through cognition marks the distinction between natural and animal appetite, we can assume that natural appetite arises without this mediation, as Buridan says, in its inclination to its particular good.[12]

The exact sense of "natural appetite" depends on the distinction with "animal appetite" (a more detailed discussion of the latter will follow in the analysis of "doctrinal appetite (*appetitus doctrinalis*)" in the fourth and fifth parts of this study). A human being, Buridan says, possesses a twofold nature and a twofold striving, i.e., sensitive and intellectual. They are opposed to each other. In the sensitive appetite (*appetitus sensualis*), we have a principle naturally oriented against reason. That is why no human being desires to know according to his sensitive nature (*nullus homo secundum sensum appetit naturaliter scire*). Nevertheless, it is certainly true that man naturally desires to know. He does this not in virtue of his sensitive and inferior nature,

---

[12] John Buridan, *In Metaphysicen* (Paris, 1518); Book I, q. 5, fol. 5vb: "Appetitus autem solet distingui, quia alius vocatur animalis, alius naturalis. Naturalis dicitur, quia inclinat in suum bonum ... Sed appetitus animalis est, qui inclinat ad bonum vel apparens bonum mediante cognitione illius boni ... Et non oportet credere quod iste appetitus animalis sit omnino naturalis, quia hoc non esset bene credere. Appetitus enim animalis per ipsam animam, quae est natura ipsius animalis, non est praeter naturam et per consequens naturalis. Sed non est consuetum illum vocari naturalem, quia non est immediate ab ipsa natura, sed mediante cognitione ..."

but in virtue of his intellectual nature, which is of higher worth. To be human in absolute terms, then, is to be determined according reason (*qualis est homo secundum intellectum, talis est dicendus simpliciter*).[13]

In this connection, we ought to ask whether Buridan, in making this observation, identifies being human in the strict sense with reason or with being determined by reason. Could one not understand this in terms of the anthropological claim that man alone *qua* rational is capable of knowledge? Against this interpretation, however, is the fact that Buridan takes the position on the desire for knowledge that man is capable not only of possessing knowledge, whether he has this knowledge at his disposal or not, but also of appropriating it as such. Buridan accordingly gives an account of the practice of knowing as such. Second, he draws a normative conclusion that is in keeping with the practical nature of his reflection: "it must be said (*est dicendus*)." Third, he seems to be making a normative-practical rather than an anthropological claim, as is suggested by the fact that he uses the idea that something is absolute or unrestricted in a twofold sense: first, in relation to man being determined by reason (*qualis est homo secundum intellectum talis est dicendus simpliciter*); second, in relation to the truth of this assertion (*ista conclusio est simpliciter concedenda, debet homo simpliciter denominari*). This is how the normative-practical form of the idea that man is determined by reason is articulated. If the assertion were of an anthropological nature, it could not at the same time be said that it must be conceded (*simpliciter concedenda est, debet denominari*). Since Buridan is speaking in a twofold sense here, he emphasizes the difference between the assertion that being human is characterized solely according to reason, and the strictly binding sense of this assertion.

A further reason for this interpretation is given in Buridan's argument here for the renunciation of knowledge. This argument claims that, although it is willed, such a renunciation stems from a natural desire (*desiderium naturale*). Specifically, it is willed because it is based on a conception proper to animal appetite (*appetitus animalis*), which itself gives rise to the first cognition that makes the

---

[13] John Buridan, *op.cit.*, I.5, fols. 5vb-6ra: "… homo quodammodo compositus est ex duplici natura, scilicet ex natura sensuali et natura intellectuali. Ideo etiam duplex est appetitus, scilicet intellectualis et sensualis, et isti appetitus aliquomodo obviant sibi invicem. … nullus homo secundum sensum appetit naturaliter scire. … ista conclusio est simpliciter concedenda 'Omnis homo naturaliter appetit scire', quia a superiori et nobiliori parte et non ab inferiori debet homo simpliciter denominari. Natura enim intellectualis est superior et nobilior et natura sensualis inferior. Ergo qualis est homo secundum intellectum, talis est dicendus simpliciter."

will free (*voluntas est libera mediante cognitione prima*).[14] Since there is a distinction between a cognition that makes the will free and a cognition belonging to animal appetite (*appetitus animalis*), which renounces the value of knowledge, only the former cognition can serve as the basis of natural appetite (*appetitus naturalis*).

That this cognition is of an original and reflexive nature is further confirmed in Buridan's claim that the will is free by virtue of its first cognition. Because it is a cognition that makes the will free, it is the cognition of practical reason, concerning what is to be done or left undone, chosen or not chosen. And because it is the first cognition, it cannot allow the will to choose between one or another particular good: either the first cognition would choose a particular good in relation to which the will would not be free, or else the will is not involved with any such first cognition. If the will is free by virtue of a first cognition, then its choice can only be directed to the good in general, and its freedom could consist only in determining itself to willing as such by virtue of the first cognition. As Buridan says elsewhere in the *Questions on the Metaphysics*,[15] the will can move from a state of indifference, or not willing (*non velle*), to a state of willing for (*velle*) or willing against (*nolle*). This indifference is neither caused by something other than the will, nor does it exist prior to (*aliis ante positis*) the act of willing for (*velle*) or willing against (*nolle*). For if it were caused by something other than the will, then the will would not be free relative to the transition from indifference to the actualization of volition. Insofar as there is a temporal order in which the indifference precedes the will's actualization of volition, there could be a cognition embracing this transition, except that it would have to be of either a theoretical or a practical nature. If it is theoretical, the transition in question could not occur spontaneously. But if it is practical, the cognition comprehending the temporal nature of the transition from indifference to willing would occur on the basis of the volition, and so it would not be a first cognition.

---

[14] John Buridan, *op.cit.*, I.5, fol. 6ra: "... si aliquis desiderat aliqua ignorare, hoc non est desiderio naturali, sed animali. Unde multotiens voluntas quae est libera mediante cognitione prima, inclinatur contra illud, ad quod appetitus inclinatur naturaliter."

[15] John Buridan, *op.cit.*, VI.5, fols. 36vb-37ra: "... si voluntas quiescit tam ab actu volendi quam ab actu nolendi ita quod neutrum producit, aliqua alia est causa quietis, quam oportet removeri antequam agat. Dico quod non, quia ista sola per suam libertatem aliis ante positis est causa sufficiens et ad agendum et ad non agendum, etiam ad agendum volitionem et ad agendum nolitionem."

If the will is free by virtue of a first cognition, then its freedom consists in the ability to go from not willing (*non velle*) to willing (*velle*) as well as to nilling (*nolle*). This means that it is capable, on its own, of being a source of volition. If the will is free by virtue of a first cognition, then only to the extent that this cognition belongs to the will as such – i.e., to the extent that it characterizes and brings about the constitutive moment of the will – does the will actually will. The insight that serves as the basis of freedom is therefore the original and purely rational self-comportment of the will.

The final argument to be articulated here for this interpretation (viz., that Buridan identifies what it is to be a man in the normative-practical sense with being determined exclusively by reason), results from the aforementioned foundation of the will in the *Questions on the Nicomachean Ethics*. Buridan speaks of an immediate judgment of reason that does not come about through any discursive process, just as he speaks of natural appetite (*appetitus naturalis*) – i.e., which proceeds from the nature of the intellect without any prior method of proof.[16] This is further supported by the fact that any such proof determining what ought to be done is the sole constituent of the absolutely universal end.[17] In other words: the unrestricted obligation of the necessary and universal form of knowledge and science follows from reason's exclusive determination of the will. To this extent, what it is to be human is in the true sense determined by being determined exclusively by reason.

This last argument makes clear the agreement between the foundation of the will in the *Questions on the Nicomachean Ethics* and that of natural appetite (*appetitus naturalis*) in the *Questions on the Metaphysics*. If we begin by noting that Buridan is also committed in the *Metaphysics* to the original self-determination of the will, we can conclude provisionally that Buridan traces the desire for knowledge back to the human will and not to nature in the aforementioned way. The next section considers the meaning of this desire for knowledge as regards the absolutely first principles of knowledge. To what extent does Buridan base knowledge and science as such – i.e., the necessary

---

[16] Cf. n. 3 above (John Buridan, *Quaestiones super decem libros Ethicorum Aristotelis* (Paris, 1513), VI.5, fol. 121vb): "... intellectus recte iudicat de fine saltem quoad communem rationem per eius naturam absque ratiocinatione."

[17] Cf. n. 3 above (*Ibid.*): "Tertio ex eo quod ad finem tendit appetitus, intellectus movetur ad ratiocinandum ... Per appetitum igitur finis determinatur intellectus ad ratiocinandum."

and universal form of knowledge and science – on the aprioricity of reason?

3. *The Original and Reflective Nature of the Cognition of the First Principles*

The following remarks on Buridan's defense in the *Questions on the Metaphysics* of the absolutely first principles of knowledge and science are divided into three parts: first, I will examine his claim about the natural desire for knowledge (1); second, I will analyze his statements about our natural possession of principles (2); and finally, I will investigate a further passage from this work more closely (3).

(1) Buridan claims that the intellect is inclined to knowledge and truth on the basis of natural appetite (*appetitus naturalis*), as against animal appetite (*appetitus animalis*), since it is inclined to recognize the truth of first principles before it cognizes a particular. For this reason, the intellect assents to these principles, or does not assent to them, on the basis of a cognition that precedes the act of consent (*non consentit in virtute aliquorum praecognitorum*).[18]

For the interpretation of this text, it is crucial to note that Buridan excludes the possibility that the intellect's inclination toward knowing and truth belongs to the animal appetite (*appetitus animalis*) and arises through the preceding cognition. In ascribing this inclination to natural and not to mental striving, he is excluding the possibility of infinite regress. Thus, the process of assent is also shown to be original. Second, this interpretation is supported by the fact that the intellect's desire in relation to its substance is the self-cognition of reason. Of course, it might be objected that this inclination precedes any cognition. This suggests that we are not concerned here with any inclination of the animal appetite (*appetitus animalis*); that natural appetite (*appetitus naturalis*) is not to be explained through cognition means that it does not arise through a discursive process. Moreover, this natural inclination is substantially concerned with the assent of reason to first principles. Insofar as Buridan rejects the idea that the intellect's assent to these principles

---

[18] John Buridan, *In Metaphysicen* (Paris, 1518), I.5, fol. 6ra: "... intellectus sic est inclinatus ad scientias et veritates per appetitum naturalem et non animalem appetitum, prout in isto casu animale distinguitur contra naturale, quia inclinatus est ad assentiendum veritatibus priorum principiorum antequam aliquid cognoscat. Ideo non consentit in virtute aliquorum praecognitorum."

is due to a prior cognition, one is led to believe that it is attributable to the determination of reason that serves as the foundation for the natural appetite (*appetitus naturalis*). Thus, reason is agreeing with itself when it assents to principles. Therefore, the natural inclination of the intellect toward knowledge and truth is substantially reflexive in nature. When Buridan claims that the intellect is naturally inclined to agree with the truth of principles, he is also making explicit the original and reflexive determination of reason that grounds the will.

To this extent, this text agrees with the results of the preceding reflection on the natural appetite (*appetitus naturalis*). But it raises another question, leading us further, about the precise sense of the first principles recognized by reason. Let us investigate a passage in the paragraph preceding the passage just discussed. Here Buridan explains that the intellect could agree with any principle, just as it could with its opposite, if this principle were not previously known (*praecognivisset*). Further, he emphasizes that we cannot begin with the fact that all men immediately (*statim*) consent to first principles, and to the negation of their opposites, and explain this cognition by appealing to a temporally prior cognition.[19]

In keeping with these claims, Buridan hopes to explain our cognition of first principles in terms of an assent not based on temporally prior knowledge, and also as not excluding the possibility of error or contradiction relative to any principle. Thus, it is tempting to assume that the desired explanation involves assent to the principle as principle, i.e., the cognition of the form of necessity and universality. In this way, it is conceivable that even the principle of non-contradiction could be simultaneously recognized and assented to, and denied or doubted. That is because such a contradiction or error presupposes the truth of the principle. Thus, human apprehension would itself be the ground and cause of the truth of its own principles, or their necessity and universality.

Second, this interpretation fits with what Buridan further claims, viz., that the agreement is self-explanatory given that man by virtue of his reason naturally desires to know, and thus at the same time to assent to the truth of first principles, as well as being inclined by cog-

---

[19] John Buridan, *op.cit.*, I.5, fol. 6ra: "Si intellectui proponatur aliquod principium, si numquam praecognivisset illud principium, ipse non magis assentiret illi quam suo opposito. Igitur quia videmus quod omnes homines assentiunt statim primis principiis et dissentiunt oppositis, oportet concludere quod hoc sit, quia alias intellectus cognoverat haec principia. Hoc erat argumentum Platonis, quod non potest solvi nisi dicendo quod intellectus noster naturaliter inclinatur ad scire et ad veritatem."

nizing them to agree to the truth of the conclusions.[20] This supports the proposed interpretation, first insofar as it traces the cognition of first principles back to the natural desire for knowledge and thus to the original and reflexive relation of reason; and second, to the extent that in recognizing the ground of first principles, it simultaneously legitimizes the truth of the conclusions. For if one assumes that the ground of first principles does not involve the truth of the conclusions included in them, but only their form as conclusions, Buridan shows why the method of demonstration is valid. This claim is thus in agreement with the fact that he deduces the unrestricted obligation of the method of proof as a process of determining concrete action from the original self-determination of the will.[21] When he concludes the passage in question by claiming that we assent to first principles without having already recognized them,[22] one can understand, in view of the interpretation presented here, that the truth or validity of first principles follows from reason's assent, or agreement with them. Understood in this way, this claim makes the aprioricity of reason and knowledge explicit.

(2) When Buridan articulates his position on our natural possession of principles, he distinguishes between natural appetite (*appetitus naturalis*) and animal appetite (*appetitus animalis*), just as he did when discussing our natural desire for knowledge. He says that animal appetite arises from cognition (*mediante cognitione*), which is not the case with natural appetite (*sine cognitione mediante*). Accordingly, he further differentiates between natural assent (*assensus naturalis*) and doctrinal assent (*assensus doctrinalis*); it is through the former that we naturally assent to principles (*naturaliter pincipiis assentimus*). Natural assent emerges from the nature of the intellect without reaching a conclusion by necessity through prior reasoning (*ex natura intellectus sine ratiocinatione praevia necessario concludente*). It is on the basis of such prior reasoning that doctrinal assent (*assensus doctrinalis*) arises.[23]

---

[20] John Buridan, *op.cit.*, I.5, fol. 6ra: "Et sic concedendum est quod homo secundum intellectum naturaliter desiderat scire et ad assentiendum veritatibus primorum principiorum et etiam est inclinatus ad assentiendum veritatibus conclusionum mediante tamen notitia principiorum."

[21] Cf. n. 3 above.

[22] John Buridan, *op.cit.*, I.5, fol. 6ra: "Ideo consentit illis et non oppositis, quamvis illa numquam praecognoverit."

[23] John Buridan, *op.cit.*, II.2, fol. 10ra: "Ideo dicendum est quod sicut appetitus naturalis distinguitur contra appetitum animalem – ex eo quod appetitus naturalis dicitur, qui est sine cognitione mediante et animalis vocatur, qui est mediante cogni-

With regard to natural assent (*assensus naturalis*), we can say first of all that Buridan again appeals to his conception of the original and reflexive relation of reason that is the foundation of the will. The following remarks will concern Buridan's more detailed explanations of the principles to which we assent in this relation of reason.

Let us begin with Buridan's account in the body of the Question. Here he speaks of "all modes of indemonstrable principles" (*habetur omnes modos principiorum indemonstrabilium*), from singular principles (*principia singularia*) to those principles conceded on the basis of neither previous experience nor proof (*sine experientia praecedente conceduntur et etiam absque demonstratione*). In addition to the principle of non-contradiction, the latter category includes judgments about the species-genus or the species-species relationship of a class of objects.[24] Determining their relationship appears to be the reason why Buridan includes the latter judgments and the principle of non-contradiction in the same group of principles. This is quite understandable to the extent that foundational logical relationships are at issue. But it is also beyond doubt that for Buridan, the content we use to characterize individual objects – e.g., that fire is hot (*iste ignis est calidus*) – is known on the basis of experience (*accepta per sensum*). Likewise, no one should doubt that experience alone teaches us that a single object belongs to a class of objects – e.g., every fire is hot (*omne ignis est calidus*).[25] The scholarly literature tends to focus only on what the arguments are for Buridan's claim that such judgments are universal. Technically speaking, the question is: On what basis does Buridan establish the universality of inductive conclusions?[26] Though judgments about species-genus or species-species relationships and the principle of non-contradiction are not based on experience, the context of experience, the particular principles,

---

tione – ita distinguimus assensum naturalem contra assensum doctrinalem. Vocatur enim assensus naturalis, qui provenit ex natura intellectus sine ratiocinatione praevia necessario concludente. Et sic intendit Commentator et alii quod principia insunt nobis a natura, id est eis naturaliter assentimus. Sed assensus doctrinalis vocatur, qui provenit ex ratiocinatione praevia de necessitate concludente."

[24] John Buridan, *op.cit.*, II.2, fol. 9vb: "... ut quod omnis equus est animal, quod omne ferrum est metallum, quod nullum calidum est frigidum et sic etiam consimiliter primum principium."

[25] *Ibid.*: "... sciendum est ... quod aliqua sunt principia indemonstrabilia accepta per sensum, ut quod iste ignis est calidus ... etiam sunt aliqua principia universalia, quae propter experimenta in multis singularibus consimilibus conceduntur ab intellectu propter naturalem inclinationem intellectus ad veritatem sicut quod omnis ignis est calidus."

[26] Cf. the works by P. King, J. M. M. H. Thijssen, and J. Zupko, cited in n. 1 above.

and also the universal principle reached through induction, suggest that Buridan sees such judgments as being related to the principle of non-contradiction. In other respects, the question of the justification of inductive conclusions can also be connected to efforts to define how judgments about species-genus or species-species relationships and the principle of non-contradiction are related to each other, since the perspective of universality also contains a relational moment.

Buridan's account suggests that principles that are not conceded on the basis of experience or proof involve basic logical relations such as the principle of non-contradiction. This accords with what we saw above concerning Buridan's explanation of the cognition of first principles in the context of his position on our natural desire for knowledge: when our assent to first principles actually concerns the aprioricity of reason and knowledge, it has only the logical form as its object.

This analysis of Buridan's exposition of our natural possession of principles is further supported by the final argument in the series of introductory arguments in this Question. Buridan uses it to support the thesis that principles are not possessed by nature. The reason is that otherwise, principles could not be denied or assented to. He establishes this by arguing that the principle of non-contradiction or the principles foundational to natural science (*scientia naturalis*) have been negated just like practical principles. In the former case, the dissent stems from someone used to hearing the opposite, whereas we dissent from practical principles because of wickedness.[27]

With this background, the question about first principles as the object of natural assent (*assensus naturalis*) can be formulated as follows: how can these principles be valid, given that they cover foundational logical principles, such as the principle of non-contradiction, and principles fundamental to natural science (*scientia naturalis*), as well as practical principles? And how is the foundation of these principles to be understood when natural assent (*assensus naturalis*) is the original and reflexive cognition of reason?

---

[27] John Buridan, *op.cit.*, II.2, fol. 9va: "... si principia essent nobis naturaliter habita, sequeretur quod non possemus ea negare neque eis dissentire. Modo consequens est falsum, igitur. Consequentia videtur de se nota, et falsitas consequentis patet quarto huius, quia quidam negaverunt primum principium, contra quas disputat ibi Aristoteles. Et hoc etiam dicit Commentator in secundo huius, scilicet multi negaverunt principia supposita in scientia naturali propter consuetudinem audiendi opposita. Et similiter in sexto ethicorum dicit Aristoteles quod multi propter malitiam non possunt principiis practicis assentire."

To answer this question, let us begin with the aforementioned relation between the principle of non-contradiction on the one hand, and those principles fundamental to natural science (*scientia naturalis*), as well as practical principles, on the other. This relation must not be understood as consisting solely in our ability to deny the respective principles. Rather, let us begin with the relation between these three principles, or groups of principles. This relation can be found in the very nature of the principles as such. It constitutes the form of necessity and universality, and thereby the very nature of these principles as principles, allowing them to function as such.

This interpretation can be established by two arguments. First, the formal relation makes clear that a class of objects is being assumed with regard to the groups of principles, i.e., the principle of non-contradiction on the one hand and judgments as regards the species-genus or species-species relation on the other.[28] This is evidence for the interpretation we have been discussing here insofar as there is no other common characteristic of principles considered with regard to the form of knowledge except the unity of their logical form. Because of this unity, the principle of non-contradiction and judgments of species-genus or species-species relation can serve as the basis for natural science (*scientia naturalis*), as well as representing practical principles. Second, Buridan regards natural appetite (*appetitus naturalis*) and its corresponding natural assent (*assensus naturalis*) as original relations of reason. Insofar as human action is here strictly governed by the application of the method of proof,[29] it shows that the necessary and universal form (of knowledge) guarantees the unity of science in view of the practical principles. That is why Buridan also appears to accept that the form of necessity and universality is recognized through reason's original determination of action. It further suggests that the a priori foundation of the necessary and universal form of knowledge and science simultaneously arises from this determination of reason.

This interpretation of the original and reflexive determination of reason in terms of the aprioricity of reason and knowledge finds additional support in Buridan's admission that we have the power to deny the principle of non-contradiction and the founding principles of natural science (*scientia naturalis*), as well as practical principles. He traces this possibility back to habit or wickedness, noting that although we naturally assent to principles of this sort (*huiusmodi*

---

[28] Cf. the explanation in view of the text quoted in n. 24.
[29] Cf. the explanation in view of nn. 3–4.

principiis), our assent is valid only when there is no impediment. On its own, nature always acts correctly and completely, but men, due to impediment, occasionally make mistakes in the realm of nature and with regard to natural operations. For this reason, men are sometimes also hindered in assenting to principles. More specifically, impediments concern the habit acquired from having heard people assert the opposite of principles (*consuetudo audiendi opposita*), or with wickedness (*malitia moralis*). If these impediments are removed, however, we assent to the principles in the way described.[30]

It is important to note that when Buridan says that it is in our power to deny the principle of non-contradiction, he is considering that our cognition of first principles might stand in need of explanation. According to this explanation, the assent given to these principles is not based on previous knowledge, and the possibility of error or contradiction with regard to any principle is also not excluded.[31] In view of this, it was proposed that the desired explanation involves assent to a principle as principle, i.e., cognition of the form of necessity and universality. This proposal would be confirmed if it could be shown that Buridan understands the original and reflexive cognition of reason in terms of its aprioricity. As we have seen, he attributes the possibility of denying the principle of non-contradiction, or the principles of natural science (*scientia naturalis*) and practical principles, to habit or wickedness. Furthermore, he describes this denial as an "impediment," the removal of which makes it possible for us to assent to the principles in question. Accordingly, the original and reflexive cognition of reason can be understood in terms of a clarification of the precise sense of these impediments and their removal.

There is a second point that needs to be emphasized: Buridan draws the distinction between natural and animal appetite when he considers our natural desire for knowledge, as well as when he discusses our natural possession of principles.[32] It is on the basis of this distinction that he further distinguishes between natural assent

---

[30] John Buridan, *op.cit.*, II.2, fol. 10ra: "... quamvis naturaliter assentiamus huiusmodi principiis, hoc tamen est intelligendum, si non occurrat impedimentum. Natura enim quantum est de se semper agit recte et perfecte, sed aliquando per impedimentum accidit peccatum in natura sive eius operatione. Ideo etiam aliquando homines impediuntur ut non assentiant principiis, scilicet vel propter consuetudinem audiendi opposita vel propter malitiam moralem. Sed impedimentis remotis non assentimus eis modis praedictis."

[31] Cf. n. 19.

[32] Cf. nn. 12 and 23 above.

and doctrinal assent.[33] As we shall see, the precise sense of "doctrinal assent" is also to be explained by the relation of reason involved in natural appetite and natural assent.

(3) Before moving on, let us look once more at Buridan's discussion of our natural desire for knowledge, examining more closely an example presented there. With reference to Plato's teaching of anamnesis, this example compares human learning to a master's efforts to catch his fugitive slave. If the search is to have any success at all, the master must know which one of his slaves has run away from him.[34] With the explanation provided thus far of the original and reflexive meaning of natural assent – and especially considering that Plato himself illustrates the doctrine of recollection with the knowledge not of the master, but of the slave[35] – we can see that the original meaning of the intentional origin of knowledge is articulated in the relation of mastery. Seen in this way, it is clear that Buridan from the outset grasps the problem of learning, or more generally, the possibility of human knowledge, from the standpoint of intentional self-determination, i.e., from the standpoint of freedom. From a practical perspective, it is necessary, in the interest of freedom, to attain certainty in relation to the claim of learning or knowledge. Going back to the example, we can ask: Over which slave does the master have a claim of mastery? In order to identify this slave, the master must actually know him. Likewise, in the discussion at hand, we can say that in order to be able to judge knowledge claims, we must first understand the nature of knowledge as such.

4. *The Aprioricity of the Cognition of the First Principles*

Turning to Buridan's discussion of the possibility of error as regards the first principle, let us begin with the distinction he draws as regards the possibility of error in the mind (*mente*). Error can arise, he says, because one assents to a false proposition or denies a true proposition, one of whose terms supposits for the first principle (*pri-*

---

[33] Cf. n. 23 above.
[34] John Buridan, *op.cit.*, I.5, fol. 6ra: "... dicebat Plato quod nostrum addiscere non est nisi quoddam reminisci. Et hoc probat per exemplum de servo fugitivo, quia si dominus eius quaerat eum et non prius cognoverit ipsum, non magis capiet illum quam istum."
[35] Cf. Plato, *Meno*, 82a8–84b1.

mum principium). Thus, one can be in error in many ways. Buridan even points to his own inquiry as an example.[36]

The doctrine of supposition has to do with the capacity of spoken expressions to stand for their propositional content or meaning. That Buridan situates it on the "mental" level[37] indicates that he understands the concrete thoughts or representations in which the propositional content or meaning is comprehended as terms performing this function. In so doing, he makes thoughts or representations equivalent to the expressions of written or spoken language. Hence, by pointing to his own investigation of the concept grasped by means of the Latin expression *primum principium*, he is considering its ability to stand for the content in question. One is led to believe that in itself, this content possesses no being of its own. In any case, it can be distinguished from the reality of the concrete thoughts in which it is considered.

The second way in which Buridan admits the possibility of error with regard to the first principle (*primum principium*) concerns the mental level of the content expressed by the term *primum principium* (*circa primum principium est ipsi dissentire vel eius opposito assentire*). In this case, Buridan agrees with Aristotle that deception is impossible, "at least according to nature" (*saltem naturaliter*).[38]

When Buridan claims that we cannot be deceived "by nature" with regard to the objective content of the *primum principium*, he is suggesting that it does not follow from the fact that reason assents by nature to the content of the first principle that it also assents to the proposition in which this principle is expressed. When reason assents by nature to the content of the first principle, it is not assenting to the

---

[36] John Buridan, *op.cit.*, IV.12, fol. 21[va]: "... errare circa primum principium est mente assentire propositioni falsae vel dissentire propositioni verae. Et hoc potest esse dupliciter. Uno modo quia nos assentimus propositioni falsae vel dissentimus propositioni verae, in qua terminus aliquis supponit pro primo principio. Et sic adhuc sine dubio possumus circa primum principium errare multipliciter sicut in praesenti quaestione ..." This text, as well as the one following this Question, are quoted from the edition by L. M. de Rijk in his "John Buridan," p. 300. De Rijk corrected the 1518 edition on the basis of two manuscripts.

[37] Buridan admits that with regard to a spoken expression (*ore*), one can affirm the contrary. See John Buridan, *op.cit.*, IV.12, fol. 21[va]: "Dicendum est breviter quod 'circa primum principium errare' potest intelligi dupliciter. Uno modo ore affirmare contrarium, et ⟨hoc modo⟩ est valde possibile, prout quilibet potest experiri." Quoted from De Rijk, "John Buridan," p. 300.

[38] John Buridan, *op.cit.*, IV.12, fol. 21[va]: "Alio modo errare circa primum principium est ipsi dissentire vel eius opposito assentire. Et de hoc dico cum Aristoteli quod impossibile est sic circa primum principium errare, saltem naturaliter." Quoted from De Rijk, "John Buridan," p. 300.

complex or propositional expression of the first principle (*primum principium complexum*). Thus, the natural assent of reason (*assensus naturalis*) to the first principle, understood in terms of its content, represents our cognition of the form of necessity and universality.

Moreover, it is also true that one cannot deceive oneself by nature with regard to the objective content of the *primum principium*. Buridan notes that Aristotle proves this in the right way (*bene*), i.e., in the sense of an "explanatory demonstration (*demonstratio propter quid*):" by not assenting to the first principle or affirming its opposite, there would be simultaneous convictions in the intellect contradicting each other in their content (*est habere simul in intellectu opiniones contradictoriorum*). But this is impossible, for convictions contradicting each other in this way are contraries, and contraries cannot exist in the same subject or intellect (*non possunt esse simul in eodem subiecto vel intellectu*).[39] We can express this idea by considering the following two sentences: (1) "Buridan is of the conviction that one can affirm the objective content of the *primum principium* as such;" and (2) "Buridan is of the conviction that one can deny the objective content of the *primum principium* as such." Taking the contents of (1) and (2) together, we can say: (3) "Buridan is of the conviction that one can affirm and deny the objective content of the *primum principium* as such." Thus, we have a possibility that cannot be denied on the grounds that one cannot be deceived as regards the objective content of the first principle as such, i.e., a possibility referring to those propositions thought by Buridan when he considers the possibility of error with regard to the *primum principium* as such, having earlier denied this possibility. The two convictions do not oppose each other in the mind as contradictories, but contraries. This is so for two reasons. First, each conviction stands for itself in affirming the objective content of the *primum principium* as such. Insofar as each is actually thought, it is based on an act of will. But because this act of will is in turn based on the natural assent (*assensus naturalis*) or original determination of reason, including the a priori cognition of first principles, each conviction affirms the objective content of the *primum principium* as such. Second, each conviction stands to the other as its contrary. In the case of the conviction that the objective con-

---

[39] John Buridan, *op.cit.*, IV.12, fol. 21[va]: "Tamen bene Aristoteles demonstrat propter quid hoc est, scilicet quia ei dissentire vel eius opposito assentire est habere simul in intellectu opiniones contradictoriorum. Et hoc est impossibile, cum opiniones contradictoriorum sint contrariae, et contraria non possunt esse simul in eodem subiecto vel intellectu." Quoted from De Rijk, "John Buridan," p. 300.

tent of the *primum principium* as such can be affirmed, the conviction and the objective content agree, whereas if the opposite conviction is affirmed, they do not. The two convictions are contraries because the intellect finds itself simultaneously affirming and denying the objective content of the first principle as such. Thus, insofar as we are concerned with the relationship of the intellect to the objective content of the first principle, we are concerned with the relationship of the intellect to itself. That these convictions are contraries means that the intellect simultaneously affirms and denies itself. But this is impossible (*contraria non possunt esse simul in intellectu*).

Should we agree with Buridan that it is impossible to deceive oneself as regards the objective content of the *primum principium* as such in this way, i.e., in terms of an explanatory demonstration (*demonstratio propter quid*)? One can only make this claim if one has first admitted that the objective content of the *primum principium* as such refers to the form of necessity and universality or the form of knowledge. Furthermore, the aforementioned proof counts as a priori if the original and reflexive knowledge of the form of necessity and universality is recognized and presupposed in the relation. Under this assumption, the intellect cannot simultaneously affirm and deny itself. The proof is a priori because the intellect's self-comportment is a priori. In other words, the aprioricity of the proof stems from the aprioricity of the self-cognition of reason.

This confirms the view that in thinking about the original and reflexive determination of reason, Buridan "discovers" the aprioricity of reason and knowledge. Moreover, Buridan claims that eliminating the impediments obstructing our assent to principles permits us to assent to them in the ways described (*nos assentimus modis praedictis*).[40] The two ways of assent being referred to here are natural assent (*assensus naturalis*) and doctrinal assent (*assensus doctrinalis*). The a priori aspect of the proof involves the removal of impediments and the twofold assent obtained thereby: first, relative to the object, insofar as the proof-procedure has the self-cognition of reason as an object (to this extent, the removal of impediments is the same as the cognition of the aprioricity of reason or the form of knowledge and science); second, from a practical or normative standpoint, insofar as the assent realized in the self-cognition of reason results in an act of assent from the process of argumentation. To this extent, the removal of impediments is equivalent to understanding the strictly

---

[40] Cf. n. 29.

binding function of the form of knowledge. Accordingly, the aforementioned proof of the aprioricity of reason presupposes natural assent (*assensus naturalis*) as well as doctrinal assent (*assensus doctrinalis*), if it is conceded that reason cannot affirm and deny itself at the same time.

Given the precise sense of doctrinal assent (*assensus doctrinalis*), there remains the task of explaining how habit and wickedness, in relation to the self-cognition of reason, make error possible. They appear to correspond to the two ways in which impediments can block the self-cognition of reason. Thus, habit leads to error when it prevents cognition of the aprioricity of reason, and wickedness, when it does not permit action according to the binding function of the form of knowledge.

5. *The Fact of Reason*

The aforementioned explanation should be seen in light of Buridan's view that the mistaken possibility of error relative to the objective content of the *primum principium* as such cannot be demonstrated in terms of a "factual demonstration (*demonstratio quia*)" (*non potest probari quantum ad quia est*). Buridan touches on this in connection with his claims about the a priori possibility of proving the aprioricity of reason.

Let us begin with his claim that cognition of the fact of the *primum principium* as such represents an a priori cognition of the form of knowledge or reason as reason. On this assumption, Buridan admits that one cannot prove this cognition as a "fact that" (*quia*), i.e., according to its reality. Furthermore, he admits that one cannot prove this reality of reason. Rather, one can only assume it and thus be only subjectively certain of it. In other words, that the reality of reason can only be presupposed is acceptance of the fact of reason.

A broad definition of the possibility of error can be given in Buridan's terms as follows: relative to one's own action, there is the possibility of evident appearance by virtue of which one can be morally certain, even though it is possible that the judgment is false when viewed objectively because of unavoidable circumstances.[41] For ex-

---

[41] John Buridan, *op.cit.*, II.1, fol. 9ra: "... est adhuc alia debilior evidentia quae sufficit ad bene agendum moraliter, scilicet quando visis et inquisitis omnibus circumstantiis factis, quas homo cum diligentia potest inquirere, si iudicet secundum exigentiam huiusmodi circumstantiarum, illud iudicium erit evidens evidentia suffi-

ample, Buridan remarks that it is possible for a judge to act in a morally good and meritorious way in condemning a saint to death because it appears to him sufficiently certain, from witnesses and legal documents, that this person is a murderer.[42]

So error and subjective certainty are possible with respect to one's own actions. The judge could not rule out an objective error, since he would accept that the person being judged might be a saint. He could acknowledge that from another perspective, i.e., from which one knows the holiness of the judged, the act that led to the judgment has a different appearance. But from the judge's perspective, the holiness would be irrelevant, i.e., when viewed objectively from the standpoint of the function of judging. Hence, the judge can be certain of the rationality of his conduct as judge because he practices it objectively under given conditions, although the possibility of objective error from the other standpoint would render his actions only subjectively certain. In other words, although he cannot be absolutely certain of the rationality of his actions, he can be factually certain, and thus be factually certain of his reasoning.

Our inability to provide an a priori proof of the aprioricity of reason suggests that objective error and subjective certainty can be simultaneously present in our actions as well. There is the example Buridan mentions on numerous occasions, in which he questions some old women about the truth of the principle of non-contradiction. He turns to the old women and asks them if they have the ability to sit and not sit at the same time. They answer immediately that this would be impossible. He then asks them if they believe that God could do such a thing. They reply that they don't know, for "God can do everything, and it must be believed that he can do the impossible."[43]

In general, Buridan argues that one can be in doubt (*formido*) about the principle of non-contradiction, although no one would

---

ciente ad bene agendum moraliter, etiam licet iudicium sit falsum propter invincibilem ignorantiam alicuius circumstantiae."

[42] *Ibid.*: "... possibile esset quod praepositus bene et meritorie ageret suspendendo unum sanctum hominem, quia per testes et alia documenta secundum iura sufficienter apparet ipsi quod ille bonus homo esset malus homicida."

[43] Buridan also uses this example in Questions addressed to the natural possession of the principles and the possibility of error in relation to the *primum principium*. In the former context (John Buridan, *op.cit.*, II.2, fol. 9vb), he states: "... petivi enim a pluribus vetulis, utrum scilicet crederent quod simul possent sedere et non sedere. Statim dicebant quod erat impossibile. Et nunc petivi ab eis: 'Nonne creditis quod Deus posset hoc facere?' Statim responderunt: 'Nescimus, Deus potest omnia facere, et quod impossibilia deum posse facere credendum est.'" The latter context can be found in John Buridan, *op.cit.*, IV.12, fol. 21va.

deny this principle in his mind (*mente*).[44] There is objective error in this doubt to the extent that the women doubted the truth of the objective content of the first principle. Moreover, one can admit that the women were not in the position to prove the a priori self-cognition of reason, and that unavoidable circumstances of habit led to the error in question. Nevertheless, the women could be certain of the rationality of their actions insofar as they could be certain of their doubt of the absolute truth of the principle of non-contradiction: because they believe God can do anything.

Whereas the case of the old women concerns objective error about a fact of reason, the second type of error occurs relative to the self-cognition of reason, concerning the wickedness that impedes our assent to practical principles (*propter malitiam non possunt principiis practicis assentire*).[45] In the aforementioned case, one could only speak of wickedness if the circumstances impeding the women are not given. The wickedness would then consist in the demand for absolute certainty in the cognition of principles. That is because the self-confirmation of reason as reason – i.e., the theoretical self-confirmation of reason – is possible only when proof of the aprioricity of reason is possible given the fact of reason. However, the reality or fact of reason cannot itself be proved. Hence, there cannot be any appearance that is evident without restriction as far as reason is concerned. This means that relatively evident appearance provides the sufficient, subjective foundation for all cognition of principles, including the foundation of first principles. The demand for unrestricted certainty in the cognition of principles is theoretically impossible to satisfy. In practical terms, it forces us to question the subjective certainty of reason. It is wicked first because it is objectively false, and second because viewed subjectively, there are no inescapable circumstances impeding it. Thus, the demand for unrestricted certainty in the cognition of principles is unfulfillable.

There is further support for this interpretation of the effect of moral wickedness. First, there is a passage in which Buridan argues for the possibility of physics and moral philosophy based on relatively evident appearance against the objection that absolutely evident appearance is needed for these sciences.[46] He reproaches the objec-

---

[44] John Buridan, *op.cit.*, II.2, fol. 9vb: "… quamvis nullus mente negaret primum principium, tamen potest de eo habere formidinem."
[45] Cf. n. 26 above.
[46] John Buridan, *op.cit.*, II.1, fol. 9ra: "Ideo conclusum est correlarie, quod aliqui valde male dicunt volentes interimere scientias naturales et morales eo quod in

tors for having a morally corrupt will (*male dicunt volentes*), charging them not merely with objective error, by holding natural science to an unattainable standard of absolute certainty, but also with subjective error, by questioning the basis of his own objective claim, viz., the subjective certainty of reason. Second, there is the question of the historical significance of this text. According to the standard interpretation, Buridan is attacking Nicholas of Autrecourt,[47] and Buridan is suggesting that by demanding that all scientific principles be traced back to the principle of non-contradiction, Nicholas is jeopardizing the subjective foundation of his own scientific activity.

In addition, there is support for this interpretation of wickedness in Buridan's concession that one can only be certain of the *primum principium* in a restricted way. There are three points here. First, our grasp of the principle of non-contradiction, like all other principles, is through experience and thus "in" time.[48] The experience consists in our cognizing the terms from which the first principle is constituted in its spoken as well as in its propositional forms (*primum principium complexum*). Second, the scholarly literature assumes that Buridan is committed to the principle of non-contradiction being "absolutely evident (*evidentia simpliciter*)."[49] According to the text usually cited here, there is a third mode of certainty that proceeds from evident appearance, which is said to be absolutely evident relative to a proposition. In this mode, a man is forced or compelled to assent to a proposition in such a way that it is not part of the nature of his intellect to dissent from it. According to Aristotle, this kind of evident appearance is consistent with (*conveniret*) the propositional expression of the first principle (*primum principium complexum*).[50]

---

pluribus earum principiis et conclusionibus non est evidentia simplex, sed possunt falsificari per casus supernaturaliter possibiles, quia non requiritur ad tales scientias evidentia simpliciter, sed sufficient praedictae evidentiae secundum quid sive ex suppositione."

[47] Cf. the works cited in n. 1 by P. King, J. M. M. H. Thijssen and J. Zupko.

[48] Cf. John Buridan, *op.cit.*, II.2, fol. 9vb: "... debitis scire quod adhuc circa huiusmodi principia indigemus habitu vel determinatione ad assentiendum quantum ad illa principia, quae indigent experimentis praeviis vel saltem indigent habitu confirmante assensum, ut aliquo casu possit circa ea contingere formido. Unde quamvis nullus mente negaret *primum principium*, tamen potest de eo habere formidinem."

[49] Cf. the works cited in n. 1.

[50] John Buridan, *op.cit.*, II.1, fol. 8vb: "Tertio modo firmitas assensus provenit ex evidentia et vocatur evidentia propositionis simpliciter, quando ex natura sensus vel intellectus homo cogitur sive necessitatur ad assentiendum propositioni, ita quod

With regard to absolutely evident appearance, Buridan on the one hand speaks of the evident appearance that applies to a proposition when one is forced to assent to it and cannot deny it. This supports our interpretation of Buridan's position, for when the objective content of the *primum principium* is recognized a priori, no proof is needed of the aprioricity of reason. One is compelled to assent to the conclusion in question. But on the other hand, it runs contrary to the suggestion that such an appearance is called *evidentia simpliciter*, since Buridan does not describe the appearance under discussion in this way. Most importantly, however, he does not say that he considers an appearance of this sort to be evident without restriction. When Buridan speaks of Aristotle's belief that the *primum principium complexum* has unrestricted certainty, he puts it in a hypothetical way. Thus, when he speaks of the *primum principium complexum*, he is considering the principle of non-contradiction in its spoken form, i.e., as accessible to experience and in terms of the genesis of its cognition. Moreover, by speaking hypothetically, he makes it possible to take his own position on the question under discussion.

Buridan develops his position by first acknowledging that absolutely evident appearance is not a feature of experience, only that which is sufficient as a foundation for physics. This permits us to conclude that he is speaking here of relatively evident appearance. He continues by noting that there are also principles whose evident appearance is not based on any experience, but given by what is included in the terms contained therein. Buridan includes the *primum principium* among these principles.[51] That he does not exclude the necessity of experience for the *primum principium* does not run contrary to the fact that he declares it to be relatively evident in terms of the genesis of its cognition, because experience is not the basis of its truth or worth. Instead, Buridan introduces the necessity of experience as regards the *primum principium* by considering it in relation to our cognition of the terms from which it is constituted in its spoken form. From this perspective, he determines the prin-

---

non potest dissentire. Et huiusmodi evidentia secundum Aristotelem conveniret primo principio complexo."

[51] John Buridan, *op.cit.*, II.1, fol. 9rb: "Et potest concedi, quod huiusmodi experientiae non valent ad evidentiam simpliciter, sed valent ad evidentiam, quae sufficit ad scientiam naturalem. Et cum hoc etiam alia sunt principia ex inclusionibus vel repugnantiis terminorum vel propositionum, quae non indigent experientiis, sicut est de primo principio."

ciple of non-contradiction to be relatively evident. This is the only interpretation of the text that makes sense.

6. *Conclusions*

The considerations presented in this essay can be summarized in two points. First, let us recall the provisional conclusion of the second part: that for Buridan the desire for knowledge has its foundation and origin in the human will. To the extent that this desire is originally self-determined, it does not result from nature, which is independent of and prior to the understanding. Its source is rather in the human will.

Second, as we saw in the previous section, Buridan does not believe that there is any explanatory or *propter quid* demonstration of the claim that we cannot be mistaken about the objective content of the *primum principium* as such.[52] Accordingly, he views the propositional expression of the first principle (*primum principium complexum*) as certain in a relative sense.[53] It is thus clear that for Buridan, the actual self-certifiability of reason is the basis for all cognitions of reason. The actual realization of the cognition of reason is the precondition from which the possibilities of cognition and knowledge, or science, are determined and adjudicated.

This condition can also be defined so that evident appearance is given priority over truth. For no cognition of reason can ultimately be doubted unless it is actually realized and insofar as it is certain. Of course, there is always the possibility of the theoretical self-certification of reason as reason, i.e., the demonstration of the aprioricity of reason and knowledge, or, in Buridan's terms, the explanatory or *propter quid* demonstration of the impossibility of our being in error about the objective content of the *primum principium* as such.[54] Yet this demonstration depends on an actual cognition

---

[52] John Buridan, *op.cit.*, IV.12, fol. 21va: "... errare circa primum principium est ipsis dissentire vel eius opposito assentire. ... Et ista conclusio non potest probari quantum ad quia est ..." Cf. nn. 37 and 40 above.

[53] Cf. n. 51 above.

[54] John Buridan, *op.cit.*, IV.12, fol. 21va: "... errare circa primum principium est ipsi dissentire vel eius opposito assentire. ... bene Aristoteles demonstrat propter quid hoc est, scilicet quia ei dissentire vel eius opposito assentire est habere simul in intellectu opiniones contradictoriorum. Et hoc est impossibile, cum opiniones contradictoriorum sunt contrariae, et contraria non possunt esse simul in eodem subiecto vel in intellectu." Cf. nn. 37–38 above.

of reason. For even doubts about reason as such require its self-certification, to the extent that such doubts are actually realized. As Buridan puts it, we can be as certain of the activity of reason as we are of our own actions, even when the judgment (i.e., doubt in relation to the first principle) is false from an objective standpoint.[55]

---

[55] John Buridan, *op.cit.*, II.1, fol. 9ra: "… illud iudicium erit evidens evidentia sufficiente ad bene agendum moraliter, etiam licet iudicium sit falsum propter invincibilem ignorantiam alicuius circumstantiae." Cf. n. 41 above.

JOHN BURIDAN ON INFINITY*

JOHN E. MURDOCH &
JOHANNES M. M. H. THIJSSEN

There were three main routes by which discussions of the infinite entered late medieval thought. The first route originated from *De caelo*, Book I, chapters 5–7 and *Physics*, Books III, chapters 4–8 and VI, in which Aristotle analysed the notions of infinity and continuity. The second route by which the infinite was discussed had to do with arguments about the eternity of the world. Since an eternal world would imply an infinite past and future, a number of arguments allowing or refuting the thesis of an eternal world relied on the infinite. Discussions of the world's eternity usually occurred in a theological context, either in commentaries on the *Sentences*, or in separate questions. They also occured, however, in the context of book VIII, chapter 1 of the *Physics*, in which Aristotle presented his proof for the eternity of motion. As a result, even thinkers such as John Buridan, who never became a theologian, had an opportunity to discuss the question of the world's eternity. A third route which fostered discussions of the so-called intensive infinite was the infinity of God himself.[1]

In what follows, we shall concentrate our attention almost entirely upon Buridan's *Quaestiones super octo Physicorum libros Aristotelis, secundum ultimam lecturam*. There, in book III, in *quaestiones* 14 to 19, he presents the most central aspect of his treatment of infinity,[2] which is most characteristic about his approach in natural philosophy and metaphysics in general, namely, the further development

---

* This article relies on material that was previously discussed in Thijssen, *Johannes Buridanus over het oneindige*, a Ph.D. thesis that was co-directed by John E. Murdoch, and in an unpublished manuscript of a lecture delivered by the latter in 1979 as part of the conference "Infinity, Continuity, and Indivisibility in Antiquity and the Middle Ages," organized by Norman Kretzmann at Cornell University. A revised version of part of this lecture appeared as Murdoch, "William of Ockham." Since in the past both authors have collaborated so closely on this theme, it seemed only fitting that they co-author this article. Thijssen's research for this paper was financially supported by a grant from the Netherlands Organization for Scientific Research (200-22-295).

[1] See Sweeney, *Divine Infinity*, and Davenport, *Measure of a Different Greatness*.

[2] John Buridan, *Tractatus de infinito* [Thijssen]. A new edition in their entirety of books III and VI of *Buridan's Quaestiones super octo Physicorum libros Aristotelis, secundum ultimam lecturam* is being prepared by Thijssen.

of Aristotelian notions and doctrines in terms of medieval logic and semantics.

In contrast, Buridan's treatment of infinity in *quaestiones* 13 to 17 of Book I of his *Quaestiones super libris quattuor De caelo et mundo*, with the exception of q. 17 (see below), mainly give a resumé of views and arguments developed in Aristotle's own *De caelo* I, chapters 5–7, to the effect that no infinite body exists.[3] *Quaestio* 13, for instance, recapitulates Aristotle's argument that an infinite body that moves in a circle implies an infinite heaven, which is considered impossible (274b29–33); *quaestio* 14 argues that an infinite body cannot act, nor be acted upon; since, however, it is characteristic of a natural body that it possesses the power of acting or of being acted upon, a natural body cannot be infinite. *Quaestiones* 15 and 16 develop the idea that a body cannot be infinite for the reason that it cannot move, either in a circle (q. 15), or in a straight line (q. 16). These arguments are to be found in Aristotle's *De caelo*, I, 5–7, 275b12–276a17.

1. *The Aristotelian Background*

The central theme of the debate on infinity is, whether an infinite magnitude can exist.[4] Aristotle's most lasting contributions to this debate were his distinction between the actual and the potential infinite, and his view that the infinite can only exist in potentiality. The paradigm case of a potential infinity in Aristotle's discussion is the infinite division of a continuum. According to Aristotle, a continuous magnitude of finite size can be infinitely divided into parts which, in turn, can be further divided. Moreover, the parts arrived at by this processs of division can be added and will thus constitute another potential infinite (Aristotle, *Physics*, III, 206b3–6). Furthermore, the unending successive division of the continuum affords Aristotle with the infinity of natural numbers (Aristotle, *Physics*, III, 207b1–5). Note that Aristotle's notion of the infinite relies on the assumption that any continuous magnitude is composed of always further divisible parts, a thesis that he would only establish in book VI of the *Physics*.

---

[3] John Buridan, *Quaestiones super libris quattuor De caelo et mundo* [Moody], pp. 57–82. Another edition of the same text, though with few differences, is now available in John Buridan, *Expositio et questiones in Aristotelis De caelo et mundo* [Patar], pp. 292–323, which also contains Buridan's literal commentary of the *De caelo*.

[4] See Maier, "Diskussionen," and Murdoch, "Infinity and continuity" for a survey of the medieval debates.

Both in the process of dividing a given quantity into divisible parts and of numbering those parts "... it is always possible to take a part which is outside a given part." (*Physics*, III, 207a3). In other words, the process of dividing or of numbering is never completed, and in this sense the infinite exists merely potentially. Averroes explained the potential character of an infinity by pointing out that it was a potentiality mixed with an actuality (*actus permixtus*).[5] Part of the infinite exists actually, namely that part whose division or addition has been completed, but there always remains some part beyond that potentially can be still further divided or numbered. The only permissible infinite is an infinite *in fieri*, that is, an infinity-in-becoming that will become completed only successively. Alternative medieval expressions of the potential and the actual infinite were *non tantum quin maius* and *tantum quod non maius*, respectively. That is to say, the potential infinite was a quantity that was not so great but that it could be greater, whereas the actual infinite was defined as a maximum value, as a quantity so great that it could not be greater. In similar fashion, a potentially infinite number was defined as a number that was not so many but that it could be more (*non tot quin plures*), whereas an actually infinite number was so many that it could not be more (*tot quod non plures*).

During the fourteenth century, another important distinction was introduced, namely that between the categorematic and syncategorematic uses of the term "infinite" or "infinitely many" in different propositions. Anneliese Maier believed that it was merely a matter of terminology that a categorematic infinite corresponded to an actual infinite, whereas a syncategorematic infinite was equivalent to a potential infinity.[6] Nowadays, the generally accepted view is that the distinction between the categorematic and the syncategorematic infinite is at the heart of the logic of the infinite. It is part of a new approach in natural philosophy which focuses on propositional analysis. Until now, the examples used to illustrate this approach have been mainly taken from the works of William Ockham or the Mertonians and have largely neglected Buridan.[7] The latter, however, was one of the most consistent practitioners of this logico-semantic approach towards natural philosophy.

---

[5] Averroes, *Aristotelis opera cum Averrois commentatoris* (Venezia, 1562–1574), vol. IV, fol. 112vb (*In Arist. Physicam*, Book III, text 57).

[6] Maier, "Diskussionen," p. 44, n.7.

[7] See, for instance, Murdoch, "William of Ockham," and Sylla, "William Heytesbury."

2. *Categorematic and Syncategorematic Infinites:*
The Semantic Background

The distinction between categorematic and syncategorematic terms was basically a grammatical distinction.[8] The categorematic terms are those that can function as subject or predicate, or part of a subject or predicate, in a proposition, that is, the nouns, adjectives, pronouns, and verbs. Syncategoremata are all the non-categorematic terms, that is, the conjunctions, adverbs, and prepositions. Although these terms are not part of the subject or predicate, they do affect the meaning of a proposition, and for this reason had drawn the attention of medieval logicians. Syncategorematic terms were considered to have no signification on their own (unlike categorematic terms), but yet, to modify the signification of the categorematic terms adjoined to them. Towards the end of the thirteenth century, the class of categorematic and non-categorematic terms became both narrower and broader than this grammatical classification. It became narrower in that many syncategorematic terms were omitted in treatises on syncategoremata, because they were considered not to be of enough interest to the medieval logician to be discussed. It became broader in that some words which were categoremata from a syntactical perspective, came to be treated as syncategorematic terms, because they too appeared to affect the semantic interpretation of a proposition. Among this latter class are terms like "all" (*omnis*), "whole" (*totus*), "another" (*alter*), and "infinite" (*infinitus*). In some cases these gramatically categorematic adjectives can function as a quantifier (*signum*) in a proposition, and on this ground can be considered as syncategorematic terms. In other words, on the basis of the semantic function which some categorematic terms could exercise in a proposition, they were classified as syncategorematic terms. In any case, it is noteworthy that late-medieval authors usually distinguish between the categorematic and the syncategorematic *use* of the term "infinity" in a given proposition: *categorematice sumpto* or *syncategorematice sumpto* is one of the recurrent formulas of this distinction.

Interpreters of late-medieval theories of the infinite have often claimed that the categorematic use of the term "infinite" is equivalent to an actual infinite, and its syncategorematic use is equivalent to a potential infinity. Buridan's discussion – and that of other

---

[8] The following is based on Braakhuis, *De 13de eeuwse tractaten*, and Kretzmann, "Syncategoremata," which are still the most fundamental studies for our understanding of the medieval theory of syncategoremata.

fourteenth-century authors, for that matter – indicates, however, that the situation is slightly more complex. The different occurrences of the terms "infinite" or "infinitely many" in a proposition yield different referents of these terms, which in turn affect the truth-value of the proposition. One use or the other of the term "infinite" could imply either an actual or a potential infinite. From the truth-value which an author attributed to a particular proposition, one can infer which kind(s) of infinity he found permissible. Much depended on how a particular use of the term was defined. Buridan distinguished two ways by which the syncategorematic use of "infinity" could be defined. Thomas Bradwardine, Gregory of Rimini, and Albert of Saxony too indicated that the term "infinity" used syncategorematically could be expounded in several different ways.[9]

Furthermore, fourteenth-century authors held divergent views as to when a term was used categorematically or syncategorematically. As we will see below, John Buridan maintained that when a term is used categorematically, it makes no difference whether it precedes or follows the term it qualifies. In addition, he did not agree that "infinite" could only be used as a syncategorematic term if referred to the subject (*a parte subjecti*). Even in the predicate position, the term "infinite" was still used as a syncategorematic term with respect to the term it modified, since it still functioned as a distributive sign (*signum*).[10]

3. *Categorematic and Syncategorematic Infinites*

In Book III, question 18 of his commentary on the *Physics*, Buridan discusses the types of infinity that are involved in infinitely divisible continuous magnitudes: Are the parts in a continuum infinite in number? The reply to this question depends on whether one

---

[9] See Thijssen, *John Buridan over het oneindige*, pp. 271–273, and the texts cited there.

[10] John Buridan, *Tractatus de infinito* [Thijssen], p. 61$9^{-16,\ 22-26}$: "Sed aliqui obiciunt dicentes quod omnis dictio posita a parte predicati tenetur categoremaatice et non sincategorematice, et ideo in ista propositione 'linea est infinite partes' hec dictio 'infinite' non potest teneri sincategorematice, propter quod male dicebatur quod ista propositio esset vera capiendo 'infinitum' sincategorematice ... Et ita est in proposito, nam licet hoc totum 'infinite partes' teneatur categoremaatice ad istum sensum, quia hoc totum est predicatum, tamen non debet negari quod hec dictio 'infinite' respectu istius dictionis 'partes' retineat suam significationem sincategorematicam et quod debeat exponi secundum expositonem predictam." See Thijssen, *Johannes Buridanus over het oneindige*, p. 268.

takes the term "infinite" categorematically or syncategorematically. According to the categorematic use of the term "infinite" the two propositions "Infinite in number are the parts of a continuum" ("*Infinitae secundum multitudinem sunt partes in continuo*") and "The parts of a continuum are infinite in number" ("*Partes in continuo sunt infinite secundum multitudinem*") are equivalent. Both propositions are false, according to Buridan, since the categorematic use of the term "infinite" implies that there is a single particular number which is infinite. However, neither these two specific parts of a continuum are infinite, nor these three, nor these four, etc. And even if one were to take *all* the parts of a continuum collectively (*omnes simul sumpte*), it would be false to maintain that all the parts of a continuum were infinite in number, because there is no number of parts of the continuum which constitutes *all* parts. One cannot determinatedly and properly say how many parts constitute *the* set of all parts in a continuum. Used categorematically, the term "infinite" exercises *suppositio determinata*, that is, it should refer to some one infinite thing. The fact that Buridan considers the proposition false when taken in this interpretation, indicates that he believes that there is no such referent.[11]

If this, then, is the case, what are we to maintain in order to account for the admitted infinite divisibility of continua? Buridan's answer is that we must appeal to the syncategorematic infinite, provided we give a proper exposition of it. When used as a syncategorematic term, Buridan distinguishes two different definitions of an infinite magnitude and multitude, respectively. First, "infinitely large" and "infinitely many" are defined in the usual way as "a certain quantity and not so great but that it could be greater" (*aliquantum et non tantum quin maius*) and "a certain amount and not so many but that it could be more" (*aliquota et non tot quin plura*), respectively. The

---

[11] John Buridan, *op.cit.*, p. 49$^{16-26}$; p. 50$^{7-11}$: "Et tunc etiam ego credo esse ponendum quod nec infinita secundum multitudinem sunt aliqua nec aliqua sunt infinita secundum multitudinem, quia si aliqua essent infinita secundum multitudinem, hoc maxime esset verum de partibus continuorum, ita quod hoc esset vera: 'partes vel alique partes linee B sunt infinite'. Sed hec esset falsa, scilicet quod partes linee B sunt infinite. Probatio igitur quod hec sit falsa: 'partes linee B sunt infinite,' quia quereres 'que sunt ille?'. Et oportet hoc dicere, cum ille terminus 'partes' stat determinate. Et hoc non potest dici, nam nec iste tres sunt infinite, quia numerando veniremus ad ultimam istarum, nec iste decem, nec iste centum, et sic de aliis ... Sic autem nulle partes continui sunt omnes, quia quecumque partes continui habent iterum alias partes in quas sunt divisibiles, et omnino preter quascumque sunt adhuc alique alie que non sunt alique istarum. Et sic etiam dicendum est de medietatibus proportionalibus columne B, scilicet quod nulle sunt omnes."

definition of infinite magnitude is further explained by the expression that "for any magnitude B there exists a greater B" (*omni B est B maius*). Thus, "Infinitely great is B" should be expounded as "for any magnitude B, there is a greater magnitude B." Buridan observes that the term "B" has different referents in this proposition. In the first part of the proposition ("For any magnitude B"), the term "B" has distributive supposition (*suppositio distributiva*), meaning that given any quantity no matter how great, one can always find a greater one. The second term "B," however, ("there is a greater magnitude B") has merely confused supposition (*suppositio confusa tantum*), meaning that there is not any particular magnitude B which happens to be greater than any magnitude B chosen in the initial situation.[12] Buridan claims that according to this reading, the proposition "Infinitely many in number are the parts in a continuum" ("*Infinitae secundum multitudinem sunt partes in continuo*") is false. Since there are only two parts in any continuum, namely the two halves generated by the first division of that continuum, there are not more parts for any given number of parts, as the definition stipulates. Any other parts are merely subdivisions of the original two halves.[13]

In addition, Buridan provides a second definition of the syncategorematic use of "infinity," which he clearly prefers. According to the latter definition, the proposition "that B is infinitely many" ("*Infinita esse B*") means that B is two, and three, and a hundred, etc., without

---

[12] John Buridan, *op.cit.*, p. 54$^{3-10}$: "Sequitur dicere de 'infinito' sincategorematice sumpto. De quo notandum est quod diversis modis solet exponi illud nomen 'infinitum' sincategorematice sumptum. Uno modo in magnitudinibus, quia: aliquantum et non tantum quin maius, et in multitudine, quia: aliquota et non tot quin plura. Et videtur mihi quod equivalens expositio datur sub verbis manifestioribus et brevioribus, scilicet quod infinitum esse B significat quod omni B est B maius, et infinitum esse B secundum longitudinem significat quod omni B est B longius, et sic de infinito secundum velocitatem vel tarditatem vel parvitatem etc." and p. 60$^{10-21}$: "Quinto dubitatio est circa totam predictam expositionem: quomodo supponit iste terminus 'B' in ista propositione 'infinita est B' vel 'infinite longum est B'? Et sic de aliis. Respondetur non solum de ista propositione, sed de omni alia que indiget exponente vel exponentibus: si aliquis terminus semel tantum capiatur in exposita et indigeat capi pluries in exponente vel exponentibus et in illis pluribus acceptionibus supponit diversis suppositionibus, in tali casu videtur mihi dicendum quod ille terminus in exposita non supponit unica suppositione sed illis pluribus. Et sic ego dico quod in ista: 'infinite longius est B', hec dictio 'B' supponit suppositione distributiva et etiam suppositione confusa tantum, quia cum dico 'omni B est B longius', primum 'B' supponit distributive et secundum 'B' confuse tantum."

[13] John Buridan, *op.cit.*, p. 55$^{16-19}$: "Quarta conclusio de infinito secundum multitudinem, scilicet quod hec est falsa: 'infinite secundum multitudinem sunt partes in continuo', quia si in continuo sunt due partes, tamen non sunt in eo plures partes quam iste due, quoniam ille due sunt et centum et mille, sicut ante dictum est."

end (*et sic sine statu*). Thus, there are infinitely many parts in a continuum, because two, three, a hundred, etc., or, more generally, a finite number, and a finite number greater than that finite number, and a finite number greater than that greater finite number.[14] Of course, such a view fit neatly with Aristotle's contentions that a permissible infinite was not such that there was nothing beyond, but was such that there was always something beyond (*cuius secundum quantitatem accipientibus semper est aliquid accipere extra*).[15] According to this latter interpretation of the term "infinite" used syncategorematically, the proposition "Infinitely many in number are the parts in a continuum" is true. At bottom, the infinity of parts involved in the infinite divisibility of a continuum is not opposed to the finite. It is an infinite that is rather like an "unending finite."

4. *The Potential Infinite*

Aristotle's observations on the potential infinite still puzzle interpretaters today.[16] Is the potentiality involved in a potential infinite one that will be actualised in the plenitude of time just like any other potentiality in Aristotle's theory? Or is the potentiality of the potential infinite of a different order and will never be completed? Medieval authors usually assumed the latter when they started to investigate Aristotle's notion of the potential infinite. A particularly interesting new way of doing this was opened by William Ockham and John

---

[14] John Buridan, *op.cit.*, pp. 60[23]–61[8]: "Alia expositio infiniti sincategorematice sumpti est per carentiam status in rationibus numeralibus. Ideo primo exponuntur 'infinita' secundum multitudinem, scilicet quod infinita esse B significat: duo esse B et tria esse B et centum esse B et mille et sic sine statu. In aliis autem exponitur: certa quantitate accepta per multitudinem numeralem tante quantitatis sine statu, ut infinite longum esse B significat quod data longitudine alicuius B, verbi gratia pedali, tunc est B duorum pedum et est B trium pedum, et est B centum pedum et sic sine statu; ita etiam, si infinitum velocem dicamus esse motum, hoc significat quod dato motu alicuius determinate velocitatis est motus dupliciter velocior et est motus tripliciter velocior et motus centupliciter velocior et sic sine statu. Et secundum istam expositionem ponuntur conclusiones communiter concesse. Prima est quod infinite sunt partes continui secundum multitudinem, quia due, tres, centum, et sic sine statu; immo hec linea est infinite partes, quia due, tres, centum etc.;" p. 63[2–5]: "His visis videndum est que sunt proprietates huius nominis 'infinitum' sincategorematice sumpti. Et est prima proprietas quod isti termini 'finitum' et 'infinitum' non opponuntur ad invicem, sicut nec isti termini 'omnis homo' et 'homo'."

[15] Aristoteles Latinus, *Physica. Translatio vetus* [Bossier & Brams], 207a7–8.

[16] See, for instance, Hintikka, "Aristotelian Infinity," Lear "Aristotelian Infinity," and Bostock, "Aristotle's Potential Infinites."

Buridan.[17] They analyzed the infinite's potential being (*esse in potentia*) in terms of propositions involving predication by means of the verb "can" (*potest*). But not only the potential infinite, but also the possibility of an actual, completed infinity was examined by means of propositions in which the modal operator "possible" occurred. In Buridan's hands the potential infinite divisibility by means of modal propositions becomes the central concern of *quaestio* 19 of Book III, which in its opening words explicitly refers to difficulties of the infinite *quantum ad propositiones de possibili*.[18]

An important ingredient of these discussions is the semantic distinction between the divided sense (*sensus divisus*) of a proposition and its composite sense (*sensus compositus*). As was the case with the categorematic/syncategorematic distinction, Buridan takes the semantic theory behind the *divisus/compositus* distinction for granted. It has to be culled from logical works by him or by his contemporaries.

The distinction between *sensus divisus* and *compositus* has its roots in Aristotle's *De Sophisticis Elenchis* (166a22, and following). The medievals developed this concept and clearly show that the purpose of the distinction was to establish the scope of the modal operators "necessary" and "possible" in ambiguous propositions: did the modal operator affect the entire proposition in which it occurred, or only a certain part of it? For instance the ambiguous proposition "That every man is an animal, is necessary" ("*Omnem hominem esse animal est necessarium*") can either be taken in its divided sense and is, then, equivalent to "Every man is of necessity an animal" ("*Omnis homo de necessitate est animal*") or it can be taken in a composite sense and then means "This [statement] is necessary: every man is an animal" ("*Hec est necessaria: omnis homo est animal*"). Both propositions, however, have different truth value.[19]

In Buridan's and many other fourteenth-century scholars' discussion, the distinction between divided and composite senses had evolved into a distinction between composite and divided propositions (*propositiones compositae/divisae*). If the position of the modal operator is at the beginning or end of a proposition it is composite. In that case, the modal operator affects the entire proposition.

---

[17] Ockham's treatment of the potential infinite in terms of modal propositions is discussed in Murdoch, "Ockham," esp. pp. 190–196.

[18] John Buridan, *Tractatus de infinito* [Thijssen], p. 66[1-2]: "Adhuc restant difficultates de infinito quantum ad propositiones de possibili. Et ideo formetur decima nona questio: Utrum possibile est infinitam esse magnitudinem et in infinitas partes lineam esse divisam."

[19] See Kretzmann, "*Sensus compositus, sensus divisus*."

If, however, the modal operator (*modus*) intervenes between what is being expressed by the subject and the predicate, it is a divided proposition. In the latter case, the modal operator only modifies the copula. The proposition "That a man runs is possible" ("*Hominem currere est possibile*"), for instance, is a composite proposition, whereas the proposition "For a man it is possible to run" ("*Hominem possibile est currere*") is divided.[20]

How does Buridan apply this semantic technique to the debate of the potential and the possible infinite? In effect, according to Buridan, the composite proposition "This is possible: God separates and separately conserves all the parts of line B" ("*Hec est possibilis: Deus separat et separatim conservat omnes partes linee B*") is false. The reason is that a division in *all* proportional parts would be impossible. The division of line B, or of any continuum for that matter, is never completed. God *can* divide, however, all the parts of line B. The path that leads to this conclusion is not a very direct or rapid one, but we can encapsulate what is involved by starting from a thesis that directly concerns the infinite divisibility of a continuum.

Buridan claims that all the parts of line B God can separate and separately conserve (*omnes partes linee B Deus potest separare et separatim conservare*).[21] He argues the truth of this statement on the basis of two rules that govern the analysis of divided propositions. Both rules are exemplified by propositions parallel to those involved in the infinites and the infinite divisibility asserted by his conclusions, namely the divided proposition: "I can see every star" ("*Omne astrum possum videre*") and the related composite proposition: "This is possibile: I see every star" ("*Hec est possibilis: omne astrum video*").

The first rule is that every possible can be reduced to being (*omne possibile potest poni inesse*). This rule allows one to move from the *de possibili* proposition to the (possible) affirmative assertoric proposition. Buridan notes, however, that the universal divided proposition should not be translated into the universal composite proposition. In other words, although this proposition is true "I can see every star,"

---

[20] See also Van der Lecq, "Buridan on Modal Propositions." The same distinction was made by William Heytesbury, whereas Paul of Venice defined the distinction between divided and composite propositions in a different way. See Kretzmann, "*Sensus compositus, sensus divisus*," and Van der Lecq, "Paul of Venice," p. 321.

[21] John Buridan, *Tractatus de infinito* [Thijssen], p. 75$7^{-11}$: "Quarta conclusio est quod omnes partes linee B Deus potest separare ab invicem et separatim conservare, quia et istas duas et istas centum et sic de singulis; nulle enim sunt de quibus posset dari instantia, nisi daretur de omnibus collective; et dictum est prius quod omnes non sunt, capiendo 'omnes' collective."

it is not possible: that "I see every star." Instead, the divided proposition "I can see every star" has to be analysed into various singular propositions. One can move from this divided proposition to "This is possible: that I see this star," and "This is possible that I see that star."[22]

Buridan brings his example of the stars that can be seen back to bear upon his thesis dealing with the infinite divisibility of a line. The divided proposition "God can separate and separately conserve all the parts of line B" should be expounded into the composite singular propositions "This is possible: God separates and separately conserves these parts of line B," and "This is possible: God separates and separately conserves those parts of line B," etc.[23]

The second rule which governs the analysis of divided propositions is that the predicate appellates its form (*predicatum appellat suam formam*).[24] By *appellatio formae* or *appellatio formalis*, Buridan, and other medieval logicians for that matter, indicated the property that was signified by a connotative term within a proposition. In the proposition "Peter will be white," for instance, the connotative term "white" was said to appellate its form, that is, the property of being white, as inhering in the thing which is its material significate, that is, Peter, according to the tense of the copula, that is, according to the future.[25] The rule now stipulates that if propositions in which there occurred a connotative predicate term were to be resolved into equivalent propositions, the predicate term could not change, but had to be preserved in its entire form.[26]

---

[22] John Buridan, *op.cit.*, p. 76[1, 4–8, 14–17]: "Sed tu dices quod possibili posito inesse nullum sequitur impossibile ... Dico quod universali de possibili in sensu diviso non oportet correspondere universalem de inesse possibilem, sed sufficit quod quelibet singularis de inesse sit possibilis. Verbi gratia: hec est vera 'omne astrum possum videre et sine miraculo', et tamen sic non est ista possibilis: 'omne astrum video'. Sed quelibet singularis est possibilis ... Dico quod hoc non sequitur, sed bene sequitur quod omnes singulares possunt esse simul vere. Tamen impossibile est quod omnes sint simul vere; semper enim in proposito deficit consequentia de possibili divisa ad compositam stante universalitate."

[23] John Buridan, *op.cit.*, p. 78[21–25]: "Ita similiter opinor quod hec sit vera: 'Deus potest separare et separatim conservare omnes partes linee B'; quia omnes singulares quantum ad singularitatem correspondentem isti universalitati omnes partes sunt possibiles et compossibiles et possunt esse simul vere, licet non sit possibile quod omnes sint simul vere."

[24] John Buridan, *op.cit.*, p. 77[3–6]: "Ad hoc dicunt multi quod hec est falsa: 'ego possum videre omne astrum', quia predicatum appellat formam vel formalitatem. Ideo propositio de possibili, si sit vera, debet poni inesse salvato predicato secundum eius totam formam."

[25] Nuchelmans, *Late-Scholastic and Humanist Theories of the Proposition*, pp. 56–57.

[26] Cf. Thijssen, *Johannes Buridanus over het oneindige*, p. 294, which also quotes the

In what way now, does this rule affect the analysis of the divided proposition "I can see every star?" If one applies the rule, this proposition is supposed to correspond to the proposition "This is possible: I see every star," which, however, according to Buridan, is false. As we have seen above, Buridan denies that the inference of "This is possible: I see every star" from "I can see every star" is valid. Hence, the objection that he should apply the rule that the predicate remains invariant is quite central in his analysis of divided propositions.

His reply is now that perhaps there is something to be said for this objection, but that he believes the opposite carries more weight, namely that "I can see every star" is true, whereas "This is possible: I see every star" is false. The reason is that, according to Buridan, the second rule does not apply. It does not always follow that given a divided proposition, that the corresponding assertoric proposition will be true while preserving the whole form of the predicate. Or, in other words, there is a possibility of "transfer" from a divided proposition to a corresponding assertoric proposition on behalf of the predicates involved, but the predicate term involved need not come through unchanged. In this respect, Buridan draws an analogy with the subject term, which is not preserved either in the analysis of divided propositions. Thus, given the proposition "This white thing can be black" ("*Hoc album potest esse nigrum*"), it does not follow that "This is possible: this white thing is black" ("*Hec est possibilis: hoc album est nigrum*"), the reason being that the subject term ("*hoc album*") of the original divided proposition is not preserved. On the other hand, given this same initial divided proposition, it does follow "This is black" ("*Hoc est nigrum*"), because here the term "this" designates that for which the subject stands in the original divided proposition. And Buridan believes that this can be true of predicates as well.[27]

following pertinent passage from John Buridan, *Sophismata* [Scott], p. 63: "Sed in reductione propositionis de futuro ad propositionem de praesenti oportet auferre appellationem a subiecto propositionis, et oportet mutare subiectum appellativum in subiectum non appellativum ... Sed tamen non est ita de praedicato appellativo, ipsum enim in tali reductione debet manere in propria quantum ad appellationes." And also p. 64: "Et hoc est quod debemus intelligere per illud commune dictum quod praedicatum appellat suam formam et non subiectum ... Unde ponendo in esse praedicatum oportet remanere in propria forma cum sua appellatione, sed non subiectum."

[27] John Buridan, *Tractatus de infinito* [Thijssen], p. 77[10-20]: "Sed quamvis ista opinio possit forsan probabiliter sustineri, tamen magis opinor oppositum, scilicet quod hec sit vera: 'ego possum videre omne astrum' et quod Deus potest separare et separatim conservare omnes partes linee B, quoniam omnes concedunt quod non oportet ponere inesse illam de possibili salvata tota forma subiecti, etiam si propositio sit sin-

But there is a further reason why some divided propositions are exempted from this second rule. In the example "I see every star," neither the predicate term "every" (*omne*), nor the operative verb "to see," connected with it, connotes simultaneity in time.[28] In brief, they are not connotative terms at all.[29] According to Buridan, this explains why the predicate does not remain invariant in the assertoric universal propositions that can, wrongly, be inferred from the divided proposition. On the other hand, from the proposition "I can see every star" one can veridically infer that I can see this star, and I can see that star, etc., although successively, and not all at the same time.[30]

Similarly, although Buridan does not specify which assertoric propositions would be possible under the divided proposition "God can separate and separately conserve all the parts of line B," he does tell us that all of the singulars falling under the "all parts" (*omnes partes*) of the line that God can separate and conserve *can* be simultaneously true, although it is not possible that all *are* simultaneously true.[31] Buridan ends his argument with these words, but it would seem fair to infer that he intended the following conclusion. The analysis of the above proposition into the proposition "This is possible: God separates and separately conserves all the parts of line B" –

---

gularis, particularis vel indefinita. Non enim sequitur, si verum est quod hoc album potest esse nigrum, quod hec est possibilis: 'hoc album est nigrum', vel 'album est nigrum', sed sufficit quod ponatur inesse per pronomen demonstrativum demonstrato eo pro quo subiectum supponebat, aliis que in subiecto implicabantur circumscriptis. Et ita credo de predicato esse …" See also John Buridan, *Tractatus de consequentiis* [Hubien], pp. 75–76, which discusses the same example.

[28] John Buridan, *op.cit.*, p. 78[11–15]: "Et hoc ex alio apparet, quia licet predicatum appellat suam formam, tamen iste terminus vel hec oratio 'omnes partes linee B' nec significat tempus, nec connotat; nomen enim significat sine tempore; ideo nec significat nec connotat simultatem temporis."

[29] See also Thijssen, *Johannes Buridanus over het oneindige*, p. 296, and the passage quoted there from John Buridan, *Sophismata* [Scott], p. 65: "Deinde quantum ad alias dubitationes videtur mihi dicendum quod ille terminus 'omnis' non proprie est appellativus … Et sic reducendo illas de preterito vel de futuro vel de possibili ad illas de presenti vel de inesse, non oportet remanere talia signa distributiva, sive in subiectis, sive in predicatis. Sed sufficit ex utraque parte reducere totum per singulares …" Note that for Buridan, appellative terms served a connotative function. See De Rijk, "On Buridan's Doctrine of Connotation."

[30] John Buridan, *Tractatus de infinito* [Thijssen], p. 78[17–21]: "Ideo nullo modo sequitur propter predicatum appellare formam, quod si ego possum videre omne astrum, quod ego possum videre omne astrum simul. Sed sufficit quod ego possum videre hoc astrum et quod ego possum videre illud, et sic de aliis, licet successive unum post alterum."

[31] See the text quoted in note 23.

which, indeed, means that God conserves all the parts of a continuum simultaneously – is incorrect. The divided propositions ought to be expounded in the following way: "This is possible: God separates and separately conserves this part of line B." and "This is possible: God separates and separately conserves that part of line B...," and so on. These latter propositions entail that the infinite divisibility of a continuum is realized successively, rather than simultaneously.

5. *The Possibility of Infinites*

Another way of conducting an inquiry into the possible infinite was to invoke God's absolute power. Many medieval thinkers acknowledged Aristotle's arguments against the physical possibility of a completed infinite, but they would go beyond him in asking whether God in his absolute power could accomplish certain tasks, such as creating a stone of an infinite size or completing the infinite division of a continuum. Buridan too, frequently appeals to the concept of God's absolute power to do anything short of a logical contradiction. Thus, he admits that he believes that God could create other worlds and other spheres and also other finite magnitudes, however large he might wish, magnitudes in every proportion of finite to finite. But perhaps one need not believe that he could create an actually infinite magnitude, because, Buridan avers, if he did so, then he could not create a yet greater one. No effect is proportional to the infinite power of God.[32] However, since some have made an effort to prove that God could have created an infinite body, Buridan will once more

---

[32] John Buridan, *op.cit.*, pp. 16^{20}–17^{6}: "Primo credendum est fide quod Deus posset ultra istum mundum formare et creare alias spheras et alios mundos et omnino alias magnitudines finitas, quantascumque vellet, ita quod omni finita creata posset creare maiorem in duplo, in decuplo, in centuplo, et sic de omni alia proportione finiti ad finitum. Sed forte non oportet credere quod Deus posset creare magnitudinem actu infinitam, quia ista creata non posset creare maiorem: repugnat enim quod actu infinito sit aliud maius. Et tamen inconveniens est quod Deus posset facere creaturam potentie sue proportionatam sic quod non posset maiorem et perfectiorem facere. Et mihi videtur simile de magnitudine sicut de perfectione, quamvis enim omni creatura formata et formabili Deus posset creare perfectiorem, tamen non posset creare aliquam infinite perfectionis; illa enim esset eque perfecta vel non minus perfecta quam Deus. Nam si esset minus perfecta, ipsa non esset infinite perfectionis, sed posset creari perfectior. Et ita videtur mihi de magnitudine et velocitate et intensione caliditatis et huiusmodi." The same argument occurs in John Buridan, *Quaestiones super libris quattuor De caelo et mundo* [Moody], p. 79$^{27-33}$, and John Buridan, *Expositio et questiones in Aristotelis De caelo et mundo* [Patar], p. 319$^{64-71}$.

examine what could transpire under God's omnipotence. Suppose that God at every proportional part of an hour would create a body, for instance a stone, of one foot long. At the end of the hour, the infinitely many one-foot bodies would constitute an infinitely large body.[33]

According to Buridan, however, this argument entails two contradictions. First, it implies that an hour does and does not have a last proportional part. There seems to be no last proportional part, because the hour can continue to be divided into further proportional parts. However, there must be a last part, since one stone will be the last created stone at the end of the hour. Second, the supposition also entails the contradiction that shortly before the last instant of the hour has passed, the number of created stones is and is not infinite. The number of stones seems to be infinite, because the time that has passed before the last instant of the hour that has passed can be divided into infinitely many proportional parts. At each of those parts, God could have created a stone. On the other hand, the number of stones cannot be infinite, since the number of proportional time-parts that have passed before the last instant is finite: there still is a period of time left that has not yet been divided into proportional parts.[34]

In another context, Buridan claims that the following propositions are impossible: "In any proportional part (taken according to a double proportion) of the present day God may create a one-foot stone," and "In any such proportional part of the present day he will create a one-foot stone."[35] He could not do so, according to Buridan, because this would result in any number of contradictions or impossibilities, one of them being that God would thus indicate the *last* proportional part of the day.[36]

For a similar reason, Buridan had already concluded that even God could not create an infinitely long curved line (*linea gyrativa*) by

---

[33] John Buridan, *Quaestiones super libris quattuor De caelo et mundo* [Moody], pp. 79^{34}–80^9, and John Buridan, *Expositio et questiones in Aristotelis De caelo et mundo* [Patar], p. 319$^{71-81}$.

[34] John Buridan, *op.cit.*, pp. 80^{10}–81^5, and John Buridan, *Expositio et questiones in Aristotelis De caelo et mundo* [Patar], pp. 320–321^{14}.

[35] John Buridan, *Tractatus de infinito* [Thijssen], p. 70$^{10-13}$: "Secunda conclusio est quod hec est impossibilis: 'in qualibet medietate proportionali huius diei Deus creat unum lapidem pedalem' vel etiam ista: 'in qualibet medietate proportionali huius diei Deus creabit unum lapidem pedalem' …"

[36] John Buridan, *op.cit.*, p. 72$^{21-23}$: "Item. Si possibile esset Deum sic facere, sequeretur quod esset dare ultimam medietatem proportionalem huius diei; quod est falsum, sicut dicebatur quod non est ultima medietas columne proportionalis."

dividing a cylinder into proportional parts.[37] The assumption behind this case is that if God could create an infinitely long length, he would also be able to create the other dimensions of a body – breadth and depth – in an infinite size, and hence, could create an infinite body. However, even under God's absolute power an infinite curved line is not permissible, according to Buridan, because no curved line will run through *all* parts (*nulla est protracta per omnes*). The reason is that no proportional part of the cylinder is its *last* part. It is true, though, that through every proportional part runs a curved line (*per omnes protracta est una linea*). The semantic intricacies surrounding the use of the term "all" (*omnes*) when discussing the parts of an infinitely divisible continuum have been discussed above. But the bottom line is, that even God is unable to take *all* parts.[38]

From the foregoing, Buridan draws the conclusion that it is not possible under the assumption of any power that there be an infinite magnitude (*non est possibile per aliquam potentiam esse magnitudinem infinitam*).[39] The argument of the creation of stones in each proportional part of the day by God, or of a curved line in each proportional part of a cylinder, which would have been major arguments in support of the possibility of an infinite magnitude, had now been overturned (*magna ratio ad probandum quod hoc sit possibile destructa est*).[40] But what if, an objection reads, in place of creating stones in each proportional part of a day we grant the existence of an eternal

---

[37] John Buridan, *op.cit.*, p. 31$9^{-15}$. The connection between this passage in q. 16, which discusses "utrum linea aliqua girativa sit infinita, et semper accipio infinitum categorematice" and the preceding argument is made on p. 72$^{21-23}$. See the preceding note.

[38] John Buridan, *op.cit.*, p. 30$^{2-5, 15-16}$; p. 31$^{2-5}$: "Ex hoc sequitur decima quarta conclusio, scilicet quod si sit aliqua una linea girativa protracta per omnes medietates columne B modo predicto et non ultra omnes, ita oportet esse unam rectam protractam per omnes istas medietates proportionales et non ultra omnes ... Quintadecima conclusio est quod nulla est linea recta una protracta per omnes istas medietates, nisi sit protracta ultra omnes ... Ex hac conclusione et precedente sequitur sextadecima conclusio, quod nulla est linea girativa una protracta dicto modo per omnes medietates proportionales columne B, quia nulla talis est protracta ultra omnes, ut apparet per casum."

[39] John Buridan, *op.cit.*, p. 73$^{12-14}$: "Tertia conclusio sive sit vera sive sit falsa videtur apparenter sequi, sive apparentia fuerit probabilis sive sophistica, scilicet quod non est possibile per aliquam potentiam esse magnitudinem infinitam." See also John Buridan, *Quaestiones super libris quattuor De caelo et mundo* [Moody], p. 79$^{27-29}$, and John Buridan, *Expositio et questiones in Aristotelis De caelo et mundo* [Patar], p. 319$^{64-66}$: "Postea ego opinor quod non est possibile etiam per potentiam Dei esse corpus infinitum secundum magnitudinem nec esse etiam effectum infinitum secundum perfectionem."

[40] John Buridan, *Tractatus de infinito* [Thijssen], p. 73$^{15-17}$: "Probatur primo quia

world and then have God create them, one each day, throughout that eternity? Since God would then have been able from all eternity to have created stones (and could have conserved the product of all his handiwork), would we not then have the possibility of an actually infinite magnitude? In reply, Buridan returns to the traditional refuge of the distinction between divided and composite modal propositions and claims that although, on any given day among the infinity of days entailed by an eternal world, God could have created a stone, he could not have done so on all of the infinity of past days taken collectively.[41]

However, there is what Buridan terms a difficult objection which can be opposed to all of this. Just as nature can create things out of something, so God can out of nothing (*ex nihilo*). Hence, if the world were eternal, on any past day nature has made an amount of water in excess of "one foot," (as is clear from the fact of the generation of rivers), and therefore it is possible (even factually true) that, on every past day, God has made and has conserved a one-foot stone.[42]

To this one must respond that there is no relevant potency with respect to the past (*potentia ad preteritum*) and that therefore it is not possible, even by divine power, for the past not to have been. But, applied to the case at hand, this means that, although on every past day God has been able to create a one-foot stone, since he has not done so, it is not possible that he could do so. And Buridan again urges the significance of the lack of a potency with respect to the past by closing his argument with the assertion that just as it is impossible that, if God at some time has made B, for him not to have made B,

---

magna ratio ad probandum quod hoc sit possibile destructa est, scilicet quod possibile est Deum in qualibet medietate proportionali hore vel diei creare lapidem pedalem."

[41] John Buridan, *op.cit.*, pp. 73^{18}–74^3: "Sed diceret aliquis quod alia est ratio maior, scilicet quod, si Deus fecit mundum de novo, tamen ab eterno potuisset fecisse eum, et Aristoteles ponit quod eternaliter fuit mundus. Posito igitur quod eternaliter fuit mundus, constat quod Deus in quolibet die preterito potuisset creasse unum lapidem pedalem et conservare; igitur nunc essent infiniti lapides pedales ex quibus esset magnitudo infinita; igitur hoc est possibile. Respondeo quod si fuerit mundus eternus, tunc infiniti fuerunt dies, et in quolibet potuit Deus creare unum lapidem pedalem. Tales enim de possibili in sensu diviso concedende sunt, sicut post dicetur. Sed tamen in sensu composito non oportet concedere quod hec sit possibilis: 'Deus in quolibet die creavit unum lapidem pedalem reservando eos semper'."

[42] John Buridan, *op.cit.*, p. 74^{4-10}: "Sed adhuc contra hoc est obiectio difficilis, quia sicut natura potest facere aliquid ex aliquo, ita Deus posset illud facere ex nihilo et illud semper conservare; sed si mundus fuit eternus, natura in quolibet die preterito fecit aquam maiorem quantitatis quam pedalem, ut apparet ex generatione fluviorum; igitur hec est possibilis, immo de facto vera, quod omni die preterito fecit Deus lapidem pedalem et reservavit."

so it is impossible, if he at some time has not made B, for him to have made B; all this, even though previously it was possible that he could still make B.[43] In sum, then, the restrictions that Buridan saw fit to place upon God's absolute power relative to the production of actual infinites are paralleled by his views of the infinity of parts in infinitely divisible continuous magnitudes.

6. *Eternity and Infinity*

From the beginning of the thirteenth century, Aristotle's theory of the infinite played a central role in discussions of the world's eternity. The issue to be debated was, whether the world *could* have been eternal, if God had so willed. Or, in other words, was the concept of the world without a beginning consistent? Many authors believed that the beginning of the world could be proved by (demonstrative) arguments, namely by arguments demonstrating the impossibility of an eternal world. A number of such arguments rested upon assumptions concerning the nature of the infinite.[44] An eternal world was held to imply the existence of an infinite series of past events. The contradictions that allegedly arose from this assumption were considered reasons why an eternal world was impossible. Most of the arguments were drawn from Aristotle's widely accepted theory of the infinite, but were here employed in a new context to substantiate the un-Aristotelian conclusion that the world was not eternal but had a beginning. Especially the following generally accepted Aristotelian rules were considered to be violated by the idea of a beginningless world: that it is impossible to add to the infinite (*De caelo* 283a9–10); that it is impossible to traverse what is infinite (*Metaphysics* 994a1–

---

[43] John Buridan, *op.cit.*, pp. 74[11–13, 19]–75[4]: "Ad illud potest responderi per illud quod dicitur primo Celi, scilicet quod non est potentia ad preteritum. Ideo dicitur quod non est possibile etiam per divinam potentiam preterita non fuisse ... Sed adhuc obicitur quia ex quo omni die Deus potuit creare unum lapidem pedalem et post semper conservare, queritur si ita fecisset, quid modo esset, nonne modo essent lapides infiniti? Respondeo quod si ita fecisset lapides, nunc essent infiniti. Sed huius antecedens et consequens sunt impossibilia. Immo posita mundi et temporis eternitate semper fuit impossibile quod Deus omni die preterito creavit unum lapidem semper post conservando, quia semper fuit verum dicere quod ita non fecit. Et tamen, cum non sit potentia ad preteritum, impossibile est, si Deus aliquando fecit B, ipsum non fecisse B, et similiter impossibile est, si aliquando non fecit B, ipsum tunc fecisse B, licet tamen ante fuerit possibile quod ipsum faceret."

[44] See Maier, "Diskussionen;" Bianchi, *L'errore di Aristotele*; Dales, *Medieval Discussions*.

19); that it is impossible for the infinite to be grasped by a finite power (*Metaphysics* 999a27); and that it is impossible that there be simultaneously an infinite number of things (*Physics* 204a20–25; *Metaphysics* 1066b11). In addition, the theory of a possible eternal world seemed to clash with self-evident principles such as all infinites must be equal.

Now how does Buridan fit within this framework of the eternity discussion? Book VIII of Aristotle's *Physics* brought medieval thinkers in the arts faculty routinely into contact with arguments proving the eternity of motion and of the world. In question 3 of Book VIII, Buridan discusses Aristotle's complex proof of an eternal motion. Aristotle's proof is based on the following counterfactual: If motion or change had an absolute beginning, the first motion would either (1) had to have come into existence, or (2) have been eternal. Next, Aristotle sets out to reject both alternatives. He concludes that since motion cannot have had a beginning, it must be eternal (*Physics*, 251a17–28).

Buridan finds Aristotle's arguments reasonable, and hence infers a number of probable theses "which one need not even deny according to faith."[45] The first thesis is that there never is, has been, or will be a motion that is infinite according to duration.[46] The second, and most important, thesis is that motion *could* be infinite, and time too, though only the future time and not the past time. Buridan passes over a discussion of the possible past eternity of time, because it violates the principle that there is no potentiality with regard to the past: if time never had been eternal, then neither *could* it have been eternal.[47] As all medievals, Buridan did not discuss whether motion or time had actually been eternal, but whether they could

---

[45] John Buridan, *Quaestiones super octo Physicorum libros Aristotelis, secundum ultimam lecturam*, MS København, Kongelige Bibliotek, Cod. Ny kgl. Saml. 1801 fol., Book VIII, q. 3, fol. 154va: "De ista questione sunt alique conclusiones mihi probabiles, quas concessissent Aristoteles et Commentator et quas etiam non oportet negare secundum fidem nostram."

[46] John Buridan, *op.cit.*, VIII.3, fol. 154va: "Prima est quod nullus est vel fuit vel erit motus perpetuus seu infinitus secundum durationem et nullum est vel fuit vel erit tempus perpetuum seu infinitum; immo etiam quod impossibile est esse tempus perpetuum vel motum perpetuum."

[47] John Buridan, *op.cit.*, VIII.3, fol. 154vb: "Secunda conclusio est quod eternus vel infinitus potest esse motus et sic de tempore, saltem a parte post. Et non curo dicere nunc a parte ante propter hoc quod dicitur non esse potentiam ad preteritum. Propter quod posset dici quod si non eternaliter fuerunt tempus et motus, non eternaliter possunt fuisse motus et tempus." That God cannot make the past not to have been, is also invoked in Book III; See above note 43.

have been eternal. In other words, whether the idea of their eternity was consistent.

Apparently, Buridan thought it was. He is, however, aware that some authors held the opposite view and, consequently, deals with arguments against eternal motion. At the beginning of his *quaestio* he lists ten such arguments (*rationes quod non*). Three of these have a bearing on the infinite. So although the infinite does not play a role in Aristotle's own proof of an eternal motion, it thus appears in Buridan's discussion.

Since the three infinity arguments are all stock arguments against the eternity of the world, Buridan does not spell them out completely. The first of these arguments relies on Aristotle's thesis that the infinite cannot be traversed. However, an eternal motion implies an eternal past time, which, in its turn implies that an infinite number of days and an infinite number of years have been traversed. The traversal of an infinite series of events is, however, considered impossible. Consequently, the present moment never could have been reached. Aristotle's principle is employed to draw the un-Aristotelian conclusion that motion is not eternal.[48]

The second argument against an eternal motion is not strictly Aristotelian. It relies on the assumption that the infinite is a maximum value which cannot conceivably be exceeded. Consequently, all infinites are equal. Yet, given an eternal past, the infinite number of days which have passed is greater than the infinite number of years which have passed. The paradox of equal yet unequal infinites leads to a rejection of the antecedent that motion or time are eternal.[49]

The third argument, finally, relies on Aristotle's thesis that a completed infinite multitude is impossible. However, since the human intellect is immortal, the assumption of an eternal time and motion implies the existence of infinitely many intellects.[50] As a re-

---

[48] John Buridan, *op.cit.*, VIII.3, fol. 154ra: "Item, seqeretur quod infinita essent pertransita, scilicet infiniti dies et infiniti anni; immo etiam infinitum tempus esset pertransitum, quia totum tempus praeteritum esset pertransitum, et tamen esset infinitum. Et hec videntur impossibilia."

[49] *Ibid.*: "Item, seqeretur quod infinitis secundum multitudinem essent aliqua plura, quod videtur impossibile. Consequentia probatur, quia plures sunt dies quam anni, cum in quolibet anno fuerunt dies multi. Et tamen fuerunt infiniti anni, si tempus sit perpetuum, vel etiam motus."

[50] John Buridan, *op.cit.*, VIII.3, fol. 154^{ra-rb}: "Item, seqeretur quod nunc essent actu infiniti intellectus humani, quia infiniti fuerunt homines, si tempus et motus fuerunt perpetui, et quilibet homo habuit suum intellectum. Et cum illi intellectus non sunt corruptibiles, ideo sic nunc sunt omnes, et sic sunt infiniti."

sult of this inconsistency, the claim of an eternal motion has to be rejected.

As mentioned before, Buridan finds Aristotle's proofs for an eternal motion convincing. Consequently, he rebuts the ten arguments that had been adduced to disprove the eternity of motion. Again, we will only concentrate on the three arguments having to do with infinity. Note that Buridan presents the rebuttals as if Aristotle could have given them in this way, although it is quite clear that none of them are Aristotelian (*quomodo ipse credidisset solvisse rationes ad oppositum*). Starting with the easiest one, against the contention that an eternal motion entails an infinite number of intellects, Aristotle could have replied, according to Buridan, that there is only one intellect for the entire human kind, "such as the Commentator claimed," or that human intellects are corruptible "such as Alexander (of Aphrodisias) claimed."[51]

The other two rebuttals are based on a distinction between the categorematic and the syncategorematic infinite, which Buridan had explained in Book III of his commentary on the *Physics*. The upshot of these rebuttals is that only one infinite, namely the infinite in its syncategorematic sense, is permissible, since it is really only a finite multitude. Thus, syncategorematically speaking, an infinite number of days and years have been traversed, although no time has been traversed, except a finite one.[52] In similar fashion, the number of days in such a finite stretch of time are greater than the number of years.[53]

---

[51] John Buridan, *op.cit.*, VIII.3, fol. 155ra: "Ad aliam diceret Aristoteles, ut credo, quod est idem intellectus in numero omnium hominum, sicut posuit Commentator, vel quod intellectus humani sunt corruptibiles, sicut posuit Alexander."

[52] *Ibid.*: "Ad aliam concedendum est quod capiendo 'infinitum' vel 'infinita' categorematice, non est possibile pertransitum esse infinitum, vel pertransita esse infinita. Sed sincategoremance loquendo diceret Aristoteles quod infiniti dies et anni pertransiti sunt vel infinitum tempus pertransitum, licet nullum sit tempus pertransitum nisi finitum."

[53] John Buridan, *op.cit.*, VIII.3, fol. 155ra: "Ad aliam conceditur similiter quod loquendo categorematice impossibile est aliquo infinito aliquid esse maius, vel etiam aliquibus infinitis secundum multitudinem esse plura. Sed tamen sincategorematice loquendo infinito motu est motus finitus maior et infinitis diebus vel annis secundum multitudinem sunt dies vel anni secundum multitudinem plures. Et sic etiam, cum omne tempus preteritum fuerit finitum, ego concedo quod in omni tempore in quo fuerunt dies et anni, fuerunt plures dies quam anni. Sed esset ita, et non esset ita, si esset infinitum loquendo categorematice, sicut dictum fuit in tertio huius." The last remark is an allusion to John Buridan, *Tractatus de infinito* [Thijssen], p. 51$^{1-12}$, and pp. 53$^{4-7, 11-15, 19}$–54^{2}: "Nec tempus, si esset infinitum, contineret plures dies quam annos. Et hoc est rationabile, quia sicut infinito secundum magnitudinem non est maius, ita infinitis secundum multitudinem non sunt plura. Et tamen tempus

In sum, Buridan denies that the thesis of an eternal motion entails an actual and completed infinite. Note that Buridan in his discussion of the eternity of motion does not fully enter into any of the new developments regarding the puzzle of the equality of infinites.

Thus, Buridan supports Aristotle's arguments in favor of eternal motion or at least rebuts the arguments raised against Aristotle. At the end of his *quaestio*, he notes, however, that the thesis that motion is eternal, contradicts faith. More precisely, he indicates that "... although Aristotle has said these things, we – on the basis of pure faith and not through a demonstration having evidence arising from the senses – must hold the opposite, namely that before this day no motion was either eternal or infinite."[54] And then, without any further comment, Buridan sets out to refute Aristotle's arguments for an eternal motion. This ending seems rather disappointing, yet should be seen in its proper perspective. Buridan clearly distinguishes the realm of philosophy from that of faith. The method of philosophy is based on reasoning and on empirical evidence. It is within this context that Buridan examines and discusses Aristotle's proofs for an eternal motion and finds them convincing. However, Aristotle's

---

infinitum, si esset, contineret annos infinitos secundum multitudinem, ideo non contineret plures dies quam annos. Et hoc iterum probatur, quia si poneremus tempus preteritum infinitum, nulli dies preteriti essent plures quam anni preteriti, nisi acciperentur a parte ante omnes dies sic quod nulla dies eos precessisset. Sed hoc est impossibile, quia nulli essent omnes dies preteriti, sicut nulle medietates proportionales linee B sunt omnes eius medietates proportionales ... Oppositum arguitur quia in omni tempore finito preterito in quo fuerunt anni et dies, fuerunt plures dies quam anni; sed omne tempus preteritum est tempus finitum preteritum quo fuerunt dies et anni ... Respondeo quod hec est concedenda: 'fuerunt plures dies quam anni' et similiter ista: 'dies fuerunt plures quam anni', quia iste terminus 'fuerunt' connotat tempus preteritum et nihil precedit illud verbum quod sit distributivum illius connotationis; igitur illud verbum 'fuerunt' stat pro tempore preterito indefinite ... Ideo hec est neganda: 'non fuerunt plures dies quam anni'. Et sic etiam hec est concedenda: 'quandocumque fuerunt dies et anni, plures fuerunt dies quam anni'. Sed si fuisset tempus infinitum, ista esset falsa: 'quandocumque fuerunt dies et anni, plures fuerunt dies quam anni', quia instantia esset pro isto tempore infinito, et iste etiam essent concedende: 'aliquando sic fuerunt plures dies quam anni' et 'aliquando non fuerunt plures dies quam anni'; ideo ista esset neganda: 'non fuerunt plures dies quam anni', quia equivalet isti: 'nunquam fuerunt plures dies quam anni', contra quam esset instantia pro temporibus finitis."

[54] John Buridan, *Quaestiones super octo Physicorum libros Aristotelis, secundum ultimam lecturam*, MS København, Kongelige Bibliotek, Cod. Ny kgl. Saml. 1801 fol., VIII.3, fol. 155rb: "Sed tamen quamvis Aristoteles ita dixerit, nos fide pura et non per demonstrationem habentem ex sensibus ortum et evidentiam debemus tenere oppositum, scilicet quod nec eternus, nec infinitus fuit motus; immo, tantus fuit motus ante hanc diem quod ante hanc diem nullus fuit maior."

position is in fact mistaken. On the basis of God's revelation, which is considered a fortunate complement to empirical knowledge, motion had a beginning. Revelation provided an occasion to discover the truth and to revise Aristotle's arguments.[55]

7. Conclusion

Most characteristic of Buridan's discussion of the infinite is the predominance of propositional analysis. These examples dealing with infinity and continuity can be multiplied by many other examples from other areas of Buridan's (natural) philosophy. Buridan's own semantic theories, displayed in his logical works and usually not explained *expressis verbis* in his natural philosophy, metaphysics, or ethics, provide the method for almost all the moves he makes in those other areas.

The reason why the metalinguistic or propositional approach takes flight in his work may have had something to do with how Buridan understood the task of philosophy. A major task of medieval academic (natural) philosophy was to interpret Aristotle's text and those of his commentators. In certain thinkers, among whom Buridan is one of the foremost, this situation fostered the analysis of verbalized concepts and other linguistic and logical formulations. The exposition of Aristotle's text compelled him to reflect about the precise meaning of inherited expressions and notions, and to this end he applied all the techniques he had learned during his previous training in medieval logic and semantics. In this respect, Buridan's idea of the function of philosophy may have been like that of many contemporary analytic philosophers.

In the fourteenth century, Buridan's analytic approach was equalled only by his slightly older contemporary William of Ockham, that other champion of propositional analysis.[56] Perhaps Buridan got the idea of applying semantics to all other corners of his philosophy from him. But if he did, he certainly gave his own twist to the endeavor.

---

[55] See further Edith Sylla's contribution in this book for a discussion of the intrusion of theological considerations in Buridan's philosophy, in particular as exemplified by the controversy over the world's eternity.

[56] See, for instance, Brown, "A Modern Prologue," and Murdoch, "*Scientia mediantibus vocibus*" and "William Ockham."

BURIDAN'S CONCEPT OF TIME.
TIME, MOTION AND THE SOUL IN JOHN BURIDAN'S QUESTIONS ON ARISTOTLE'S *PHYSICS*[*]

Dirk-Jan Dekker

1. *Introduction*

The *Questions on Aristotle's Physics* is one of the most important of John Buridan's works about the philosophy of nature.[1] This commentary, the last version of which was probably written between 1350 and 1357,[2] was very influential at the university of Paris. It was widely read all over Europe for more than two hundred years.[3] Centuries later the commentary would even give its author a reputation as one of the precursors of Galileo.[4] Despite all the original ideas that Buridan developed in his natural philosophy, however, he cannot be regarded as a precursor of the Scientific Revolution. Modern research therefore no longer investigates him from the point of view of the Early Modern period but rather tries to investigate his natural philosophy in its own context.

In this article, I would like to present one topic of Buridan's natural philosophy: his concept of time[5] as it is found in the fourth book of the aforementioned *Questions on Aristotle's Physics*. In reconstructing this concept of time we will have to make do with the *Physics* commentary as the only source. Although John Buridan was a prolific writer of logical treatises no natural philosophy can be found in those.[6] Because Buridan remained a master of arts and never became

---

[*] Research for this paper was made possible through financial support from the Netherlands Organization for Scientific Research (NWO), grant 200-22-295. I would like to thank Marlies Veerkamp for correcting my English; of course, I remain responsible for any mistaken usage.

[1] Until the new edition of this work – which is being prepared at Nijmegen – becomes available, the most accessible printed version of the text is John Buridan, *Quaestiones super octo Physicorum libros Aristotelis* (Paris, 1509).

[2] John Buridan, *Tractatus de infinito* [Thijssen], pp. xx–xxi.

[3] See Michael, *Johannes Buridan* and Schönberger, *Relation als Vergleich*, pp. 3–29.

[4] For instance by Pierre Duhem: see Murdoch, "Pierre Duhem."

[5] Anneliese Maier devoted little attention to John Buridan in her *Metaphysische Hintergründe*, pp. 89–90 and 135–136.

[6] Biard, "Le statut du mouvement," p. 141.

a theologian he did not write a commentary on the *Sentences* of Peter Lombard, which otherwise is an important source for late scholastic concepts of time.[7]

In the following paragraphs I would like to focus on two themes that are essential to Buridan's concept of time: firstly, the relation between time and motion, and secondly, the relation between time and the activity of the intellective soul. Although these two themes were not the only topics under discussion in late scholastic thought about time[8] they are the only topics dealt with by John Buridan in the final questions of book IV of his *Physics* commentary. In fact, almost all of this commentary in the *Tractatus de tempore* – as that set of questions is sometimes known – deals with what we now refer to as the "metric" of time, i.e., the function of time as a measure. Other philosophical topics such as the unicity and continuity of time are only briefly touched upon. Theological subjects such as the time of angels are not dealt with at all, because Buridan was not a theologian.

Buridan's concept of time can be summarized in the following three statements:

1. Time is a successive thing (*res successiva*) and is as such identical with motion.
2. The term "time" signifies the same as the term "motion" and connotes in addition that the *suppositum* is applicable as a measure.
3. The existence of time does not depend on an activity of the intellective soul.

It is clear that the first two statements deal with time and motion, whereas the third one deals with time and the soul. Let us begin with time and motion.

2. *Time and Motion*

Aristotle defined time as "the number of motion in respect of the before and after."[9] As Anneliese Maier noted,[10] late scholastic philosophers widely agreed that the Aristotelian definition was correct. How-

---

[7] Cf. Porro, *Forme e modelli*, p. 92.
[8] An impression of the scope of the discussion can be obtained from Zimmermann, *Verzeichnis*, pp. 146–301.
[9] Aristotle, *Physics*, IV.11, 219b2.
[10] Maier, *op.cit.*, p. 47.

ever, in contrast to the consensus that *tempus est numerus motus secundum prius et posterius*, there was no common opinion about the reality of time. Explaining the ontological status of time generally came down to explaining the relation between time and motion.[11]

John Buridan also sided with his contemporaries in defining time as "the number of motion in respect of the before and after."[12] Yet he also held that time is identical with motion.[13] For him time is first of all a successive thing, a *res successiva*.[14] The idea that it is successive seems obvious in our everyday speech about the duration of temporal things: a man lives for seventy years and the travel from Paris to Avignon takes twelve days. But it is less obvious to call time "a thing," a *res*, thus suggesting that it exists independently from other entities.[15] To explain Buridan's expression I shall briefly turn to his explanation of motion.

In book III of his commentary on the *Physics*, Buridan dealt with the question, "Whether local motion is something distinct from place and distinct from the mobile thing."[16] This particular question is well-known, not in the least because in it he explicitly rejected William of Ockham's identification of local motion with the mobile thing. Buridan showed that it is not correct to maintain that local motion can be explained as a mobile thing acquiring one place after the other continuously. For let us imagine a local motion of the outermost heavenly sphere. This sphere is not in any place in its Aristotelian sense because it lacks an *ultimum continentis*, i.e., a surface of the surrounding body. Still, it must be accepted that God can move the entire universe in local motion in a straight line. Denying this would be heretical because of the well-known 49th proposition,[17]

---

[11] Trifogli, "La dottrina del tempo," p. 249.
[12] John Buridan, *Quaestiones super octo Physicorum libros Aristotelis* (Paris, 1509), Book IV, q. 13, conclusio 5: "... ista [i.e., the Aristotelian definition] est bona descriptio temporis ..." In this article Latin quotations from the forthcoming edition of Buridan's complete *Physics* commentary are used. They are based upon MS København, Kongelige Bibliotek, Cod. Ny kgl. Saml. 1801 fol.
[13] John Buridan, *op.cit.*, IV.12, conclusio 3: "... tempus propriissime acceptum est motus primus ..."
[14] John Buridan, *op.cit.*, IV.12, conclusio 1: "... tempus est res successiva cuius partes non sunt simul, sed alia prius et alia posterius, et ... priori existente nondum est prior et posteriori existente non amplius prior est."
[15] Cf. also John Buridan, *op.cit.*, IV.16, ad 1: "... dicendum est quod tempus est res extra animam et praeter operationem animae nostrae existens."
[16] John Buridan, *op.cit.*, III.7: "Utrum motus localis sit res distincta a loco et ab eo quod localiter movetur."
[17] Denifle e.a. (eds.), *Chartularium Universitatis Parisiensis*, n. 473, p. 546: "Quod

condemned at Paris in 1277, that "God could not move the entire universe in a straight line because if He did, a vacuum would be left behind." But if God can move the entire universe in a straight line, then He surely can simply rotate it as one solid body including the earth at the center. This rotation obviously constitutes a motion, but how can it be accounted for in terms of mobile and place – for no place is involved in it? Apparently, local motion cannot be explained as a mobile thing acquiring one place after another, and thus something else must be involved. For Buridan this "something else" is a purely successive thing, intrinsic to the moving body. So for him local motion is a successive quality inhering in the mobile thing as a passion in a subject. Because time is motion it is also a successive thing: and that is exactly what we find when we reflect on time.[18] It has parts that occur one after the other in a way similar to motion. When the present exists the future does not exist yet; when the present exists the past exists no longer. Like motion, the succession of time occurs continuously. Not only Buridan's definition of motion "*continue aliter et aliter se habere*"[19] can be applied to such a continuum, but also Aristotle's definition, "*actus entis in potentia secundum quod in potentia.*"[20]

Since there are countless motions around us, we might ask ourselves which one is time. The motion of the outermost heavenly sphere combines three properties by virtue of which the function of time should be attributed to it. First, according to an epistemological principle of Aristotle's, knowledge of posterior things through prior things is more properly knowledge than that of prior things through posterior ones. If possible one should prefer a prior motion as time to a posterior one. Secondly, it is desirable that the motion used as time be regular. The most regular of all motions is that of the outer sphere of heaven. Finally, the astronomers ultimately use the first motion as time in their calculations of the positions of heavenly bodies.[21]

Deus non possit movere celum motu recto. Et ratio est, quia tunc relinqueret vacuum." Cf. Murdoch e.a., "The Science of Motion," esp. pp. 217–218.

[18] For a critical evaluation of this argumentation see my forthcoming dissertation.

[19] I.e., "... continuously being different and different." About Buridan's theory of motion see Biard, "Le statut du mouvement," pp. 141–159.

[20] I.e., "... the fulfilment of what exists potentially, in so far as it exists potentially" (Aristotle, *Physics*, III.1, 201a10).

[21] John Buridan, *op.cit.*, IV.12, conclusio 3: "... tempus propriissime acceptum est motus primus. Quia de ratione temporis est quod sit mensura motuum, ideo magis proprie tempus debet esse ille motus qui magis proprie dicitur esse mensura aliorum.

Buridan agrees that in everyday life, many inferior motions are used as time. Above all, ordinary people use the motion of the sun because it is most perceptible to their senses, whereas they do not perceive the motion of the outer sphere.[22] And yet other motions can be used, for instance the rotation of the hand of a watch. And a workman who knows how much work he completes in a day will know from his remaining workload when it is lunchtime.[23] But these motions are less regular than, and secondary to, the motion of the outer sphere. Because only the first and most regular motion is properly called "time," it is only the motion of this outer sphere that is time in the first and most proper sense.

3. *The Definition of Time*

If time is the motion of the outer sphere then how does Buridan explain the Aristotelian definition of time as a number? To answer this question I shall follow Buridan in analyzing and explaining this definition part by part.

Three properties of time can be deduced from the preceding discussion. Firstly time is successive because it has prior and posterior parts not existing simultaneously. Secondly time is continuous, not only because it is divided proportionally to motion, but also because it is infinitely divisible. Finally, as motion relates to the mobile thing as a passion to a subject, time relates to motion as a passion to a subject, too.[24]

Sed ille est primus motus propter hoc quod in unoquoque genere rationabilius est quod primum sit mensura aliorum quam e converso. ... Et etiam mensura debet esse regularis, et primus motus est regularissimus in succedendo. ... Et hoc etiam manifestum est per astrologos, qui in mensuratione motuum recurrunt finaliter ad primum motum tamquam ad primam et maxime proprie dictam mensuram omnium aliorum."

[22] John Buridan, *op.cit.*, IV.12, conclusio 4: "... apud vulgares motus compositus solis ex diurno et proprio magis est tempus quam aliquis alius motus, ... Sed maxime vulgares mensurant per dictos motus solis ... quia ille motus est eis notissimus quia maxime sensui apparens, et non est eis notus motus simplex diurnus."

[23] John Buridan, *op.cit.*, IV.12, conclusio 5: "... saepe operatores mechanici utuntur sua operatione per modum temporis, ... ex quantitate operationis concludunt quod est hora tertia et tempus comedendi. Et etiam ecclesiastici horologio utuntur per modum temporis."

[24] John Buridan, *op.cit.*, IV.13: "... quod tempore mensuramus motus finitos et quietes ad sciendum quanti sunt cum hoc fuerit nobis dubium. Deinde, quod tempus est successivum habens prius et posterius, ... Deinde ... quod tempus est motus et quod est motus primus, ... Deinde etiam quod tempus est continuum, ... Deinde

Having established these properties, Buridan argues that it is essential to time that its parts be distinguished from one another. This *discretio partium* is the basis for its being a measure. Since the motions that are to be measured are finite, the first motion (i.e., time) has to be divided into parts to measure them. Besides, it is also clear from everyday life that if we wish to measure a duration we divide heavenly motion into years, days, hours etc. A measure in which such a *discretio partium* or "a distinction of parts" is found, is a number. Therefore Aristotle correctly defined time as a number.[25]

Because time is a successive thing, it cannot be the measure of permanent things insofar as they are permanent. Although rest can be measured in time simply by measuring the amount of time coexisting with it, something purely permanent, a *res pure permanens*, cannot be measured in time. Therefore, time is the number of motion.[26]

According to Aristotle, the parts of motion can be distinguished in two different ways: either according to the division of the mobile thing, or according to the division of the distance that is traveled.[27] In the first way we cannot speak of "time" because the parts of a mobile thing exist simultaneously. If we divide motion in the second way, we can speak of "time" because the parts of the distance covered are acquired successively. So, in order to make clear that the first motion should be divided in the latter way, the words "in respect of the before and after" must be added to the definition of time. I shall return to this later because a difficulty will arise when applying this division to a hypothetical motion of the entire universe. But for now, by thus analyzing Aristotle's definition of time, Buridan not only shows how it explains the signification and all connotations of the word "time," but also that it is convertible with the *definitum*. Therefore the definition is correct.

---

etiam ex dictis sequitur quod iste terminus 'tempus' et iste terminus 'motus' se habent sicut passio et subiectum."

[25] John Buridan, *op.cit.*, IV.13, conclusio 1-2: "... de ratione temporis sive huius termini 'tempus' est discretio inter partes illius motus qui est tempus. ... Ergo ad hoc quod ille primus motus sit mensura aliorum motuum finitorum, necesse est discernere inter partes eius ad accipiendum partes motibus mensurandis proportionabiles. ... tempus est numerus quia est mensura, sicut dictum est, et non sine discretione inter partes eius quod est tempus."

[26] John Buridan, *op.cit.*, IV.13, conclusio 3: "... tempus est numerus motus, quia cum sit de ratione temporis quod sit mensura discreta, sicut dictum est, tamen non est mensura permanentium secundum quod permanentia, quia dictum est quod tempus est res successiva; et successivum non est mensura rei permanentis ea ratione qua est permanens."

[27] Aristotle, *Physics*, VI.4, 235a14-18.

4. Theory of Measuring

From the preceding paragraphs we have seen that time is motion, applicable as the number of other motions. Does this also mean that time is the number of any motion? To show which motions are measurable in time Buridan starts off by providing a theory of measuring – something Aristotle did not do.[28] Buridan begins by explaining that measuring is nothing other than the establishment of the unknown quantity of a measurable thing by the known quantity of a measure. This can be achieved in several ways, with which different kinds of measure correspond. Some measures are intrinsic to the measured thing. This is the case when we measure a whole – whether a multitude or a magnitude – by its parts, counting them successively and adding them up. Other measures are extrinsic. By comparing a magnitude of a known quantity with one of unknown quantity we can find that they are equal. This may be repeated if a measure is smaller than the measurable thing. There are still other methods of measuring. One should be singled out because it is important with regard to time: measurement according to a proportional division of the measure and the measurable regardless of any difference in their size or nature. An example can clarify this. Suppose someone wants to find out the circumference of the earth. He divides the sky into 360 degrees and walks some distance to observe how many miles it takes to see the Pole Star rising one degree in the sky. Now he only needs to multiply that number by 360 to determine the circumference of the earth.[29] The idea behind this example is that

---

[28] John Buridan, *op.cit.*, IV.14: "Ista quaestio est iudicio meo bene difficilis et multas implicans in se difficultates, quas aggrediendum notandum est primo quantum ad quid nominis, quod mensurabile mensurari aliqua mensura est scire quantitatem eius prius dubiam per quantitatem illius mensurae prius notam. ... Secundo etiam notandum quod plures et diversi sunt modi mensurandi. Primus modus est per mensuram intrinsecam ... Secundus modus ... in continuis supponens praecendentem modum. ... tertius modus in magnitudinibus invenitur: per suppositionem magnitudinis notae ad magnitudinem ignotae quantitatis. ... Quartus modus est compositus ex tertio et primo. ... Et adhuc alii modi mensurandi sunt ..."

[29] John Buridan, *op.cit.*, IV.14: "Sic enim mensuramus circuitum terrae quot leucas contineat per ipsum caelum, dividendo caelum in trecentos et sexaginta gradus, deinde proportionabiliter dividendo terram et considerando in terra quot leucae correspondeant uni gradui in caelo; ut quia forte triginta, quia cum per triginta leucas in plana patria ambulemus de austro ad septentrionem, inveniemus polum arcticum elevatum nobis in uno gradu. Tunc concludemus circuitum terrae esse totiens trecentas et sexaginta leucas quot leucae correspondent uni gradui in caelo, scilicet trecesies trecentas et sexaginta."

in order to measure there is no need for equality in size or even in nature between measure and the measurable.

How does Buridan apply this whole theory to time? Of course, time can be measured in the first way mentioned, i.e., by an intrinsic measure. We measure three days by counting three times one day and we also measure a week by counting seven days. It is not possible to measure time with an extrinsic measure because time is successive: there is so to speak nothing we can lay over a period of time and find it equally sized to that period of time.

Most important and perhaps most interesting is the fifth way of measuring: by proportional division. In this way it is possible to measure time with a watch, on the face of which a certain angle of the hand corresponds with a period of, for instance, one hour. This measuring can be applied in a number of other ways too. Suppose that it takes twenty days to walk from Paris to Rome. On the first day someone walks ten miles. If his walking is regular then one knows that the total distance is two hundred miles. One can easily correct this if the walking is irregular.[30] Another example is that of a mobile thing moving past a certain point, for example a boat passing a pole. If we know the length of the part passing the point in one hour and we know in how much time the whole mobile thing moves past it, then we know by multiplication the total length of the mobile thing.[31] This last example also shows that time is not the measure of motion according to the division of the mobile thing, but according to the division of the distance traversed. Similar accounts can be constructed for alterations very easily.[32]

---

[30] John Buridan, *op.cit.*, IV.14, conclusio 5: "De motu locali hoc primo declaratur quantum ad magnitudinem et divisionem motus secundum magnitudinem et divisionem spatii pertransiti, quia scire quantus est motus tali quantitate est scire quantum est spatium pertransitum, sed per tempus hoc scimus, verbi gratia cum viderimus quod ambulatio de Parisius ad Romam perficitur in viginti diebus, posito quod illa ambulatio fuerit regularis, et si sciverimus quod in primo die transimus decem leucas, sciemus spatium totale pertransitum usque ad Romam esse decies viginti leucarum, scilicet ducentarum leucarum."

[31] *Ibid.*: "Secundo etiam declarabitur conclusio quantum ad magnitudinem et divisionem mobilis, scilicet signando aliquod signum in spatio quod pertransitur. Tunc illud mobile pars eius post partem continue pertransit illud signum. ... Si ergo signaverimus tempus in quo mobile pertransit illud signum, ... et sit illud tempus una dies naturalis vigintiquattuor horarum, si appareat quod pars pedalis illius mobilis transit illud signum in una hora, et motus sit regularis, concludemus quod mobile est longitudinis vigintiquattuor pedum."

[32] *Ibid.*: "Primo quantum ad magnitudinem intensionis qualitatis in eodem subiecto et eadem parte eius. ... nam si tempore decem horarum intenditur qualitas in aliquo subiecto continue et regulariter, quanta intensione est qualitas acquisita in

But what happens when there is local motion without the mobile thing changing its place, for instance, when God would rotate the entire universe? I mentioned this difficulty earlier. Can one in this case really speak of the division of a traversed distance? Such motion seems measurable only in terms of the dimensions of the mobile thing. But that is problematic. Indeed it is true that motion, like any quality inhering in the mobile thing, is extended along the dimensions of the mobile thing. Therefore, it could be measured by these dimensions. The motion of half the mobile is half the motion of the whole mobile thing. This measuring clearly does not give us any insight with regard to time because the parts of the mobile thus distinguished move simultaneously. But there is another, more useful, way of speaking about the magnitude of motion: in terms of intension. Just as the intension of heat in a small spark can be greater than that in a large amount of air, so the intension of succession can be greater in a small mobile thing than in a larger.[33] Buridan also explains why he stated earlier that time should be measured by the distance traveled. The reason is of an epistemological character: one has to perceive a mobile being different with respect to place in order to find that is has moved. One can subsequently determine the quantity of motion by the quantity of the covered distance.[34]

According to Buridan, principally every motion is measurable because no motion is of such nature that it cannot be compared with any other motion. Of course, practical circumstances often prevent us from carrying out the measurement. Buridan also notes that a motion often cannot be measured with mathematical precision so that a certain margin of error has to be taken into account.[35]

---

una hora, decies tanta intensione est acquisita in totali tempore decem horarum. Et similiter etiam de quantitate et divisione alterationis secundum quantitatem et divisionem alterabilis."

[33] *Ibid.*: "Sed alio modo sic est magnitudo successionis sicut est in qualitate magnitudo intensionis. Hoc enim modo dicimus caliditatem in parva scintilla ignis esse in duplo vel in decuplo maiorem intensione quam in magno aere, et sic in motu quem vocamus localem est magnitudo successionis et fluxus, et sic alius motus parvi corporis est motu alterius magni corporis maior."

[34] John Buridan, *op.cit.*, IV.14: "... ille motus non perciperetur nisi mobile perciperetur aliter se habere ad locum vel spatium vel aliquod quiescens, ideo etiam istam quantitatem fluxus oportet cognoscere et arguere ex quantitate spatii, vel eius quod pertransitur, vel eius quod imaginatur pertransiri, quoniam motu maiori secundum quantitatem successionis maius spatium pertransiret si ille motus fieret commensurative spatio. In pertranseundo spatium tamen non esset maior aut minor secundum huiusmodi successionem, licet nullum esset spatium aliud a magnitudine mobilis."

[35] John Buridan, *op.cit.*, IV.14, conclusio 7: "... nulli motui repugnat secundum

5. *Time and the Intellective Soul*

The last question about time discussed by Buridan concerns the way time depends on the intellect. Aristotle's position is that, since time is a number, time cannot exist if nothing can exist which has the ability to count.[36] This means that time for its actual existence is dependent on the intellective soul. In late thirteenth and early fourteenth century Paris this topic was much debated.[37] But strangely enough not a trace of any debate can be found in Buridan's commentary. It is also striking that Buridan deals with the aforementioned question in a much more semantical way than with his other questions concerning time.[38] In contrast to the other questions about time, more use is made here of tools such as supposition, connotation, and logical rules.

When Buridan discusses the question he first presents arguments against Aristotle's – and his own – position. One of these is a thought experiment which runs as follows. Suppose that God destroys all intellective souls but saves the heavenly motions, then there would still be time for two reasons. Firstly because time *is* the motion of the outermost heavenly sphere. The second reason is of a more logical nature. After God's intervention some motion would still be faster than another, slower one. "Fast" and "slow" are defined by time. According to Buridan's logic, if the *suppositum* of a defined term exists, then that for which the terms of the definition supposit also exists. In the case of the terms "fast" and "slow," the terms of their definition include the term "time." Therefore, time will exist even after the destruction of all intellective souls if fast and slow still exist.[39]

---

rationem eius specificam vel generalem quin possit mensurari tempore, ... Sciendum est quod multis motibus repugnat propter circumstantias particulares quod possint a nobis mensurari, ut quia sunt in loco ad quem homo non potest naturaliter pertingere. Ideo nullus potest naturaliter istos motus percipere ad mensurandum eos, licet alibi possint similes in specie percipere. ... Notandum est etiam quod non possumus motus naturales omnino praecise et punctualiter mensurare, scilicet secundum modum mathematicae considerationis. Non enim possumus per stateram scire si praecise libra cerae sit librae plumbi aequalis: ... Sed sufficit saepe mensuratio ad prope, iuxta illud quod 'de modico non est curandum'."

[36] Aristotle, *Physics*, IV.14, 223a21–29. Cf. Sorabji, *Time, Creation and the Continuum*, pp. 89–97.

[37] See for example the discussion between John of Jandun and Thomas Wylton: Trifogli, "Il problema."

[38] For fourteenth-century semantical approaches to philosophical problems see Murdoch, "*Scientia mediantibus vocibus*" and Murdoch, "The Analytic Character."

[39] John Buridan, *op.cit.*, IV.16, ratio 2: "... quamvis annihilaretur omnis anima

Buridan also gives a logical counterargument. According to his theory of supposition, if something connoted by a connotative term no longer exists then the connotative term no longer supposits for anything. For instance, if the word "white" connotes whiteness and supposits for a man, then it will no longer supposit for anything if there is no whiteness any more. Similarly the word "time" connotes numerability and supposits for the motion of the outermost sphere. Aristotle stated that if there is no intellective soul, there will neither be counting nor numerability. Hence, time would lose its *connotatum* and therefore its *suppositum*.[40]

Buridan's own conclusions bear the same logical flavor as the arguments. He starts off with the remark that if there were no intellective soul at all then there would indeed be nothing because it would imply the non-existence of God and *ad impossibile sequitur quodlibet*. So firstly one has to exclude God from the cognizing beings mentioned in the question. Buridan accomplishes this by an appeal to God's unique way of knowing: because God knows everything perfectly, He also knows evidently the quantity of every measurable thing even if no measure would exist to measure it.

According to Buridan the word "number" does not connote actual counting but only potential counting.[41] He clarifies this once again by means of a thought experiment. If all people were asleep at the same time, there would obviously be no more actual counting, but only potential counting. Still, two people would be *two* people, which is only possible when the number "two" exists. Apparently potential counting suffices for a number to exist.[42] It follows, then,

---

intellectiva, tamen adhuc posset Deus salvare motus caelestes, ergo adhuc esset tempus. Consequentia probatur: primo quia tempus est primus motus, qui salvaretur. Secundo quia adhuc esset unus motus in caelo velocior alio, et alter tardior. Et tamen remanente veloci et tardo oportet tempus remanere, quia velox et tardum definiuntur tempore. Et tamen remanente eo pro quo definitum supponit, oportet remanere ea pro quibus termini definitionis supponunt, aliter definitio non esset convertibilis cum definito."

[40] John Buridan, *op.cit.*, IV.16, oppositum 2: "... si terminus pro aliquo supponit et aliquid etiam appellat, deficiente appellato ille terminus amplius pro nullo supponit. ... Modo verum est quod iste terminus 'tempus' supponit pro motu primo, sed appellat numerationem vel numerabilitatem. Et ideo hoc appellatum deficeret si non esset anima intellectiva ... Ergo ille terminus 'tempus' pro nullo supponit, ideo nihil esset tempus."

[41] John Buridan, *op.cit.*, IV.16, conclusio 2: "Sed apparet mihi quod iste terminus 'numerus' non connotat actualem numerationem, sed numerabilitatem. Quod potest declarari, quia si connotaret actualem numerationem, sequeretur quod nullus esset numerus nisi quando res actu numerarentur."

[42] John Buridan, *op.cit.*, IV.16: "Et hoc non videtur esse verum, quod si omnes

that there would still be number in case all cognizing beings except God were annihilated. And even if nothing would actually be counted then God could still create a soul to perform the actual counting.[43]

In only one case number – and hence, time – could not exist: if no counting subject can possibly exist. In this case time would neither exist actually nor potentially. However, the antecedent "no counting subject can exist" is false because it is impossible for God not to exist. Therefore, the logical rule *ad impossibile sequitur quodlibet* can be applied here and the proposition "... if no counting subject can exist, number – and hence, time – cannot exist" is true.[44]

6. *Conclusion*

We have now worked our way through John Buridan's treatise about time. We have seen that it is limited to two topics: first, the relation between time and motion, and second, the relation between time and the intellective soul. For Buridan time is a successive thing. Time is motion that can be divided into prior and posterior parts and can subsequently be applied as a measure to other motions. Buridan therefore agrees with Aristotle's definition "the number of motion in respect of the before and after." In everyday life, people use many motions as time, but speaking most properly, it is the motion of the outer sphere of heaven. For Buridan the term "time" signifies motion and connotes that it is applicable as a measure. To explain how time is used as a measure Buridan develops a theory of measuring. Finally,

---

dormirent et quod nullus numeraret, tamen ego et tu essemus duo homines, et non essemus duo sine dualitate vel binario. Ergo adhuc esset dualitas sive binarius, et per consequens numerus, quia necesse est si est aliquis binarius, quod omnis binarius sit numerus: ibi enim est praedicatio generis de specie, quae semper est universaliter vera si species pro aliquo supponit."

[43] *Ibid.*: "... quamvis omnia cognoscentia praeter Deum essent annihilata, manentibus aliis scilicet non cognoscentibus, adhuc res essent numerus; quia quamvis non essent numeratae, tamen adhuc essent numerabiles; et possent numerari, quia Deus posset statim creare homines et animas quae possent illas res numerare. Et hoc sufficit ad hoc quod illae res sint numerus secundum praedicta."

[44] *Ibid.*: "... si non posset esse numerans, non posset esse numerus, sicut bene arguebat ratio Aristotelis prius narrata. Nihil enim esset possibile numerari si nihil esset potens numerare. Et haec etiam conclusio patet per istam regulam quia 'ad impossibile sequitur quodlibet'; et antecedens dictae conclusionis condicionalis est impossibile simpliciter. Hoc enim est impossibile, scilicet quod non potest esse numerans."

the existence of time does not depend on an activity of the intellective soul. Even if there were no intellective soul besides God there would still be time because the motion of the outer sphere would still be applicable as a measure.

ON CERTITUDE

Jack Zupko

One of the most central and at the same time obscure notions in John Buridan's account of human knowledge is that of certainty or certitude (*certitudo*). Its centrality emerges in the *Summulae de dialectica*, where it is presented, along with evidentness (*evidentia*), as the first of three main differences between knowledge and opinion.[1] Its obscurity is a product of the different uses to which Buridan puts the term. He speaks not only of "certitude," but also of "the certitude of truth," "the certitude of assent," "the certitude of sense," and "certain judgment," as well of processes, such as inductive inference, which are said to help "certify" demonstrative knowledge. A representative, though by no means exhaustive, sample would be as follows:

> ... every [act of] knowledge has to occur with certainty and evidentness, as is clear by its nominal definition, but it is not possible for opinion to be like this (*Summulae. De demonstrationibus*, 4.4; cf. also John Buridan, *In Metaphysicen*, Book II, q. 1, fol. 8vb)

> ... certainty and evidentness are required for knowledge; and two other things besides: the certainty of truth and the certainty of assent (*Questions on the Posterior Analytics*, Book I, q. 2)

> ... since intellect is more subtle and powerful than sense, it does not seem that intellect, based on the certitude of sense, should assume this certitude for itself (*Questions on the Nicomachean Ethics*, Book VI, q. 11)

> ... the false judgment of sense gives way to the intellect's certain judgment, and yet what appears to sense does not give way to the intellect's certain judgment, as is obvious by experience (*Questions on the Prior Analytics*, Book II, q. 18)

> ... therefore, it must be understood that induction enters into a demonstration not because it certifies [the conclusion of such a demonstration] absolutely, but instrumentally (*Questions on the Nicomachean Ethics*, Book VI, q. 11)

---

[1] John Buridan, *Summulae. De demonstrationibus* [De Rijk e.a.], cap. 4, pars 4. The other two are that knowledge, unlike opinion, must be of a true proposition, and that we can acquire opinions about first principles by dialectical argumentation, although we cannot know them demonstratively, by means of a proper proof. For "evidentness" as the proper translation of "*evidentia*," see Zupko, "Buridan and Skepticism," n. 30.

> ... an enunciation is called a *question* on account of its doubtfulness, and for this reason someone who doubts it and wishes to get certain about it propounds it in the style of a query (*Summulae. De fallaciis*, 1.3)[2]

It is not Buridan's style to determine the proper use of a term without first examining the different contexts in which it is used, and "*certitudo*" is no exception. For modern interpreters of Buridan's thought, however, the question is made even more complex by the powerful association that has evolved in modern philosophy between the notions of certainty and objective warrant, so that if I am certain of something – say, that "here is one hand," to borrow the rather notorious example of G. E. Moore[3] – I am thereby understood to have legitimate (indeed, perhaps absolute) grounds for believing it. The stage is thus set for epistemology, the branch of philosophy concerned with the theory and practice of justifying knowledge claims. And if this were not enough, there is historical evidence that *certitudo* was a doctrinally charged concept even in Buridan's time. It figures prominently in two Articles of the Condemnation of 1277, a document Buridan knew well and quoted from on several occasions,[4] as well as in a list of propositions from the writings of Nicholas of Autrecourt, condemned at Paris in 1346 when Buridan was at the height of his philosophical career:

---

[2] John Buridan, *Summulae. De demonstrationibus* [De Rijk e.a.], 4.4, p. 72: "... oportet scientiam esse cum certitudine et evidentia, ut patet per quid nominis, sed non est possibile talem esse opinionem;" *Quaestiones in Posteriorum Analyticorum libros* [Hubien], Book I, q. 2, p. 10: "... ad scientiam requiritur certitudo et evidentia; et adhuc duo requiruntur: certitudo veritatis et certitudo assensus."; *Quaestiones super decem libros Ethicorum Aristotelis* (Paris, 1513), Book VI, q. 11, fol. 126vb: "... cum intellectus subtilior sit et potior sensu non videtur quod intellectus ex certitudine sensus debeat sibi certitudinem assumere."; *Quaestiones in Priorum Analyticorum libros* [Hubien], Book II, q. 18, p. 41: "... propter certum iudicium intellectus caderet falsum iudicium sensus, et tamen propter certum iudicium intellectus non caderet apparentia ad sensus, quod patet per experientiam." ; *Quaestiones super decem libros Ethicorum Aristotelis* (Paris, 1513), VI.11, fol. 127ra: "... igitur est intelligendum quod inductio habet introitum in demonstratione non quia certificat simpliciter, sed ministraliter solum."; *Summulae. De fallaciis* [Hubien], cap. 1, pars 3, p. 2: "... enuntiatio ex eo quod dubitabilis est vocatur 'quaestio,' propter quod a dubitante volente certificari de ea ipsa sub modo quaerendi profertur." Unless otherwise indicated, all translations in this paper are my own. Translations from the *Summulae* are from Klima, *Buridan's* Summulae, forthcoming in the Yale Library of Medieval Philosophy. Professor Klima has generously allowed me to consult, and quote from, the final text of his translation.

[3] See Moore, *Philosophical Papers*, pp. 127–150. For the notoriety, see Wittgenstein, *On Certainty*, § 1: "If you do know that *here is one hand*, we'll grant you all the rest (Wenn du weißt, daß hier eine Hand ist, so geben wir dir alles Übrige zu)" (2).

[4] For discussion, see Zupko, "Buridan and Skepticism."

Condemnation of 1277, Article 3: That in order for a man to have any certitude about a conclusion, he must have it from self-evident principles – This is statement erroneous because it speaks in a general way of both the certitude of apprehension and the certitude of adherence.

Condemnation of 1277, Article 5: That a man must not be content with authority to have certitude on any question.

1346 Condemnation of Nicholas Autrecourt: I have said in the same letter [i.e., the Second Letter to Bernard of Arezzo] that, except for the certitude of faith, there is no other certitude except for the certitude of the first principle, or that which can be reduced to the first principle – I recant this as false.[5]

So in "*certitudo*" we have a term that (1) has multiple, nonequivalent uses in Buridan's own work; (2) acquired a new and unprecedented connotation in European philosophy after Descartes; and (3) was misused in a way so dangerous that it evoked public censure only a generation before Buridan. These are not the characteristics of a stable concept.

Nevertheless, I wish to clarify the meaning of "*certitudo*" in Buridan's thought, and, along the way, to explore some of the subjective aspects of his account of knowledge. In order to do this, we must first draw attention to an ambiguity in its English translation. Although "certitude" and "certainty" are both perfectly acceptable translations of the Latin abstract noun "*certitudo*," "certainty" is by far the more common rendering. Unfortunately, however, "certainty" has almost from the beginnings of the English language been used to signify the certain condition of a fact or truth, and more specifically, "the quality or fact of being (objectively) certain."[6] Now one can only infer so much from the historical usage of a term, of course. But the stakes are slightly higher with "certainty" because of the exaggerated

---

[5] Articles 3 and 5 were promulgated as Articles 150 and 151, respectively, in the original text of the Condemnation of 1277. I here follow the numbering of P. Mandonnet's thematic reorganization of the text. Condemnation of 1277, Article 3: Quod ad hoc quod homo habeat aliquam certitudinem alicuius conclusionis, oportet quod sit fundatus super principia per se nota – Error, quia generaliter tam de certitudine apprehensionis quam adhaesionis loquitur (Hissette, *Enquête*, p. 20); Article 5: Quod homo non debet esse contentus auctoritate ad habendum alicuius quaestionis (Hissette, *Enquête*, p. 22); De Rijk, *Nicholas of Autrecourt*, p. 172: "Item. Dixi eadem epistola quod, excepta certitudine fidei, non est alia certitudo nisi certitudo primi principii, vel que in primum principium potest resolvi – Revoco tanquam falsum."

[6] For example, the *Oxford English Dictionary* mentions a 1565 usage in Richard Grafton's *Chronicles of Edward I*, "The king … woulde thereunto geue no credite until he had sent thether, and receyued the certainie."

role the concept of certainty has played in the history of modern philosophy. Descartes uses "*certitudo* (Fr. *certitude*)" and its cognates for what is standardly rendered as "certainty" in English translations of his work.[7] And although commentators have differed somewhat in their reconstructions of Descartes's conception of certainty,[8] all agree that he was seeking a radical form of judgment that is discontinuous with sensory appearance because he wanted to provide a new foundation for philosophical and scientific truth. It is not difficult to see how this led very naturally to our modern understanding of certainty in epistemic terms, i.e., in terms of objective warrant. But it gives us only half of what Buridan understood by "*certitudo*." The other half is more qualitative and subjective. Happily, there is another English term, "certitude," which has this qualitative and subjective aspect as its primary meaning; "certitude," according to the *OED*, is "the state of being certain or sure of anything; assured conviction of the mind that the facts are so and so; absence of doubt or hesitation," the "unfailing quality" of an action or event. The distinction between certainty and certitude will be very useful in bringing to light the different facets of Buridan's conception of *certitudo*.[9] Without suggesting that we reform our terminology, I will use "certitude" in the ensuing discussion whenever qualitative or subjective certainty is at issue, reserving "certainty" for the objective sense. It will not be possible to distinguish these senses for all occurrences of "*certitudo*,"[10] of course, since Buridan was quite innocent of the Cartesian epistemological program which led to their distinction and which has been largely responsible for elevating objective certainty into its present position as the primary sense of certainty.[11] But trying to do

---

[7] See, e.g., in *Meditations* I-II, IV-V. There are also numerous occurrences in the *Discourse on Method*, although Descartes did not himself carry out or approve the Latin translation of that work.

[8] See, e.g., Kenny, *Descartes*, pp. 191–193, and Wilson, *Descartes*, pp. 7–9.

[9] I believe that Buridan's conception of *certitudo* was the standard medieval sense, though I will not be able to show that here. But it is obvious that Aquinas, for example, is thinking of the subjective side of *certitudo* when he speaks of "the unshakeable certitude and unadulterated truth (*fixam certitudinem et puram veritatem*) about divine things that is conveyed to men through faith" (*Summa contra gentiles* I.4, and also I.65; *De veritate*, II.1, ad 4, and VI.3).

[10] Where the original sense is undistinguished or ambiguous, I will keep the Latin term "*certitudo*."

[11] I believe that in reconstructing such debates, we would do well to embrace Mark Jordan's comment on certain modern readings of Aquinas: "Thomas does not write ontologies or epistemologies, much to the chagrin of some who would be Thomists" (Jordan, *The Alleged Aristotelianism*, p. 22).

so will help us better appreciate how Buridan understood human knowing.

I have elsewhere explored the objective side of *certitudo*,[12] i.e., the side we would now refer to as "certainty," arguing that what little Buridan does say about epistemic justification most closely approximates the contemporary epistemological theory of reliabilism, and that a cogent reply to the skeptical arguments of Nicholas of Autrecourt would have been available to him by simply appealing to the reliability of human cognitive processes, whose capacity to produce true judgments is defended by him on numerous occasions. Despite all this, however, Buridan does not actually use such arguments in order to rebut skeptical challenges. Uncharacteristically, he does not want to talk about the issues here, but prefers a strategy of negative campaigning, denouncing those who resort to skeptical arguments as "exceedingly malevolent individuals, intent upon destroying the natural and moral sciences" by insisting that any principle not *a priori* certifiable "can be falsified by supernaturally possible cases."[13] It is not just that Buridan ignores such skeptical challenges; he does not think that Autrecourt's arguments present any challenge to his account of human knowledge at all. Why else would he exempt from philosophical consideration what was obviously a controversial topic at Paris in the 1340s, given his reputation within the University, and the fact that he was not shy of controversy in other areas? Even if we can construct an effective reply on Buridan's behalf to skeptical arguments based on these "supernaturally possible cases," understanding what he actually says to Autrecourt will require looking more closely at his remarks on the genesis of certitude. These remarks are more descriptive in nature, and accordingly lend themselves more to a phenomenological interpretation than to the sort of necessary-and-sufficient-conditions analysis typical of contemporary epistemology.[14] But since Buridan is a medieval and not a modern philosopher, I

---

[12] Zupko, "Buridan and Skepticism."

[13] John Buridan, *In Metaphysicen* (Paris, 1518), Book II, q. 1, fol. 9ra: "... aliqui valde mali volentes interimere scientias naturales et morales eo quod in pluribus earum principiis et conclusionibus non est evidentia simplex sed possunt falsificari per casus supernaturaliter possibiles." Cf. *Summulae. De demonstrationibus* [De Rijk e.a.], 4.4: "... because of the aforementioned requirements demanded by the concept (*ratio*) of knowledge, some people, wanting to do theology, denied that we could have knowledge about natural and moral [phenomena]." The sarcastic tone of "some people, wanting to do theology (*quidam theologizare volentes*)" is unmistakeable. Gyula Klima correctly notes that this is an allusion to Nicholas of Autrecourt, who was associated with the faculty of theology.

[14] Most contemporary epistemology, following Descartes, takes a subject's level of

believe that we can fairly say here, "so much the worse for contemporary epistemology." Indeed, this is perhaps something we should say, both because some of the most difficult problems in contemporary epistemology stem from its failure to take the qualitative aspect of belief seriously,[15] and because we cannot, in any case, arrive at a full understanding of Buridan's conception of certitude without it.

1. *The Two Sides of* Certitudo

Certitudo is one of the features that for Buridan distinguish knowledge from mere belief or opinion. He discusses the concept in at least four different works, and although there are slight differences in emphasis appropriate to the question being discussed, the primary doctrine is the throughout. For my purposes here, I shall focus on the somewhat more comprehensive treatment given in Treatise VIII of the *Summulae*, and specifically on its largest part, which has come to be known as the "Treatise on Demonstrations."[16]

Knowledge, Buridan tells us, differs from mere belief or opinion because "every [act of] knowledge has to occur with certainty and evidentness, as is clear by its nominal definition, but an opinion cannot be such." *Certitudo* is then explicated as follows:

> I say, therefore, first "with certainty (*cum certitudine*)." For certainty (*certitudo*) requires two things. One on the part of the proposition that is assented to, namely, that it should be true; for that is not certain belief on the basis of which we assent to something false, but rather it is uncertain and deceptive; and it is clear that, taken in this way, certainty (*certitudo*) is required for knowledge, for that which is false we do not know. Another thing is required on our part, namely, that our assent should be firm, that is, without doubt, or fear of the opposite side; and this is also required for knowledge, for a doubtful and fearful assent does not transcend the limits of opinion. For if someone assents to a proposition fearing [that] the opposite [may be true], he would never say that he knows that it is true, but rather that he takes it or believes that it is.[17]

---

subjective conviction in holding a belief to be irrelevant to its warrant. But this, of course, is a product of its focus on object warrant and what it offers to beliefs, i.e., justification.

[15] A good example of this would be the debate over whether someone can know something without being in possession of the reasons that justify it, even if it concerns something immediately present to his or her senses.

[16] The *Treatise on Demonstrations* is prefaced by two shorter treatises on divisions and definitions.

[17] John Buridan, *Summulae. De demonstrationibus* [De Rijk e.a.], 4.4, p. 73: "Dico

What is interesting about this passage is that Buridan sees *certitudo* as having both an external and an internal face. On the outside, *certitudo* is required on the part of the object of our assent, i.e., the proposition. What does this mean? Two things, it appears: the proposition to which we assent should be (1) true and not false; and (2) truthful and not deceptive, or such as to elicit our assent when, properly understood, it would elicit our dissent, or at least an act of deferment if it does not appear to be clearly true or clearly false (e.g., "The number of the stars is even"). Now truthfulness and deception are ordinarily ascribed to agents, not propositions. But for Buridan, propositions depend for their existence on being inscribed or uttered or mentally formed by some agent, which means that they can also be instruments of deception. The worry here is that propositions, which are proper parts of arguments, can be sophistically arranged by unscrupulous agents so that they lead to a conclusion which merely looks true. This is the "cause of appearance (*causa apparentiae*)" characteristic of every fallacious argument, which explains its power to deceive.[18] In Treatise VII of the *Summulae*, Buridan notes that his treatment of fallacies will focus on syllogisms "not because a sophist only uses syllogisms (for he uses whatever argumentation he deems advantageous for himself in order to appear to defeat the respondent), but because, aiming above all to appear to refute, he aims above all to appear to produce an elenchus (*quia maxime appetit videri redarguere, ideo maxime appetit videri facere elenchum*)." Accordingly, there are no "certain" propositions in arguments which, as Aristotle puts it, "appear to be refutations [of the novice's reasoning] but are really fallacies instead."[19]

Now it might be objected that Buridan has made a crucial mistake here by confusing the truth of a proposition with its apparent truth, for it has been well known since Plato that the latter does not entail the former. Indeed, truth is mentioned as a criterion of knowledge in the *Theaetetus* only to be set aside, for even if "a man gets hold

---

igitur primo 'cum certitudine.' Certitudo autem requirit duo. Unum ex parte propositionis cui assentitur, scilicet quod sit vera; non enim est certa credulitas qua falso assentimus, sed est incerta et fallax; et manifestum est quod hoc modo certitudo requiritur ad scientiam, quia falsum non scimus. Aliud est ex parte nostra, scilicet quod assensus noster sit firmus, scilicet sine dubitatione seu formidine de opposito; et hoc etiam requiritur ad scientiam, quia assensus dubitativus et formidinalis non transcendit metas opinionis; si quis enim assentit propositioni cum formidine ad oppositum, numquam dicit se scire quod ita sit, sed quod credit vel putat ita esse."
[18] For further discussion, see Ebbesen, "The Way Fallacies Were Treated," p. 115.
[19] Aristotle, *De sophisticis elenchis* 1.164a20.

of the true notion of something," he cannot be said to know that thing unless he also acquires an account of it.[20] But whether or not the propositions we believe happen to be true is beyond our control, which is why Plato describes true beliefs as something we "get hold of," almost as if this occurs by accident. For similar reasons, contemporary epistemologists are more interested in the "justified" and "belief" parts of the "justified true belief" account of knowledge, since it seems obvious what is true in the world is not made so by our cognitive states, even if the latter are said to reflect it.

The answer to this objection is that certitude is not just a psychological state for Buridan. He can speak of certitude as applying to propositions as well as beliefs because certitude connotes the stability and fixity of the thing to which it is applied. The way to see this, I think, is to notice that Buridan uses the term "firmness (*firmitas*)" synonymously with "certitude (*certitudo*)," e.g., when he speaks of "the certitude and firmness of assent" being necessary for knowledge, "because we can have doubts about a proposition of the most firm and certain truth, and so not firmly assent to it, in which case we do not have knowledge of it" (*Quaestiones in Posteriorum Analyticorum libros*, I.2). So when Buridan speaks of a proposition, rather than an assenting believer, possessing "the certitude and firmness of truth (*certitudo et firmitas veritatis*)," he is drawing our attention to a quality of propositions, viz., to the fact that true propositions have a firmness and stability which reflects the firmness and stability of the natural world. True propositions look stable and firm to us because they are partly caused by the natural world. Their structure mirrors the determinate order God has instilled in nature.[21] This is a basic principle of Buridan's thought, without which the whole enterprise of natural philosophy would not be possible at all.

By describing certitude in this way, Buridan would seem to have violated an important methodological rule established by Descartes and followed by virtually every epistemologist since: do not confuse mental with the physical phenomena. And so he has. But Buridan would also reject our modern assumption that the division between material and immaterial being is of the first order metaphysically.

---

[20] *Theaetetus* 202b-c translated in *The Complete Dialogues* [Hamilton e.a.], p. 909.

[21] The notion that true propositions are stable and firm is not to be confused with the demand that whatever we know must in fact be true, which Buridan mentions as the second difference between knowledge and opinion: "… every [act of] knowledge has to be of a true opinion, but not every opinion is such" (Klima, *Buridan's Summulae*, p. 548).

In his view, immaterial mental states, including beliefs and volitions, are more or less continuous with the physical states which give rise to them because both are part of what he calls the common course of nature. This view is not original with Buridan, though he did much to popularize it through the question commentaries that were published from his lectures. His idea that states we would now differentiate as "physical" and "psychological" exist in a kind of natural harmony was the default view of virtually all philosophers during the Middle Ages and Antiquity. One of its sources would have been Plato's *Timaeus*, the first half of which was known throughout the Middle Ages in a Latin translation by Chalcidius. Consider the following passage, where Plato explains discourse in the World Soul:

> And when reason, which works with equal truth, whether she be in the circle of the diverse or of the same – in voiceless silence holding her onward course in the sphere of the self-moved – when reason, I say, is hovering around the sensible world and when the circle of the diverse also moving truly imparts the intimations of sense to the whole soul, then arise opinions and beliefs sure and certain (δόξαι καὶ πίστεις γίγνονται βέβαιοι καὶ ἀληθεῖς).[22]

Firm judgment is associated with the orderly movement of the cosmos. In fact, the firmness and certitude of our beliefs is parasitic upon the firmness and certitude of the cosmos, a matter in which Buridan places his unquestioning trust as a philosopher. So rather than building the external notion of truth into the internal notion of certitude, thereby begging the whole question of epistemic justification from the modern perspective, Buridan has approached the issue the other way around, insisting that the stability and firmness we associate with certitude be present in the propositions which are the objects of our assent, as well as in the things to which the significant terms in those propositions refer, which are the indirect objects of our assent. Hence, they, too, are part of what is required for the certitude that distinguishes knowledge from mere opinion. We can advance the metaphor by imagining such propositions as "moving around,"

---

[22] *Timaeus* 37b-c, translated in *The Complete Dialogues* [Hamilton e.a.], p. 1166. Chalcidius renders the final clause of the Greek somewhat differently: "*recta opiniones et dignae credulitate nascuntur* [then arise right opinions and worthy beliefs]." See Plato, *Timaeus a Calcidio translatus commentarioque instructus* [Waszink], p. 29. But the general point is the same. Buridan does not to my knowledge ever quote this passage, though the terminology was common. See, e.g., John Buridan, *Summulae. De demonstrationibus* [De Rijk e.a.], 4.4: "... for that is not certain belief (*certa credulitas*) on the basis of which we assent to something false, but rather it is uncertain and deceptive (*incerta et fallax*)."

sometimes in unsettling ways, as they seek to emulate the completeness of their objects. Of course, uncertain propositions might present a false appearance, making it natural to associate them with deception and error if we adhere to them too uncritically. Again, we find a nice example of this in the definition of "fallacy" in Treatise VII of the *Summulae*. If we attend to its causal definition, Buridan says, "it is correctly said that a fallacy is the deception of someone unskilled in the art of sophistry."[23] But more generally, it is propositions and syllogisms which have "a certain aptitude and propensity to deceive," so that understood actively, "the aptitude to deceive, properly speaking, is the sophistic argumentation itself, or its illusion and defectiveness, which, again, are nothing but the sophistic argumentation itself."[24] Thus, epistemic properties that were originally assigned to persons, viz., to sophists and unskilled beginners or novices, are now ascribed to propositions. But if propositions can be deceptive, or illusory, they can also be firm, or certain.

The other side of Buridanian certitude refers to the quality of the act whereby we assent to a proposition:

> Another thing is required on our part, namely, that our assent should be firm, that is, without doubt, or fear of the opposite side; and this is also required for knowledge, for a doubtful and fearful assent does not transcend the limits of opinion. For if someone assents to a proposition fearing [that] the opposite [may be true], he would never say that he knows that it is true, but rather that he takes it or believes that it is.[25]

Here, firmness or certitude refers to our level of doxastic commitment towards a proposition. But although firmness is indicative of the truth of a proposition (since Buridan assumes that we are naturally disposed to assent to what it true), it is no guarantee that those propositions to which we assent firmly are actually true. If I know something, my assent to it must also be evident:

---

[23] Buridan indicates next that this definition reprises Aristotle, *On Sophistical Refutations* 1.164b26.

[24] John Buridan, *Summulae. De fallaciis* [Hubien], 1.7, pp. 9–10: "... dicitur bene quod fallacia est deceptio imperiti artis sophisticae // quod est habilitas, seu idoneitas, decipiendi // proprie habilitas decipiendi est ipsa argumentatio sophistica, vel ejus apparentia et defectus, qui iterum non sunt nisi ipsa argumentatio sophistica."

[25] John Buridan, *Summulae. De demonstrationibus* [De Rijk e.a.], 4.4, p. 73: "Aliud est ex parte nostra, scilicet quod assensus noster sit firmus, scilicet sine dubitatione seu formidine de opposito; et hoc etiam requiritur ad scientiam, quia assensus dubitativus et formidinalis non transcendit metas opinionis; si quis enim assentit propositioni cum formidine ad oppositum, numquam dicit se scire quod ita sit, sed quod credit vel putat ita esse."

And I also say "with evidentness" so as to indicate the difference [between knowledge and] that credulity that we believers ought to have concerning the articles of Catholic faith, such as that God is triune. For that credulity has the greatest degree of certitude on the part of the proposition; for it is a maximally true proposition that God is triune. And it should also be the firmest, without any fear on our part, in accordance with the Athanasian Creed (*Symbolum*), namely, at the end: "This is the Catholic faith. Everyone must believe it, faithfully and firmly; otherwise he cannot be saved." But it is compatible with this perfect certitude that because of the lack of evidentness we do not properly have knowledge of [the content of] these articles. However, improperly speaking there *is* evidentness, because the cognitive power by its nature, along with its concurrent circumstances, is disposed to assent to the truth.[26]

On the side of the proposition, the credulity, or believability, of articles of faith can be "most certain," even if only a handful of people actually believe them. That is because the certitude of such propositions is independent of our doxastic attitudes. But Buridan also draws our attention to the quality of our act of assent, to its evidentness. And here it emerges that the trouble is not propositions which refuse to "sit still," but rather, acts of assent which are not in proportion to the stability, or lack thereof, of their propositional objects.

2. *Certitude and Evidentness*

Buridan notices that people firmly believe all sorts of things which are not known and which lack evidentness, and that they do so "without any fear (*sine ulla formidine*)" that the opposite might be true. With the approval that could only come from an experienced teacher

---

[26] John Buridan, *Summulae. De demonstrationibus* [De Rijk e.a.], 4.4, p. 73: "Dico etiam 'cum evidentia' ad differentiam illius credulitatis quam nos fideles demonstrare debemus de articulis fidei catholicae, ut quod deus est trinus et unus. Nam certissima est illa credulitas ex parte propositionis; est enim verissima propositio quod deus est trinus et unus. Et debet etiam esse certissima, sine aliqua formidine, ex parte nostra, juxta illud in symbolo "quam nisi quisque fideliter firmiter que crediderit salvus esse non poterit." Et tamen cum hac perfecta certitudine stat quod, propter inevidentiam, non habemus proprie scientiam de illis articulis. Est autem evidentia ex eo quod virtus cognoscitiva ex sua natura cum circumstantiis concurrentibus est determinata ad assentiendum veritati." I have altered Klima's text here, from "highest degree of certainty (*certissima*)" to "highest degree of certitude," and from "with this perfect certainty (*cum hac perfecta certitudine*)" to "with this perfect certitude," in view of the distinction between certainty and certitude introduced above.

of Parisian undergraduates, he cites Aristotle's remark in the *Nicomachean Ethics* that "some men are no less convinced of what they think than others of what they know."[27] He then describes the threefold genesis of what we might call "no fear" belief:

> And this firmness of assent without any fear of the opposite arises in us in three ways: first, by evidentness, and this is scientific assent; in another way on the basis of will, backed by the authority of the Sacred Scripture, and this is the Catholic faith of the saints who choose to die to sustain it; and in the third way, [the firmness of assent arises] from some false appearance, along with the will's being confined by it, as is the case with stubborn heretics, who also choose to die to sustain their false opinion.[28]

Only beliefs generated in the first way, by the evidentness of the proposition, have certainty, i.e., "*certitudo*" in the modern sense of objective warrant. But the general empirical considerations Buridan brings into play by means of the term "evidentness (*evidentia*)" are not determinants of the firmness or certitude of the religious beliefs of saints and heretics. For them, the testimony of sense, memory, and experience simply does not matter.[29] Such beliefs have certitude rather than certainty. Certitude has nothing to do with the epistemological project of justifying knowledge claims and answering skeptical arguments, since obviously, evidential considerations arc irrelevant to beliefs held so firmly that one is willing to die for them. We might think of beliefs that are certain in this sense as actively ordaining a believer's noetic structure, rather than receiving support from other parts of the structure. It seems appropriate, then, that Buridan has chosen to explain the second and third modes of certitude by pointing to saints and heretics, individuals whose lives have been utterly transformed by their beliefs. These are exceptional people whose actions testify to the power of human conviction.

---

[27] Aristotle, *Nicomachean Ethics* VII.3.1146b29-30 translated in *The Complete Works of Aristotle* [Barnes], vol. 2, p. 1811.

[28] John Buridan, *Summulae. De demonstrationibus* [De Rijk e.a.], 4.4, p. 73: "Provenit autem in nobis talis firmitas assensus sine aliqua formidine de opposito tripliciter: uno modo per evidentiam et ille est assensus scientificus; alio modo ex voluntate, cum auctoritate sacrae scripturae, et sic est de fide catholica in sanctis qui pro ea sustinenda eligunt mori; adhuc, tertio modo, ex apparentia falsa cum voluntate ad hoc determinante, sicut est de pertinacibus haereticis, qui etiam eligunt mori pro sua falsa opinione sustinenda."

[29] Cf. John Buridan, *In Metaphysicen* (Paris, 1518), II.1, fols. 8^{rb-vb}, and *Quaestiones in Posteriorum Analyticorum libros* [Hubien], I.2. For discussion, see Zupko, "Buridan and Skepticism," pp. 199–211.

But saints and heretics have something in common with ordinary folk if we measure the strength of beliefs that are firmly assented to in the second and third ways against those of the first way. For although they might not be willing to die for their beliefs, most people cling to the certitude of faith, even when confronted with the most evident and therefore certain principles of natural science:

> So then, we ought to believe that when the most firm principles are proposed to us, even though the intellect necessarily assents to them and cannot dissent from them, it sometimes happens that our adhesion [to them] is weakened and eventually a certain fear produced by sophistical arguments occurring on the other side. And I have experienced this myself in connection with the first principle, or so it seemed to me. I have asked numerous old women whether they believed that they could eat and not eat at the same time. They immediately replied in the negative. But then, I argued as follows: "You know that God is omnipotent; he can annihilate the entire world. Don't you believe, then, that God could make it such that you would eat and not eat at the same time?'. And they replied, 'I don't know."[30]

This example occurs numerous times in Buridan's writings, though with the propositions expressing the first principle altered each time – "sitting and not sitting" (*In Metaphysicen*, II.2, fol. 9vb); "running and not running" (*In Metaphysicen*, II.4, fol. 21va). But what is instructive here is that when an article of faith (that God is omnipotent) is pitted against an absolutely evident principle (the principle of non-contradiction), it is our assent or "adhesion" to the absolutely evident principle, not the article of faith, which is weakened. The result is a state of perplexity. Notice that Buridan does not say that the old women attempt to resolve the tension by rejecting the principle of non-contradiction – only a misguided theologian, or epistemologist obsessed with consistency, would try to do that.[31] Rather, in the arena

---

[30] John Buridan, *Quaestiones super decem libros Ethicorum Aristotelis* (Paris, 1513), VI.11, fol. 127vb: "Sic igitur est opinandum quod cum nobis proponuntur principia firmissima, licet eis intellectus necessario assentiat et non possit eis dissentire, tamen aliquando per sophisticationes in oppositum occurrentes contingit adhaesionem debilitari et tandem quandam formidinem generari. Et hoc expertus sum de primo principio, ut mihi videtur, quaesivi eum a multis vetulis utrum crederent quod possent simul comedere et non comedere. Et statim responderunt quod non. Tunc igitur ego sic arguebam: 'Vos scitis quod deus est omnipotens; ipse potest totum mundum annihilare. Creditis ne ergo quod deus posset facere quod simul comederetis et non comederetis?'. Et responderunt, 'Nescio'."

[31] Of course, many such conflicts are merely apparent, since someone who knows about fallacies and how to expose them will be able to dissolve the "fear produced by sophistical arguments." But it turns out that not even the most absolutely evident

of everyday life, we are simply not sure what to say when beliefs made certain by faith and beliefs made certain by evidentness come into conflict. As the old women remark in another version of the story, "We don't know. God can do everything, and it must be believed that God can do the impossible." But that is where the matter is left. Buridan rejects the idea that the whole edifice of empirical knowledge must be placed on skeptical hold just because of the strength of our belief in divine omnipotence. Thus, although our recognition of the consequences of omnipotence might weaken our level of commitment or "adhesion" to evident beliefs, it does not demolish it. The reason is fairly simple. There are different kinds of belief, and it does not belong to us – fallible, wayfaring creatures that we are – to sort out conflicts between them. What it does belong to is "the certainty and evidentness of divine wisdom, to which no created cognition can attain."[32] This strikes me as exactly the right thing to say, in view of the enormous distance that exists between the various justificational protocols of empirical knowledge and genuine religious belief.

Understandably, Buridan is more interested in exploring what he can say something about, i.e., the different kinds of certitude and evidentness found in human knowing. He begins by reiterating "no fear" belief as a mark of knowledge: "... as far as we are concerned, certainty or assent should not be called that of knowledge, unless it is firm, namely, without any fear [of falsity]."[33] He proceeds to identify two degrees of certainty, in the sense of objective warrant:

> For as far as we are concerned, certainty or assent should not be called that of knowledge, unless it is firm, namely, without any fear [of falsity]. But as far as the proposition is concerned one sort of certainty is that which pertains to a proposition so firmly true that it, or one similar it, can by no power be falsified. And in this way we should certainly

---

principle is beyond doubt. Buridan's claim that it is impossible for us to be in error concerning the principle of non-contradiction contains an important qualification: "... at least naturally (*saltem naturaliter*)" (*In Metaphysicen* (Paris, 1518), IV.12, fol. 21ᵛᵃ). And many other principles, such as the psychological principle that the human intellect is not a material form, do not admit of demonstrative proof, but only of explication in terms of readily believable (*probabiles*) arguments (*Quaestiones in Aristotelis De anima, tertia lectura* [Zupko], III.4). For discussion, see Zupko, "How Are Souls Related to Bodies," pp. 591–599, and "Buridan and Skepticism," pp. 214–216.

[32] John Buridan, *Summulae. De demonstrationibus* [De Rijk e.a.], 4.4, p. 74: "... certitudo et evidentia sapientiae divinae, ad quam nulla notitia creata potest attingere."

[33] *Ibid.*: "... ex parte nostra non debet dici certitudo scientiae, seu assensus, nisi sit firmus, scilicet sine formidine."

concede, as they have argued, that it is impossible for us to have such certainty about an assertoric categorical affirmative proposition, unless it consists of terms suppositing for God, or, perhaps, if we admit natural supposition, of which we spoke elsewhere [i.e. in *Summulae. De suppositionibus*, 3.4]. But this sort of certainty is not required for natural sciences or metaphysics, nor even in the arts or morality.[34]

The "they" in "as they have argued" are those persuaded by the skeptical arguments of Nicholas of Autrecourt, since they insist that the only knowledge worthy of the name is that which is reducible to the principle of non-contradiction in a demonstrative syllogism. Since the conclusions of such demonstrations are so firmly true that they "can by no power be falsified," they produce in us a limiting case of "no fear" belief.[35] But this fails to apply to most knowledge claims, where our epistemic task consists of reducing fear, not eliminating it entirely.

More relevant to us is the second degree of certainty:

However, another sort of human certainty on the part of the proposition is that of a true proposition which cannot be falsified by any natural power and by any manner of natural operation, although it can be falsified by a supernatural power and in a miraculous way. And such certainty suffices for natural sciences. And thus I truly know, by natural science [or knowledge, *scientia*] that the heavens are moved and that the sun is bright.[36]

---

[34] *Ibid.*: "Sed ex parte propositionis certitudo una est quia est propositionis sic firmiter verae quod ipsa vel talis per nullam potentiam potest fieri falsa; et sic bene concedendum est, sicut illi arguebant, quod impossibile est nos habere talem certitudinem de propositione affirmativa categorica et de inesse nisi sit constituta ex terminis supponentibus pro ipso deo, vel forte nisi ponatur suppositio naturalis, de qua alias dictum est. Sed haec certitudo non requiritur ad scientias naturales vel metaphysicas, vel etiam ad artes aut prudentias."

[35] By "can by no power be falsified," Buridan seems to include divine power, even though there is textual evidence that he held that God could make us assent to an evident contradiction (see n. 31 above). Ultimately, perhaps his view is that the principle of non-contradiction is on the same epistemic footing as any other evident principle with respect to its certainty, the difference being that God could actually falsify a natural principle – e.g., about the motion of the heavens – whereas God could only deceive us into thinking that the principle of non-contradiction is false. Not even God could make the principle of non-contradiction false.

[36] John Buridan, *Summulae. De demonstrationibus* [De Rijk e.a.], 4.4, p. 74: "Alia est certitudo humana ex parte propositionis quia propositio est vera et per nullam potentiam naturalem et modum agendi naturalem talis propositio potest fieri falsa, licet per potentiam supernaturalem et modo miraculoso posset fieri falsa. Et talis certitudo sufficit ad scientias naturales, et ita vere scio, scientia naturali, quod caelum movetur et quod sol est lucidus."

The vast majority of human knowledge will be of "propositions which cannot be falsified by any natural power." In this case, the firmness of our assent is tempered somewhat by the recognition that an evidently true proposition could be falsified by supernatural means, in ways we could never detect, not even in principle:

> Accordingly, it seems to me to be possible to conclude as a corollary that supernaturally it is possible for my [act of] knowledge, while it remains the same, to be converted into non-knowledge. For as long as the sun and the sky are moving in accordance with all their natural ways, the assent by which I firmly and with certainty assent to the proposition "The sun is bright," is true, evident and certain natural knowledge, endowed with the evidentness and certainty appropriate to natural science. I posit, then, that if this [act of] assent, which is knowledge at the present time, remains in me for the whole day, and at 9 o'clock God removes light from the sun without me knowing this, then that [act of] assent of mine will no longer be knowledge after 9, for it will no longer be true, nor will it have a true proposition as its object.[37]

Should the natural scientist fear the possibility that our knowledge of the sun and the sky might be false because of supernatural intervention? Not at all.[38] For although the vast majority of principles in natural science cannot be known demonstratively, i.e., by reduction to the principle of non-contradiction, we "can have opinion about them by dialectical demonstration." Furthermore,

> ... this is true of many principles which at first had been doubted, until they were made evident by the senses, memory and experience. For

---

[37] John Buridan, *op.cit.*, 4.4, pp. 74–75: "Et secundum hoc videtur mihi posse concludi correlarium quod possibile est supernaturaliter scientiam meam, manente eadem, verti in non scientiam. Quam diu enim sol et caelum manent secundum ejus motus naturales, assensus quo firmiter et certe assentio huic propositioni 'sol est lucidus' est vera naturalis scientia evidens et certa, evidentia et certitudine requisitis ad scientiam naturalem. Pono igitur quod ille assensus qui nunc est scientia maneat mihi per totam diem et quod hora nona deus removeat lucem a sole, me hoc nesciente; ille assensus meus post horam nonam non erit amplius scientia, quia nec verus nec propositionis verae."

[38] Interestingly enough, the notion that the opposite view leads to disaster, and that the theologians are responsible for advancing it, is not without precedent in the history of medieval philosophy. Cf. Averroes's attack on Ghazali in *Tahafut al-Tahafut* (*The Incoherence of the Incoherence*), pp. 324–325: "When the theologians admit that the opposite of everything existing is equally possible, and that it is such in regard to the Agent, and that only one of these opposites can be differentiated through the will of the Agent, there is no fixed standard for His will either constantly or for most cases, according to which things must happen. For this reason the theologians are open to all the scandalous implications with which they are charged. For true knowledge is knowledge of a thing as it is in reality."

these can be taken to be false, and then, later opined, on the basis of insufficient evidence, and finally evidently known, when experience has sufficiently been made complete.[39]

What we have here is a three-step description of the genesis of our empirical knowledge of principles, patterned after the three ways in which simple propositions become evident to us: first, by sensation, such as when we assent to "This fire is hot" or "Jacob is writing" when we feel this fire and see Jacob writing, respectively; second, by memory, such as when we assent to "This fire was hot" and "Jacob was writing" on the basis of our memory of this hot fire and Jacob writing; and third, by experience, when, having sensed and remembered that this fire is hot and that fire is hot, I correctly judge "on the basis of the memory of the others and the similarity (*per memoriam de aliis et propter similitudinem*)" that some other particular fire is hot, even though I have not actually sensed it. Unlike brute animals, human beings have the additional capacity to assent to:

> ... principles that are universal in form, such as that every fire is hot, and that all rhubarb purges bile. For these principles are known to us by induction on the basis of sense, memory, and experience.[40]

The immediate cause of inference in the case of empirical generalizations is our intellect's natural inclination to assent to the truth, provided that the truth is expressed in the form of propositions sufficiently evident to us, and in which no counter-instances have been found. The key word here is "natural." Buridan regards the fact that we have such a capacity as completely uncontroversial. We needn't fear our considered assent to the principles of natural science because we are reliable detectors of law-like regularities in nature. And that, for Buridan, is the most important kind of *certitudo*, both on the side of the proposition and on the side of the person assenting to it.

---

[39] John Buridan, *Summulae. De demonstrationibus* [De Rijk e.a.], 4.4, pp. 75–76: "Quod est verum de multis principiis quae a principio dubitata sunt donec facta fuerunt evidentia sensu, memoria et experientia. Haec enim possunt putari esse falsa et post, per insufficientem experientiam, possunt fieri opinata, et tandem evidenter scita, cum sufficienter fuerit expleta experientia."

[40] John Buridan, *Quaestiones in Priorum Analyticorum libros* [Hubien], II.20, p. 47: "... et illa principia sunt universalia, ut quod omnis ignis est calidus, et quod omne rheubarbarum est purgativum cholerae. Illa enim principia sunt nobis nota per inductionem supponentem sensum, memoriam, et experientiam."

3. Conclusion

It should now be clear why Buridan, possessed of an adequate reliabilist response to the skeptical arguments of Nicholas of Autrecourt, failed to use it against him. Autrecourt does not know the difference between knowledge and certitude. We can see this in his insistence that the only acceptable criterion for empirical knowledge must annihilate, rather than merely diminish, our natural fear of assenting to propositions which might turn out false. This is an error for two reasons. First, empirical knowledge is not deductively closed, as Aristotle himself certainly knew, and to insist upon absolute certainty is to ignore the process by which human beliefs are actually generated – a natural process in which objects of belief gradually become more firm or stable, and thereby, because of the connection between subjective certitude and objective certainty, more worthy of our assent. Second, and more important, Autrecourt's criterion completely omits the subjective aspect of belief. For it is not just the truth of a proposition that becomes firm through experience; the intensive quality of our assent grows as well. The trouble with Ultricurian skepticism is that its narrow redefinition of *certitudo* forces us to assume that we must be either absolutely certain and fearless about our beliefs, or else wracked with doubts about their possible falsity. And virtually everything in human experience testifies against this. Furthermore, Buridan's attention to the experience of the believing subject enables him to see that even our adhesion to absolutely evident principles can be weakened, as the case of ordinary folk and the principle of non-contradiction suggests. Buridan presents us with an organic picture of human knowledge, an inexact science with inexact subjects. Knowledge in the Ultricurian picture is a zero-sum affair, free of nuance and precise in its focus on what follows from the rules of demonstration. The contrast could not be greater.

SENSATIONS, INTENTIONS, MEMORIES, AND DREAMS

Peter G. Sobol

When Aristotle turned to the study of the soul, he found, as others had found before him and as still others continue to find, that consciousness fits only with difficulty into categories that neatly contain other natural phenomena. His intellectual heirs received a theory of the soul with problems unsolved and details unexamined, for the analysis that Aristotle had employed to parse other kinds of natural change into substantial or accidental, and to locate active and receptive participants in change, simply did not work with consciousness. Regarding the means by which sensible objects affect sense organs, he wrote next to nothing.

In his questions on *De anima* and the *Parva Naturalia*, John Buridan addressed these weaknesses in Aristotle's theory of the soul. He often couched his responses in the language of conjecture, and, equally often, he introduced them with the confession that he found the question at hand to be very difficult indeed. Nevertheless, he determined each question with a surprising fealty to Aristotle, making frequent use of his own observations of the natural world in order to aid his cause. Chief among these problems were the activity or passivity of sense, the means by which sense organs become aware of external objects, the number and location of the internal senses, and the reconciliation of Aristotle's psychology with the body of medical tradition.

1. *The Activity or Passivity of Sense*

In the Aristotelian world, all change requires that something act and that something be acted upon. In the standard example, fire has the potential to act upon combustibles and turn them into fire. Wood has the potential to be acted upon and to become fire. When wood is acted upon by fire, the potential of both is actualized and the substance that was wood becomes fire because the form of the wood is replaced with the form of fire. This process is called substantial change, and sufficed for Aristotle to explain not only combustion but all kinds of generation and corruption.

But nature teems with changes that are not substantial. The water in a shallow pond remains water even if it becomes warm during the day and cool during the night. A ripening apple remains an apple as its color changes from green to red. These kinds of change are accidental – they are changes in qualities that are not essential to what a thing is.

If substantial and accidental change cover all change in nature, then what kind of change occurs in a dog when it sees its master? It cannot be substantial change because the dog is still a dog. Nor can it be accidental change because accidents are sensible qualities – and the dog is not now slightly different in size, shape, or color. What kind of change is it?

In *De anima* II.5 Aristotle admitted a third kind of change: an actualization of potential without a change of form.[1] Averroes was so struck by Aristotle's willingness to introduce a third kind of change in *De anima* that he dismissed as insufficient the discussion of change in Aristotle's book *On generation and corruption* because the third kind of change is not mentioned there.[2] To understand the need for this third kind of change, think of an infant. An Aristotelian would say that this infant is potentially a carpenter. Now picture the infant thirty years later, no longer an infant, employed as a carpenter, but currently at home in bed asleep. Quite a lot of substantial change has occurred as this person has approached full actualization as a carpenter. But the sleeping carpenter is not a fully actual carpenter. Aristotle believed that in order for a carpenter to be a fully actual carpenter, he or she must be doing carpentry. Now some kind of change must occur as the carpenter goes from not carpentering to carpentering. But does it make sense to say that the exercise of knowledge by which a resting carpenter becomes a working, actual carpenter involves substantial or accidental change? Aristotle did not think so. The exercise of knowledge is the actualization of a potential without a change of form.

The same sequence of infant to resting carpenter to working carpenter can be repeated in animals, but with respect to sensation instead of knowledge.[3] The newborn animal undergoes much sub-

---

[1] Aristotle, *De anima*, 417a20-b10.
[2] Averroes, *Aristotelis opera cum Averrois commentariis* (Venezia, 1562–1574), text 51, fol. 76ᵛ, E: "Et intendebat hic per sermones universales librum De generatione et corruptione, et non sufficit ei, quod declaratum fuit in illo libro, quia sermo hic videtur magis proprius."
[3] Aristotle, *De anima*, 417b17 sq.

stantial change before its sense organs are fully developed. But once the organs are fully developed, the step between not sensing and sensing is like the step between not carpentering and carpentering. It, too, is actualization of potential without a change of form.

But then, what is it that acts and what is acted upon when the dog sees its master? In *De anima* II.5, Aristotle had declared that "the faculty of sensation has no actual but only a potential existence."[4] But in Book III he said that "when that which has the power of hearing is exercising its power, and that which can sound is sounding, then the active hearing and the active sound occur together."[5] So it is not surprising to find that commentators on Aristotle's *De anima* from the earliest times puzzled over whether the sensitive power in animals was an agent or a recipient. By the time Buridan came to Paris, these and other problems had become part of a standard set of questions on *De anima*.

Buridan devoted two questions in the *tertia lectura* of his commentary to the agent-or-recipient problem. The ninth question of Book II asks whether sense is active or passive. Broaching this question required Buridan to precisely define the terms involved and then to make clear just what it means to say that something is capable of sensation. In order for sensation to occur, a sense organ must receive sensible species. The reception of sensible species is thus a necessary part of sensation, but it is not sufficient. An act of some kind is needed in order for an animal to sense. But if sense is both active and passive, where are the active and passive parts located? Not surprisingly, Buridan concluded that the matter in the sense organ was the passive participant, and that the soul was the active participant. But then how does the soul, by an act of its own, becomes aware of the change produced in the matter of the sense organ? And how can the extended nature of the change in the sense organ be represented in the indivisible and unextended soul? Natural philosophy offered no explanation for this latter phenomenon. "And I answer with certainty that it is miraculous," Buridan wrote, in a stunning admission for an Aristotelian commentator, "because the human soul, neither extended nor called forth from the potential of the subject in which it inheres, inheres in the body in a miraculous and supernatural way, and yet it inheres in the whole body and in every part

---

[4] Aristotle, *De anima*, II.5, 417a7 (Aristotle, *On the Soul, Parva Naturalia, On Breath* [Hett], p. 95.)

[5] Aristotle, *De anima*, III.2, 426b30 (Aristotle, *On the Soul, Parva Naturalia, On Breath* [Hett], p. 147).

of it."[6] The dimensionless soul – unlike a dimensionless point – has no single location in space. The soul instead is wholly in every part of the body, as Buridan argued in the seventh question of Book II. But having explained how the unextended soul can become aware of the extended world, Buridan concluded the question by asserting that sensation occurred not in the soul or the matter but in the living organism, which is changed by the reception of sensible species. The soul itself, unchanged by sensible species, is nevertheless somehow aware of the change in its subject, and acts to complete the process of sensation.

The tenth question asks whether there is an agent sense. This question has an interesting history because earlier in the tradition it did not ask about the active part of sense. It derives instead from a question posed by Averroes, who, contemplating the analogous relationship between sensation and thought,[7] had been moved to search for the cause of the sensibility of sensible objects. He had reasoned as follows: If the agent intellect is necessary in order to make the potentially intelligible actually intelligible, then, by analogy, there must be an agent sense that makes the potentially sensible actually sensible. He did not discover this agent sense. Indeed, he closed his discussion of it by saying "you should look into this, because it needs to be examined closely."[8]

These two problems – the activity or passivity of sense and the cause of the sensibility of sensible objects – became conflated in the thirteenth century in the work of Albert the Great. Like most thirteenth century scholars, Albert concluded that sense was a passive potential which operated by being acted upon by sensible objects. The alternative view, that sense was an active potential was, to Albert, absurd because it meant that sensation operated on sensible objects. In good scholastic form, Albert raised arguments in favor of an active sense in order to refute them. One such argument claimed that sense must be active because sensation entails judgment and judgment is an act. To this Albert replied that there is no wholly passive potential! He

---

[6] John Buridan, *Quaestiones in Aristotelis De anima, tertia lectura* [Sobol], p. 138: "Et certe ego respondeo quod hoc est mirabile, quia mirabili et supernaturali modo anima humana inheret corpori humano non extensa nec educta de potentia subiecti cui inheret, et tamen etiam toti corpori inhereat et cuilibet parte eius."

[7] Aristotle, *De anima*, III.4, 429a16–17 (Aristotle, *On the Soul, Parva Naturalia, On Breath* [Hett], p. 165: "... as the sensitive is to the sensible, so must mind be to the thinkable.").

[8] Averroes, *Aristotelis opera cum Averrois commentariis* (Venezia, 1562–1574), fol. 81ʳ, B: "Et debes hoc consyderare, quoniam indiget perscrutatione."

allowed that sensation entailed some kind of act, such as judgment, but not an act of the soul beyond the body.

Having established, as he saw it, the passivity of sense, Albert went on to consider the question raised by Averroes regarding the agent of actual sensibility. "This is a question worth considering," Albert wrote, "and it is a question which Averroes touched on and left unsolved."[9] Worth considering, perhaps, but not for very long. To ask why sensible objects are sensible, Albert wrote, is like asking "why does light shine?"[10] Not satisfied with this refutation, Albert introduced two current theories that proposed a cause of actual sensibility. One theory claimed that the cause was visible light. Albert dispensed with this theory by pointing out that sound can be heard in the dark. The other theory claimed that the cause of actual sensiblity was in the soul. Part of his refutation of this view directs the reader back to the refutation of an active sense. I suspect that it is because Albert applied the same arguments against an act of the soul outside the body and against the cause of actual sensibility, that later scholars referred to them both under the heading "agent sense."

Buridan's question on the agent sense considers the activity of sense.[11] He used the same arguments that Albert and most other scholars used to argue against the mere passivity of sense. But where Albert had raised these points only in order to refute them, Buridan used them to argue that sense must have an active as well as a passive power. Yes, sensation appears to be passive because, in order to sense, an animal needs to receive sensible species. But sensation must be active because the sensitive powers of the soul are nobler than sensible objects and also nobler than the vegetative power of the soul, which has its own acts of nutrition, growth, and reproduction. The mere presence of species does not cause an animal to experience an external object because species abound in our memory, and yet we experience remembered objects only when their species return to the common sense.

---

[9] Aristotle, *De anima*, II.3.6; Albertus Magnus, *De anima* [Stroick], 104, pp. 48–50: "Hoc enim dignum est consideratione, et est quaestio, quam tangit Averroes et relinquit insolutam."

[10] Albertus Magnus, *op.cit.*, 106, pp. 24–26: "Et ideo frustra quaeritur, quid conferat ei illud, sicut si quaereretur, quid conferat luci lucere secundum actum."

[11] The only part of the question that touches on the cause of sensibility is his reference to Avicenna's *Dator formarum* as both agent intellect and cause of sensibility. "Hec est opinio quorundam," Buridan wrote, "que non est improbabilis." John Buridan, *Quaestiones in Aristotelis De anima, tertia lectura* [Sobol], p. 153.

But if sense makes use of an active power, what is the act? It seemed at first glance to be an immanent act, an act of the soul upon itself. Aristotle had denied that anything can act upon itself.[12] Buridan knew this and escaped the problem by saying that the soul uses species in sensation the same way it uses heat in digestion.[13] The soul is thus no more acted upon by the arrival of species than a heavy object is acted upon by whatever removes the support from beneath it and allows it to fall.[14]

2. *The Number of External Senses*

Buridan defended Aristotle's claim that we possess no more than five external senses. To do so he had to overlook the legitimate claim that touch was in fact two senses because the pairs hot and cold and wet and dry each comprised a contrariety, and Aristotle had stated that each sense must be of one genus of contraries.[15] Buridan also had to entertain the possibility that our experience of hunger and other internal pains, or the localized experience of sexual pleasure might indicate the existence of an extra, heretofore unnoticed external sense. He evaded the claim about sexual pleasure by simply asserting that the sense of touch was responsible, but he added that nature had made these sensations especially pleasurable to insure that procreation would occur.[16] He explained the sensation of internal pain by reminding his students that all organs were external to each other and that hence no pain was really sensed as pain internal to any single organ.[17]

---

[12] John Buridan, *op.cit.*, p. 156: "Sed tunc est dubitatio quomodo sit possibilis quod idem in seipso agat, et a seipso patiatur, cum Aristoteles videatur hoc negare in multis locis."

[13] John Buridan, *op.cit.*, p. 158: "Sed tunc est bene dubitatio utrum ista species requiritur tamquam dispositio passivi ad recipiendum sensationem vel tamquam dispositio activi ad agendum instam sensationem. Et de hoc apparet mihi probabile dicere quia, sicut anima utitur calore tamquam instrumento ad agendum nutritionem, ita ipsa utitur ista sensibili vel intelligibili ad agendum sensationem vel intellectionem."

[14] John Buridan, *op.cit.*, p. 156.

[15] Averroes believed that Aristotle had referred to touch as a single sense as a concession to common speech."Item notandum est quod bene dicit Commentator quod, licet sit alius sensus calidi et frigidi et alius humidi et sicci secundum predicta in precedenti questione, tamen in hac enumeratione sensuum quinque Aristoteles accipit illos duos sensus tangendi tamquam unum, secundum modum loquendi vulgarem." John Buridan, *op.cit.*, p. 336.

[16] John Buridan, *op.cit.*, pp. 348, 351.

[17] John Buridan, *op.cit.*, p. 335.

SENSATIONS, INTENTIONS, MEMORIES, AND DREAMS

As for the likelihood that animals possessed abilities to detect whole genera of sensibles undetectable to humans, he noted that no animal pursues the good and avoids the noxious by mysterious means.[18] We might guess that an animal that did so possessed a sixth sense of some kind. He thereby left the door open for the discovery of further senses should such mysterious behavior ever be detected, as it has been in the magnetic sense of migrating birds and the electrical sense of various marine creatures.

3. *The Nature of Species*

Buridan, like other commentators before him, had to flesh out the little that Aristotle had said about how sensible objects affected sense organs. To do so he relied on the concept of sensible species, adapted by thirteenth-century perspectivists – most notably Roger Bacon – from Muslim writers. Because Bacon believed that light in the medium was the model species, his conclusions about the nature of species in general relied heavily on the nature of light. Light moved imperceptibly fast. Its rate and direction of propagation were not affected by any motion of its medium. It disappeared from the medium as soon as its source was removed. Bacon believed that an effect with characteristics such as these could not be due to anything corporeal. Bacon instead believed that species had incomplete corporeal being, and that they were propagated not by local motion but by successive actualizations of the potential of the medium.

In adopting species to explain vision, however, scholars implicitly granted species a role in all sensations. After all, if light was the paradigm of all natural effects, then whatever was responsible for light was also responsible for other effects, and other sensations. Moreover, Aristotle had claimed, and every medieval scholar believed, that a sensible object placed directly on a sense organ was not sensed,[19] indicating that a sensible object could cause sensation only when separated from the organ by an intervening medium. Something, therefore, must cross the medium between the object and the organ. No one doubted that vision occurred when species of light and color arrived at the eye. But scholars did not agree on the

---
[18] John Buridan, *op.cit.*, p. 342: "Et tamen nullum invenimus animal quod videamus prosequi convenientia vel fugere nociva nisi propter sensus istorum sensibilium per que innati sumpmus sentire."
[19] Aristotle, *De anima*, II.7, 419a12–26.

role of species in the other senses. Some commentators on *De anima* explained that the finite speeds of sound and odor followed from the fact that their species were less intentional and more corporeal than the species of light. The sense of touch received the species of heat and cold across external media, but many touch-sensations are perceived only on contact between an object and the flesh. These instances showed that the flesh itself was also a medium for certain species, while the nerves embedded in the flesh were the actual organs of touch. Buridan may have intended his long digression on sensible species in the eighteenth question of the *tertia lectura* of Book II to bring consistency to the debate. But if sensible species were to have the same nature among all the sensible qualities, then all species must be multiplied not instantaneously but in time, and all species must persist in the medium.

Buridan's first task was then to show that light was not instantaneously propagated from its source, even if its medium did not resist the propagation. His first argument would have been stronger had he offered some examples from terrestrial natural philosophy of propagations that face no resistance and that occur in time nevertheless. As it was, his examples were the finite speed of the celestial orbs and God's ability to move an object at a finite speed even though nothing can resist God.[20] But he went on to present evidence that transparent media do indeed resist the propagation of light. The fact that the moon remains visible during a total lunar eclipse suggested that some kind of partial reflection occurred in the celestial medium, redirecting some weak rays toward the moon.[21] Hence partial reflection must occur at every point in even the most transparent of media, hence transparent media must resist the multiplication of species, hence that multiplication must occur in time.

Buridan's conclusion that the speed of light is finite contradicted Aristotle's claim at *De anima* II.7, 418b21–26, that light is propagated instantaneously. Buridan pointed out, however, that we may not see the sun as soon as it rises, as Aristotle had claimed,[22] but only as soon as its rays reach our eyes, during which time the sun may have moved above the horizon.

Buridan defended the persistence of light in the medium by drawing attention to the phenomena of after-images. The glowing tip of a stick, waved before our eyes, appears to leave a trail of light.

---

[20] John Buridan, *op.cit.*, p. 275.
[21] John Buridan, *op.cit.*, pp. 281–282.
[22] Aristotle, *De anima*, II.7, 418b24–26.

Buridan attributed this phenomenon to the persistence of species in the eye, and argued that species may well persist in the medium, too, albeit not long enough for us to notice.[23] At the very least, the species at a particular point must persist long enough to actualize the species in the subsequent point in the medium.

Having shown that the species of light are multiplied in time and persist in the medium, Buridan turned to the species of sound. Previous writers, including Roger Bacon, had likened sound to waves in the medium – created when, for example, a clapper strikes a bell – that propagate to the ear.[24] Buridan recalled the Aristotelian dictum that the material presence of a sensible object at a sense organ cannot cause sensation to support his claim that sound waves could not cause hearing. Not only that, but daily experience shows that sound does not propagate as waves in the air. Imagine a choir singing in a room with open windows across which lightweight curtains have been drawn. The choir can be heard from far away, yet the curtains do not move at all.[25] Some writers asserted that the sensation of odors was triggered by the arrival at the organ of smell of particles expelled by odoriferous bodies. But the Aristotelian dictum also told against this theory. And anyway, if scented bodies made their presence known by emitting fumes or particles, only bodies that did so would be detectable by smell from a distance. Yet even bodies that do not do so have an odor.[26] Moreover, odors are rapidly detected from a great distance, and the air would resist such rapid penetration by odoriferous fumes.[27]

The argument for the existence of species of tangible qualities – hot, cold, wet, and dry – begins with the observation that when you first enter a hot bath, the water feels intensely hot, but that after a while the heat seems less.[28] This occurs because the nerves in the

---

[23] John Buridan, *op.cit.*, pp. 283–284.

[24] Roger Bacon, *De multiplicatione specierum* [Lindberg], 6.2, 20:8 *et seq.*

[25] John Buridan, *op.cit.*, p. 296: "Item si sint cantantes in domo alta voce, et sint tele ad hostium vel ad fenestras appense valde tenues et faciliter mobiles, iste non apparebunt moveri. Ymo nec tela aranea apparebit moveri. Et tamen notabiliter moverentur si aer intra domum, impulsus a voce cantantium, deberet, mediantibus illis telis, sic velociter impellere aerem extra domum, quod impulsio duraret usque ad istum locum valde distantem in quo vox istorum notabiliter auditur."

[26] John Buridan, *op.cit.*, p. 303: "Sed ultra, sine tali fumo, multipliciter species illius odoris, quia, sicud dictum est prius, aliqua sunt multum odorifera que valde modicum sunt frangibilia vel evaporabilia."

[27] John Buridan, *op.cit.*, pp. 303–304: "Ideo videtur quod iste parve exalationes aer resisteret, quod non sic ita cito posset undique ad tantam distantiam pervenire."

[28] John Buridan, *op.cit.*, p. 163: "Consimiliter apparet de tactu, quia intrans bal-

flesh can only detect species from a heat greater or lesser than their own.[29] The moment you enter the bath, species of heat begin arriving at your nerves. But as your flesh begins to warm, and your nerves with it, because of direct contact with the hot water, their heat approaches the heat of the water and they no longer sense it as intensely.

In another example, Buridan asked his students to recall the experience of entering a chamber in which a fire burns on a cold day, and to recognize that the fire warmed them without having to warm the intervening air. But not only do the students perceive warmth, their clothing and the bench upon which they sit are also warmed, suggesting that not only do species of heat evoke the sensation of heat from nerves, but they also can make a thing warm.[30]

Buridan went beyond the realm of sensation to show that certain meteorological and speleological effects can be explained only if rays of heat and cold exist. Take, as an example, the formation of hail in summer. Buridan reported that people who ascend Mt. Ventoux in summer find it cold at the summit but not cold enough to cause the rapid freezing of water that, Buridan believed, produced hail. Yet he observed a cloud, at an altitude below that of the summit, producing large hail.[31] To explain this, he suggested that hot and cold bodies both emit species of heat and species of cold, but that the species of cold are reflected from a hot body, while the species of hot pass on through. Thus, if, in a cloud, a portion of aqueous vapor were surrounded by warm and dry exhalations, the species of cold emanating from the vapor would be reflected back into the vapor, while any species of heat would pass on through.[32] Consequently, the

---

neum iudicat aquam nimis calidam et postquam, per spacium fuerit in balneo, iudicat de calidiori non esse ita calidam." John Buridan, *op.cit.*, pp. 357–358: "Ymo, sicud alias dicebatur, tactus intense sentit caliditatem aque quando aliquis intrat balneum. Cum tamen diu fuerit in balneo, et organum sensus tactus interius fuerit realiter calefactum, tunc nec sentit ita caliditatem suam nec caliditatem carnis sibi coniuncte; ymo etiam nec caliditatem aque sicud a principio sentiebat eam, cuius causa dicebatur alias."

[29] John Buridan, *op.cit.*, p. 327: "Tactus enim, prout vult Aristoteles, cum non possit esse denudatus totaliter a calido et frigido, humido et sicco, debet consistere, quantum ad organum, in media proportione huiusmodi qualitatum, ita quod sit denudatus ab excellentiis earum us possit percipere huius excellentias."

[30] John Buridan, *op.cit.*, p. 305.

[31] John Buridan, *op.cit.*, p. 312.

[32] John Buridan, *op.cit.*, p. 313: "Sed tamen corpora calida et frigida, cum fiunt subtilia et levia vel omnino faciliter mobilia, bene experiuntur fugere ab invicem. Sed etiam sine fuga corporali, debemus ymaginari quod radii caliditatis et frigiditatis non solum habent naturam quod refrangantur a corporibus solidis et densis, ymo etiam radii caliditatis refranguntur a frigido, licet sit rarum et subtile, et radii frigiditatis

vapor would become colder and colder until it froze and fell from the cloud. Hence hail is larger in summer than in spring, because the hotter the dry exhalations, the colder the pockets of vapor will become.[33]

Buridan's willingness to attribute material effects to species here may constitute a betrayal of his commitment to consistency. If the species of heat can make a thing hot, and the species of cold can make a thing cold, then a red wall facing a white wall should turn pink. A person peering into a pond should become wet. But even if he did not treat species consistently, Buridan's extended inquiry into their nature highlights two aspects of his natural philosophy that I believe echo the discussion of impetus in Book VIII of his *Physics*. In both cases, he tried to fill in where he found Aristotle's ideas vague or inadequate, and in both cases he extended the new concept beyond its original purview, by applying impetus to natural motion and by applying species to meteorological phenomena. Not a revolution, by any means, but a respectable effort by a scientist of any era.

4. *Intentions*

The concept of intentions originated with Islamic writers, chiefly Avicenna, who coined the Arabic term (*wahm*) that eventually was translated into Latin as *estimativa*. Avicenna believed that this power explained instinctive and learned behavior in animals and humans, and that it did so by detecting intentions, in the same way that sense detected forms.[34] Buridan used the term "intention" with three different meanings. He used it most frequently to refer to species in the nervous system, in recognition of the fact that the nature of what was continuing its journey to the organ of common sense was now different than it had been in the external medium. Species of light and color in the medium required transparency and were multiplied in straight lines. Once received in the eye, however, the species wound their way along various nerves until they reached the organ of the common sense. If they no longer required either transparency or

---

a calido. Radii autem caliditatis non sic refranguntur a calido, nisi sit grossum et solidum, sed multiplicantur in ipsum."

[33] John Buridan, *op.cit.*, p. 318.

[34] Avicenna Latinus, *Liber De anima seu Sextus De Naturalibus IV-V* [Van Riet], IV.1, 8, pp. 2–3: "Usus autem est ut id quod apprehendit sensus, vocetur forma et quod apprehendit aestimatio, vocetur intentio."

a straight path, they must be different in some way.[35] Buridan did not make clear whether intentions were triggered by the arrival of species at the external sense organ or whether reception at that organ conferred something upon the received species – or abstracted something from them – that allowed them to continue in the new environment. He probably believed some form of the latter.

Buridan also used the term "intention" as it was used by most commentators to refer to a feature of a sensed object that was not sensed *per se* by any external sense or by the common sense. Buridan used the traditional example of the lamb that flees from the first wolf it has ever seen because something in the lamb detects the intention of harm in the sensible species that arrive from the wolf.[36] He used "intention" in its third sense to refer to the sense of time that must accompany a memory, so that when species return from the organ of memory to the organ of the common sense, we do not mistake the image for an object currently present to our senses.[37]

5. *The Number of Internal Senses*

Buridan denied the multiplicity of internal senses. Albert the Great had distinguished five internal senses: the common sense, imagination, fantasy (a storehouse for the imagination), estimative/cogitative powers, and memory. Late in the Renaissance, many scholars believed that they were returning to a pristine Aristotelianism by reducing the number of external senses to three: common sense, imagination and memory. But Buridan refused even to distinguish imagination from the common sense, drawing upon certain passages in *De anima* and ignoring others that suggest that Aristotle did in fact make the distinction.[38] Why did Buridan do so? I wonder if the reason may not be found by considering his position with regard to the organ of the common sense, the heart.

---

[35] John Buridan, *op.cit.*, p. 406: "Et alie et dissimiles sunt species seu intentiones que ultra multiplicantur ad organa virtutum interiorum, nec ille requirunt tales dispositiones ad sui receptionem quales requirebant priorem, ut alias dictum est."

[36] John Buridan, *op.cit.*, p. 155.

[37] John Buridan, *De memoria et reminiscentia* [Lokert], q. 1. fol. XLI[r].

[38] At II.23, Buridan went so far as to argue that Aristotle had not distinguished memory from common sense, on the basis of Aristotle, *De memoria*, 450a14, but he chose not to pursue that line of reasoning. John Buridan, *Quaestiones in Aristotelis De anima, tertia lectura* [Sobol], p. 378. Aristotle seems to have distinguished imagination from sense, as at *De anima*, III.3, 428a5 *et seq.*

6. *The Location of the Common Sense*

In the twenty-fourth question of Book II, Buridan defended Aristotle's assertion that the common sense is located in the heart. Given the heart's primacy in the body as source of vital heat, its appearance before all other organs in embryonic development, and the impression that humans have of feeling emotion and extremes of sensation in the chest, Aristotle had argued that the heart is the seat of sensation. Buridan agreed, despite the medical and philosophical custom of placing the internal senses in the cerebral ventricles. Galen and Avicenna had recognized that injuries to the front of the head damaged the ability to coordinate sense impressions, injuries to the middle of the head interfered with thinking, and injuries to the back of the head affected memory. Aristotle had stated that sleep was a condition of the common sense, yet physicians treated sleep-related maladies by applying remedies to the head.

In order to maintain that the heart was the organ of the single internal sense that fulfilled all the functions of the imaginative and estimative faculties, Buridan had to save the medical phenomena. His response was to admit that the brain was indeed the first way station of sensory input, but that then all intentions proceeded to the heart. He argued that such an arrangement shielded the heart from extremes of sensation that could cause damage.[39] But to make it work, there had to be the necessary conduits between the brain and the heart, and indeed Buridan proposed that a nerve joined the nexus of sensory nerves in the forebrain to the heart, and that another nerve joined the hindbrain – the organ of memory – to the heart.[40] Where did he learn his anatomy? Not from Avicenna's *Canon*, which describes only one cranial nerve – the sixth – as reaching the heart,[41] or from Galen's *De usu partium*. Perhaps he misread a medical text or received a poorly presented description of nervous system anatomy from an acquaintance in the medical faculty. Perhaps he invented it to suit his theory. We may never know. But his denial of other internal senses and his defense of the heart as the organ of the common sense may have come from an exaggerated desire to resolve questions in Aristotle's favor.

---

[39] John Buridan, *op.cit.*, p. 404.
[40] John Buridan, *op.cit.*, p. 408.
[41] In fact, the pneumogastric nerve is the tenth cranial nerve, and is the only nerve to reach from the brain to the heart. The heart is also supplied via cervical nerves. On Avicenna, see Shah, *The General Principles*.

7. Memory

The fact that animals possess memory indicated that memory consisted of sensible species and intentions stored in the hindbrain. These stored species and intentions did not enter into the realm of consciousness until they returned through the posterior nerve to the common sense in the heart, accompanied by an intention representative of past time, so that memories are not confused with present sensations. But humans, the sole incarnate possessors of intellect, can remember things that were never sensed, such as logical proofs, God, and universals. Buridan concluded that the storehouse for intellectual memory had to be the intellect itself.[42] The storehouse for sensitive species and intentions needed to reside in an organ different from the organ in which species and intentions were received and were manipulated. The organ of manipulation needed to be fluid, while the organ of the storehouse needed to be dry.[43] But the intellect was not housed in an organ; hence, it and its storehouse must be the same. If we are not constantly aware of our intellectual memories, it is because the mere presence of intellectual species is not enough to instigate thought. The will must also act.[44]

8. Sleep and Dreams

Granted the two nerves referred to above, Buridan could offer explanations of the varieties of sleep and dreams. Aristotle had taught that sleep was caused by an ascension of vapors from digestion, followed by a coalescence and descent, which had the effect, as Buridan saw it, of ligating the two nerves. The heart remained active and in fact became more active as it regenerated lost animal spirits and increased heat to aid digestion. But with the anterior and posterior nerves sealed, no species could arrive at the heart – even from organs below the heart – and no images could return to the heart from the organ of memory. This was dreamless sleep. But if the ligature on the posterior nerve was not perfect, some species might flow back down and become such things as dreams are made of. Several factors con-

---

[42] John Buridan, *De memoria et reminiscentia* [Lokert], q. 3, fol. XLI^v.

[43] *Ibid.* Why he then chose the brain, which, according to medieval theory, was cold and wet, is not clear.

[44] *Ibid.* The topic of intellectual memory in Buridan is treated at some length in Zupko, "What Is the Science of the Soul?"

tributed to the nature of dreams. Your memories are the raw material of your dreams. In the absence of other influences, species return to the heart during sleep and we dream of what those species represent. This much of Buridan's theory agrees with the latest theories of our era on the origins of dreaming, in which images run amok as the mind struggles to impose some narrative sequence on it all. But, for Buridan, there were other influences as well. One's bodily complexion affects the species in the posterior nerve, enhancing the similar and suppressing the dissimilar. Hence feverish persons dream of fire, of burning. Physicians take an interest in patients' dreams because a change in the predominant quality of the dream may portend a change in the patient's condition. When a fever sufferer dreams of drowning at sea, his fever will soon break.[45] Our emotions also influence our dreams. In the heart of an angry person, an image of a burning candle may become a dream of a house afire.

The dreams so far described offer no useful basis for prognostication beyond the life of the dreamer. But there is a fourth influence, which nothing terrestrial save human intellect can resist: the influence of the heavens. Our bodies receive the same influence that our environment at large receives. Consequently, our dreams can serve as indicators of the increased influence of some quality. A dream predominated by watery images might portend a flood. A dream predominated by images of dryness might portend a drought. Buridan also listed war and sedition as forecastable through dreams, but he did not reveal what complex of qualities went with each.

As an Aristotelian, he could not deny that the celestial realm affects the terrestrial, to which human bodies belong. And while allowing that prognostication by dreams can occur, he pointed out that only an experienced interpreter could hope to parse out the celestial influence from the emotional, physiological, and experiential inputs.

9. *Conclusion*

There are two features of Buridan's psychology that, I think, reveal something of his character and something of how he approached the task of teaching Aristotle. The first is his remarkable excursus on the nature of species; the second is his willingness to argue for a

---

[45] John Buridan, *De somno et vigilia* [Lokert], q. 8.

single internal sense located in the heart. The excursus on species suggests a philosopher willing to challenge Aristotle on the minor point of the speed of light in order to supplement Aristotle's own theory of sensation. His attempt to employ species in meteorology, beyond the area for which they were first proposed, suggests that they were more than an ad hoc addition to the Aristotelian world, and that their admission to Aristotelian psychology demanded that their nature be thoroughly understood.

Buridan's refusal to accept the common understanding of several internal senses located in the cerebral ventricles excused him from spending much time on the estimative and cogitative powers, which are not Aristotelian. But why would he have turned his back on the common understanding and then constructed such an elaborate scheme – complete with phantom nerve – to explain the medical evidence that the organs of the internal senses were located in the head? Was he lecturing about nature, or about Aristotle? His inquiry into species sounds like someone thinking about nature, but his defense of the cardiac common sense does not. It sounds instead like a clever defense of a position that might be abandoned when he concluded his lectures and went home for the night. It sounds like cleverness for its own sake. Perhaps – and this is a wild guess – his return to Aristotle was his way of scoffing at certain curricular requirements, of subtly rebelling against the attitude that he addressed at the end of his excursus on species. You can almost hear him sigh with impatience as he says, "And that is what I have to say about the species of proper sensible qualities, by way of a digression. And it seems to me that it is not without purpose. The arguments that were made at the beginning of the question have been answered according to what has been already been determined here and elsewhere. I have commented on the authorities according to the requirements of predetermined matters, just as anyone might wish."[46]

---

[46] John Buridan, *Quaestiones in Aristotelis De anima, tertia lectura* [Sobol], p. 321: "Et hec sint dicta de speciebus proprium sensibilium per modum quandam digressionis. Sed non inutilia, prout videtur mihi. Rationes autem que fiebant in principio questionis solvuntur secundum predeterminatam hic et alibi. Glosentur enim auctoritates secundum exigentiam predeterminatorm, sicud unusquisque voluerit."

THE NOTION OF "NON VELLE"
IN BURIDAN'S ETHICS

Fabienne Pironet

In recent literature, many scholars have discussed the position Buridan held on free will: is he an intellectualist, a voluntarist, or does his theory represent a certain kind of compromise between the two? The contribution made by Jack Zupko to this debate in his article, "Freedom of Choice in Buridan's Moral Psychology,"[1] shows very convincingly why no one should hesitate to say that Buridan is an intellectualist, and even a strict intellectualist. So I will not repeat what he said, but aim instead to continue the work he began. In particular, I will try to answer the question he asked at the end of his article: "Why did Buridan feel it necessary to appropriate voluntarist terminology in his theory of the will?"[2]

It seems clear that Buridan used some Scotistic vocabulary, the most evident of which is the distinction between the natural and the voluntary, as well as the notion of "not-willing (*non velle*)." There are many other points of comparison between Scotus and Buridan: e.g., both distinguish between the compound and divided senses of propositions such as "In willing A, the will can not will A (*Voluntas volens A potest non velle A*)." But of course, the mere fact that they discuss and even agree on the same points does not imply that they are in full agreement, or that they are using the same terms in the same senses. But how deeply Scotus could have influenced Buridan is not a question I intend to answer. I will concentrate here only on the notion of "*non velle*" and try to shed some light on the questions to which it gives rise. In order to do that, I will first review the different medieval meanings of the verb "*not-will*," and then offer a brief reconstruction of Buridan's theory of the will. Next, I will try to show how Buridan could reconcile his strict intellectualism with the idea that the will is free, in the course of which we shall see then that his notion of not-willing is quite different from Scotus's. Finally, I will discuss the extent to which Buridan's position agrees with the "Parisian Articles" he mentions from the Condemnation of 1277. My conclusion will be that one of the reasons – although not the most important reason – why "*non velle*" appears so often in

---

[1] Zupko, "Freedom of Choice."
[2] Zupko, "Freedom of Choice," p. 98.

Buridan's writings on free choice is that this "Scotistic-sounding" or "voluntarist-sounding" notion is the only way for a strict intellectualist such as Buridan to escape the charge of determinism, especially after 1277.

Before continuing, I should note that in order to avoid a confusion that the English language cannot easily resolve, I will adopt the translation used by Zupko and translate "*velle*" by "will," "*nolle*" by "nill" and "*non velle*" by "not-will."

1. *The Different Meanings of the Verb "non velle"*

As is well known, "*non velle*" can be understood in several ways.

If we look at what Thomas Aquinas,[3] Scotus,[4] Ockham,[5] and Dante,[6] have written on this subject, they all say that the will is capable of willing X, nilling X, or not-willing. Willing and nilling are considered as positive acts: "*volo X* (I will X)" means "*volo X esse* (I will X to be)," "*nolo X* (I nill X)" means "*volo X non esse* (I do not will X to be)," and "*non volo* (I do not will)" means that I stay in a passive state because I neither will nor nill. Unlike nilling, not-willing is *not* an act of the will. But in a broader sense, we could say that "*non volo*" is whatever is opposed to "*volo*." In this sense, "*non volo*" includes "*nolo*," which is the contrary of "*volo*," as well as "*non volo*" in the strict sense, which is the contradictory of "*volo*."

Because "*non velle*" can be understood in these two senses, it is always possible for a proposition containing "*non velle*" to be ambiguous. As we shall see, Buridan takes advantage of this ambiguity.

---

[3] Thomas Aquinas, *Scriptum super primum librum Sententiarum* [Mandonnet e.a.], I, dist.VI, expositio textus: "Differt nolo et non volo, quia, cum dicitur 'non volo' negatur actus, et ideo opponitur sicut negatio ad affirmationem, sed in hoc 'nolo' et in toto condeclinio eius remanet actus voluntatis affirmatus, et negatio fertur ad nolitum, unde sensus est 'nolo hoc' id est volo hoc non esse." See also Thomas Aquinas, *Summa Theologiae*, Ia–IIae, q. 6, a. 3, ad 2.

[4] John Duns Scotus, *Quaestiones super libros Metaphysicorum Aristotelis* [Andrews e.a.], IX, q. 15, pp. 152–153: "… experitur enim qui vult se posse non velle sive nolle."

[5] William Ockham, *Scriptum in librum primum Sententiarum. Ordinatio* [Gál e.a.], I, d. 1, q. 6 (OTh I: 506, ll.21–24): "Si autem accipiatur frui large pro actu appetendi, sic dico quod finem ultimum, sive ostendatur in generali sive in particulari, sive in via sive in patria, potest absolute voluntas eum velle vel non velle vel nolle."

[6] Dante, *Monarchia* [Kay], III, 2: "Sed ad non nolle alterum duorum sequitur de necessitate: aut velle aut non velle; sicut ad non odire necessario sequitur aut amare aut non amare; non enim non amare est odire, nec non velle est nolle, ut de se patet." See also Dante, *Banquet* IV, viii, 11 and 13.

"*Non velle*" can have yet another sense: when Aristotle discusses, for example, the case of the man who throws his goods into the sea during a storm,[7] it is clear that the man wills to do this absolutely (*simpliciter*) to save his life, although there is a respect (*secundum quid*) in which he does not-will it because of the financial loss. This is what Buridan calls willing against part of a judgment.

Risto Saarinen notes that "in medieval Latin, '*non volo*' refers to a passive state, whereas *nolo* denotes a case of active resistance."[8] It seems to me that this notion of not-willing does not really apply to Buridan. In *Quaestiones super decem libros Ethicorum Aristotelis* (Paris, 1513), *E* III.1, Buridan discusses some arguments about the generation of an act of motion or an act of volition. In this context, he uses the expression "*non velle*" to signify a passive state: before an instant t, the will did not will; and at instant t, it wills.[9] In most cases, however, not-willing simply means deferring the act of willing until the intellect has further deliberated.[10] Buridan thus gives a new di-

---

[7] Aristotle, *Nicomachean Ethics*, III.1, 1110a8–12.

[8] Saarinen, "John Buridan," p. 137, n. 30.

[9] John Buridan, *Quaestiones super decem libros Ethicorum Aristotelis* (Paris, 1513), Book III, q. 1, fols. 36vb–37va.

[10] See, for example, John Buridan, *op.cit.*, III.1, fol. 36^{rb-va}: "Verbi gratia, quod de Parisius ad Avinionem possum ire vel per Lugdunum vel per Dunonem, quorum utrumque praesentatur voluntati sub ratione boni, et voluntas quodcumque bonum sibi sub ratione boni praesentatum acceptare potest, et non potest illa duo simul acceptare propter incompossibilitatem; ideo libere potest se determinare ad quodlibet illorum absque alio quocumque determinante ipsam, vel etiam potest ad neutrum illorum se determinare, sed in suspenso manere donec fuit inquisitum per rationes quae via fuerit expeditior vel melior;" John Buridan, *op.cit.*, III.1, fol. 36va: "Et ita patet quod rebus sic stantibus quod ego vellem esse apud Avinionem, et scio me posse illic ire per Lugdunum vel per Dunonem, et percipio laborem viae, voluntas, absque alio quocumque determinante ipsam praeter ea quae dicta sunt, potest libere utramque viam refutare propter laborem, immo et totaliter ab eo discedere quod prius volebat, vel potest libere quamlibet viam acceptare, scilicet hanc vel illam, vel etiam per hanc omne⟨m⟩ determinationem in suspenso tenere donec ratio docuerit vel determinaverit quae melior et quae pejor sit;" John Buridan, *op.cit.*, III.5, fol. 44vb: "Prima est quod voluntas potest illud non velle quod per intellectum judicatur esse bonum. Aliter enim non esset domina sui actus. Dictum enim fuit prius quod voluntas potest differre actum volendi ut antea fiat inquisitio si bonitati apparenti fuerit aliqua malicia consequens vel annexa, potest etiam illud non velle propter annexam tristitiam vel laborem. Et eodem modo dicendum est quod voluntas potest non nolle quod intellectus judicat esse malum;" John Buridan, *In Metaphysicen* (Paris, 1518), Book IX, q. 4, fol. 58ra: "Istis judiciis et omnibus aliis de mundo similiter se habentibus et existentibus voluntas sine alio determinante potest velle accipere ex eo quod judicatum est utile vel etiam potest nolle accipere quia judicatum est inhonestum et injustum vel etiam potest dif⟨f⟩erre ita quod nullum actum volendi aut nolendi producat, sed differat et prolonget donec intellectus magis consideraverit et magis de-

mension to the verb "*non velle*:" the will is *active* when it not-wills, i.e., when it chooses to defer its act ("*non volo*" in fact means "*volo differre* (I will to defer)"), although it remains in a passive state (*manet in suspenso*) towards the external object because it neither wills it nor nills it.

2. *Buridan's Theory of Will*

When speaking about acts of the will, Buridan, like most authors, mentions only willing and nilling. But it is clear that a third choice is always possible, even if it is not always explicitly mentioned. This third choice is not-willing. To understand the significance of not-willing in Buridan's texts, we should first reconstruct his theory of will, which is based on three principles:

1. The will cannot will something bad as such, or nill something good as such, because God did not create us free so that we might sin; rather, God gave us freedom because it is noble and very useful as a guide in life, so that we might attain happiness.[11]
2. The will cannot will that which the intellect has not judged to be good in some way or other.[12]

---

liberavit ita quod intellectus non est sufficiens ad determinandum voluntatem, sed hoc habet voluntas de sua libertate."

[11] See John Buridan, *Quaestiones super decem libros Ethicorum Aristotelis* (Paris, 1513), III.3, fol. 42[va]: "... est sciendum quod libertas secundum quam voluntas potest non acceptare quod sibi praesentatum fuerit sub ratione boni vel non refutare quod praesentatum fuerit sub ratione mali, prodest valde nobis ad vitae directionem pro tanto quia in multis in quibus prima facie sunt aliquae rationes bonitatis apparentes, latent semper mille maliciae vel annexae vel consequentes ...; (*Ibid.*, fol. 42[vb]) libertas non est data nobis ut nocere possit homini ordinata, saltem bene; sed ipsa noceret si contingeret hominem nullam in bonis vel malis sibi apparentibus habere complacentiam vel displicentiam: quia tunc nullam haberet occasionem amplius inquirendi de eis an essent bona vel mala simpliciter aut acceptanda aut refutanda, propter quod saepe acceptandorum acceptatione et refutandorum refutatione privaretur;" John Buridan, *op.cit.*, III.5, fol. 44[vb]: "Ad primam, dicendum est quod voluntas non dicitur libera quia possit velle malum sub ratione mali vel nolle bonum sub ratione boni (sic enim libertas per se disponeret ad peccandum et esset utilis conditio et mala), sed dicitur libera loquendo de libertate oppositionis: quia ipsa potest in opposito: quia ipsa potest non velle quod prima facie apparet bonum et non nolle quod prima facie apparet malum et quia potest illud velle aut nolle illud in quo simul apparent rationes boni et mali."
[12] Cf. John Buridan, *op.cit.*, III.5, fol. 44[vb]: "Secunda conclusio est quod voluntas non potest velle illud in quo intellectui nulla apparet bonitatis ratio quoniam tale nullo modo esset praesentatum intellectui seu voluntati sub ratione volibilis."

3. The will can never choose a lesser good while a greater good is taken into consideration, because the lesser good is bad in comparison to the greater good.[13]

Let us now see what happens in the five possible situations that could arise, based on these principles:

1. *If X is absolutely good*, the will cannot nill it, although it can not-will it; so I will X because it is good or I not-will until the intellect has considered the matter further. But does further deliberation make sense when a judgment is absolutely certain? I cannot nill X because it has no bad aspect.
2. *If X is absolutely bad*, the will cannot will it because it is in no way willable; so I nill X because it is bad or I not-will until the intellect has considered the matter further. But again, does further deliberation make sense when a judgment is absolutely certain?

We should notice that these situations are practically impossible because we very rarely have absolute certainty in practical matters. Thus, in most cases, even if we are very prudent and consider everything that can be considered, we find ourselves in one of the following three situations, where judgments are based on appearances rather than on absolute certainty:

3. *If X appears to be good*, the will can not-will it, because the agent thinks that something bad could lie hidden in it or follow from it; so I will X because it is judged to be good (although perhaps I will realize later that it was a mistaken choice), or I not-will until the intellect has considered the matter further. And here it can make sense to proceed to further consideration. But I cannot nill X, because the only appearances it has are appearances of goodness. (If, after further consideration, no bad aspect appears, X will be considered absolutely good, and we go back to situation 1. If, after further consideration, some bad aspect appears, we will be in situation 5.)
4. *If X appears to be bad*, the will can not-nill it, because the agent thinks that something good could lie hidden in it or follow from it; so I nill X because it is judged to be bad (although perhaps I will realize later that it was a mistaken choice); or I not-will until

---

[13] This point is shown in John Buridan, *op.cit.*, III.4, and the third conclusion is clear (fol. 44[ra–rb]): "Tertia conclusio: si voluntas debeat eligere, ipsa necessario eliget majus bonum: quia aut majus bonum aut minus, sed non potest minus pro tunc; igitur oportet quod majus."

the intellect has considered the matter further And here too it can make sense to proceed to further consideration. But I cannot will X because the only appearances it has are appearances of badness. (If, after further consideration, no good aspect appears, X will be considered absolutely bad and we go back to situation 2. If, after further consideration, some good aspect appears, we will be in situation 5.)

5. *If X appears good and bad,* the will can will it, nill it, or not-will it; so I will X because it is judged to be good in some way, or I nill X because it is judged to be bad in some way, or I not-will until the intellect has considered the matter further. And here too it can make sense to proceed to further consideration.

In this connection, Buridan discusses the case of adultery, which is both pleasurable and dishonorable. As further deliberation is useful in such cases, let us say that the first reaction of the will is to not-will. If, after further consideration, i.e., by comparing these two goods (pleasure and honor) to determine which is the greater, the intellect judges that honor is a greater good than pleasure, I nill the adultery because it is bad. As pleasure is judged to be a lesser good, it is bad, according to principle 3, and, as it is bad, I cannot will it. But if, after further consideration, the intellect judges that pleasure is a greater good than honor, I will the adultery because it is good. As honor is judged to be a lesser good, it is bad, according to principle 3, and, as it is bad, I cannot will it.

We should here mention the case of equally strong judgments: what happens if, after further consideration, the intellect judges that both choices are equally good? Is this situation even possible? It seems that although this situation is not impossible,[14] the agent cannot stay in such a passive state for very long, unless, of course, he is an ass. This state is only provisional because even if the intellect is unable to rationally determine which is the greater good, it can always judge that it is not good to remain in this state, and so it will judge that making an irrational or arbitrary choice (by playing dice or by other means) is the greater good. Even if Buridan does not explicitly give this answer, it seems to follow from his theory of will.

---

[14] See John Buridan, *op.cit.*, VII.6, fol. 143[rb]: "Si quaeras cui istarum viarum magis assentiam, dicam quod potest poni differentia inter apparentiam et judicium, quoniam saepe, ut mihi videtur, expertus sum quod cum rationes viderem ad utramque partem probabiles, tamen ad neutram partem judicii determinabam me, etiam neque novis rationibus ad unam partem vel ad aliam supervenientibus sed in suspenso tenebam me."

We can conclude from all this that whatever the case, either (1) the will makes its choice according to the judgment of the intellect about the goodness, or at least about the appearance of goodness, of its object, or (2) the will defers its decision. Where, then, is the freedom of the will? Buridan answers that in the situations covered by principles 1–4 above, the will is not determined to will whatever is good or appears to be good, because it can not-will, and that in situation 5, the will is not determined to will whatever appears to be good, because it can will against a part of the judgment.

But this answer should not convince anyone who reads the texts attentively. It seems rather that the will is not free at all but is fully dependent upon the intellect. And this becomes evident if we consider that:

1. The will can decide to not-will its object if and only if the intellect has presented it to the will as something good, and even as the greater good (principles 2 and 3). This is the main point of Zupko's demonstration.[15]

2. Since it is practically impossible to remain in the state of not-willing for a long time, it is necessary for the will to make its choice at some moment or other, in keeping with the judgment of the intellect.[16]

3. The will can will adultery (or any other sin) if and only if the intellect has presented it to the will as something good, and even as the greater good (principles 2 and 3).

To the question, "How can someone will adultery when it is evident to everyone that honor is a greater good than pleasure?," Buridan replies that there are several possible explanations: (1) the intellect – deceived, ill-disposed, carried away by passion, etc. – ignores the comparison between these two goods so that, ignoring honor, pleasure would look good; (2) the intellect makes a false judgment; or

---

[15] Zupko, "Freedom of Choice," pp. 83–94.

[16] Buridan also adds that it is also possible to completely renounce the object. See John Buridan, *op.cit.*, III.1, fol. 36ᵛᵃ: "Et ita patet quod rebus sic stantibus quod ego vellem esse apud Avinionem, et scio me posse illic ire per Lugdunum vel per Dunonem, et percipio laborem viae, voluntas, absque alio quocumque determinante ipsam praeter ea quae dicta sunt, *potest libere utramque viam refutare propter laborem, immo et totaliter ab eo discedere quod prius volebat,* vel potest libere quamlibet viam acceptare, scilicet hanc vel illam, vel etiam per hanc omne⟨m⟩ determinationem in suspenso tenere donec ratio docuerit vel determinaverit quae melior et quae pejor sit" (emphasis mine). But it is clear that it is the intellect which judges that it is better to renounce to the trip. I will discuss later how the expression "(voluntas) absque alio quocumque determinante ipsam potest ..." should be interpreted.

(3) the will orders the intellect not to consider the comparison.[17] We should remember that, according to principle 2 above, the will can command the intellect not to consider something *if and only if this possibility has been presented to it as good by the intellect.* It is therefore evident that the intellect, culpable or not, is alone responsible for sin and that the will can never act directly against reason.

Buridan is accordingly a strict intellectualist for whom the will is well and truly dependent upon the intellect. His contemporaries called such a doctrine "psychological determinism." It was condemned in the so-called "Parisian Articles" that Buridan himself cites in his *Questions on the Ethics,* and indeed, cites so often that Saarinen remarks that, "… one can say without exaggeration that questions 1–5 of Book III deal more with the *Parisian Articles* than with Aristotle's *Ethics.*"[18]

At least two questions need to be answered if we want to understand why Buridan's theory of will has been interpreted in so many different ways by his contemporaries as well as by modern scholars: (1) How can Buridan reconcile this psychological determinism with the conclusion, which he explicitly claims to defend, that the will is free?; (2) Are Buridan's various gestures of fidelity to the Parisian Articles sincere? I will begin with the first question.

3. *Freedom of the Will According to Buridan*

Buridan talks about freedom of the will in terms of its "freedom of opposition (*libertas oppositionis*):" the will is free to choose either one of two opposing acts, or it can simply not-will. "It is evident," he says, "that no power of the soul is freer than the will."[19] In his *Questions on Metaphysics,* he adds that the will is indifferent, free and does not need anything else determining it to act in one way or another.[20]

---

[17] See John Buridan, *op.cit.,* III.4, fol. 44ra: "… est in potestate voluntatis imperare intellectui ut desistat a consideratione illius boni majoris, et tunc poterit acceptare minus."

[18] Saarinen, *Weakness of the Will,* p. 168.

[19] John Buridan, *op.cit., E* III.2, fol. 38ra: "Credo quod voluntas sit activa illorum actuum quorum nos primo domini sumus et quod ceteris omnibus eodem modo se habentibus nos possumus in utrumque oppositorum, quoniam nulla videtur potentia animae quae sit magis libera et magis domina illorum actuum libertate et dominio contradictionis, scilicet quibus possumus in opposita, quam voluntas, sed libertas et dominium magis videntur attribuenda agenti quam patienti."

[20] John Buridan, *In Metaphysicen* (Paris, 1518), IX.4, fol. 58ra: "Istis judiciis et omnibus aliis de mundo similiter se habentibus et existentibus voluntas sine alio de-

Some interpreters have concluded that it was to save the freedom of the will that Buridan insisted so strongly on the possibility of the will's not-willing. Buridan himself suggests as much: "... the will is not said to be free because it can will what has been judged to be bad, and do so in the face of that judgment, but because, in the face of a judgment concerning goodness or badness itself, it can defer the act of willing or nilling."[21] But we have seen why this is not really the case: not-willing does not make the will free, or freer than the intellect, because the will can not-will if and only if the intellect has judged that not-willing is the greater good at that moment. Is Buridan then a liar when he claims that the will is free? Of course not! Denying freedom of the will would lead to determinism and would have dangerous consequences for both morals and religion. So how can the will be free if it is so dependent upon the intellect?

In Book X of his *Questions on the Nicomachean Ethics*,[22] Buridan clearly says that the intellect and the will are equally free because the intellect and the will are not really distinct; rather, they both are the same thing, viz., the soul. So, if the intellect is free, then the will must be free too, and they equally free of necessity. But he also says that the soul produces an act of understanding more freely than an act of willing, because what he calls "... freedom of final ordination (*libertas finalis ordinationis*)" is nobler than freedom of opposition. Thus, the freedom of the will immediately depends upon the freedom of the intellect, and it is because the intellect is free that the will is free.[23]

Despite these intellectualist-sounding remarks, it is evident to me that Buridan's understanding of not-willing is in keeping with some of

---

terminante potest velle accipere ex eo quod judicatum est utile vel etiam potest nolle accipere quia judicatum est inhonestum et injustum vel etiam potest dif⟨f⟩erre ita quod nullum actum volendi aut nolendi producat, sed differat et prolonget donec intellectus magis consideravit et magis deliberavit ita quod intellectus non est sufficiens ad determinandum voluntatem, sed hoc habet voluntas de sua libertate."

[21] John Buridan, *Quaestiones super decem libros Ethicorum Aristotelis* (Paris, 1513), VII.7, fol. 144va: "... voluntas non dicitur libera quia possit velle quod judicatum est esse malum et hoc illo judicio stante, sed quia ipsa stante judicio de bonitate vel malitia potest differre actum volendi vel nolendi, sicut magis visum fuit in tertio libro."

[22] John Buridan, *op.cit.*, X.1, fol. 205ra.

[23] If we want to look for an author who might have influenced Buridan here, and who would have supported the same kind of view on the relation between the intellect and will as he does, we should look not at Scotus but at Siger of Brabant, who also supports a strict intellectualism. Many parallels could be drawn between Siger's and Buridan's positions, but it is not my aim here to develop this point; I would rather focus on the comparison between the notions of not-willing in Buridan and Scotus.

what Scotus has to say about the subject. In particular, Buridan draws three conclusions that seem to echo the views of Scotus. Consider, for example, the following passages from each author:

> The first [conclusion] is that the will can not-will what the intellect has judged to be good because otherwise, it would not be master of its own act. For it was stated before that the will can defer the act of willing in order to inquire first whether an apparent good has some badness consequent upon it or adjoined to it. It can also not-will it on account of the attendant sadness or effort. And in the same way, it must be said that the will can not-nill what the intellect judges to be bad. The second conclusion is that the will cannot will that in relation to which no aspect of goodness appears in the intellect, since such a thing has in no way been presented to the intellect or will under a willable aspect. And in the same way, it must be said that the will cannot nill, although it could not-will, that in relation to which no aspect of badness appears in the intellect, since such a thing has in no way been presented to the intellect under an avoidable or rejectable aspect, etc. The third conclusion is that the will can will that which has in some way been judged to be bad, and nill that which has in some way been judged to be good, just as in the case of adultery. There is an appearance of dishonesty and pleasure, [but] the will, the dishonesty notwithstanding, can will the adultery under the aspect of pleasure, or can not-will it under the aspect of dishonesty. And so it is in the case of the man who throws his goods into the sea during a storm, and enough has already been said about this above. So then, it is obvious that the will can will against part of the judgment, but not against the whole, or in addition to the whole.[24]

To the third argument, I say that in one way, even if there is in it no lack of any good(ness), nor anything bad, (so that perhaps the will

---

[24] John Buridan, *op.cit.*, III.5, fol. 44vb: "Prima est quod voluntas potest illud non velle quod per intellectum judicatur esse bonum. Aliter enim non esset domina sui actus. Dictum enim fuit prius quod voluntas potest differre actum volendi ut antea fiat inquisitio si bonitati apparenti fuerit aliqua malicia consequens vel annexa, potest etiam illud non velle propter annexam tristitiam vel laborem. Et eodem modo dicendum est quod voluntas potest non nolle quod intellectus judicat esse malum. Secunda conclusio est quod voluntas non potest velle illud in quo intellectui nulla apparet bonitatis ratio quoniam tale nullo modo esset praesentatum intellectui seu voluntati sub ratione volibilis. Et eodem modo dicendum est quod voluntas non potest nolle licet possit non velle illud in quo nulla apparet intellectui ratio maliciae: quia tale nullo modo est praesentatum intellectui sub ratione fugibilis vel refutabilis et cetera. Tertia conclusio est quod voluntas potest velle illud quod aliquo modo judicatum est esse malum et nolle illud quod aliquo modo judicatum est esse bonum, sicut in adulterium apparuerit inhonestum et delectabile voluntas non obstante inhonestate potest velle adulterium ratione delectationis vel potest non velle ratione inhonestatis, et ita est de illo qui tempore tempestatis projicit merces in mare, et haec dicta satis fuerunt prius. Sic igitur patet quod voluntas potest velle contra partem judicii, sed non contra totum vel praeter totum."

cannot *nill* it because the object of the act of nilling is what is bad or defective), it can nevertheless *not-will* that perfect good because it is in the power of the will not only to will in this way or that, but also to will or not-will, because its freedom is to act or not act. For if it can command other powers to move not only to act in this way and that, but also to act determinately or not act, it does not seem that its freedom is less, taken in itself, as far as the determination of the act is concerned. And this could be shown in that passage of Augustine from *Retractations* I, where he comes out against the first article above.[25]

Buridan agrees with Scotus in claiming that the will can not-will something that is judged to be perfectly good. But the agreement between the two authors ends there because they do not share the same notion of not-willing. For while Buridan explicitly states what happens when the will not-wills (viz., it orders the intellect to look again for any hidden badness or goodness in the object), Scotus does not make it clear what happens then. According to what Scotus says about willing an end, the will has the power to turn the intellect from its consideration of this end, after which it is simply not-willed because the will cannot be exercised as regards what is not (or no longer) known to it.[26]

Thus, although Buridan and Scotus agree that the will can give orders to the intellect, in Scotus's view the will can do so on its own, whereas for Buridan, the will can do so if and only if doing so has been presented to it by the intellect as the greater good. And although they also agree that the will can order the intellect not to consider something, Scotus's will can apparently do so even as regards a perfect good, i.e., an object in which there is no appearance of badness, whereas for Buridan, it is always impossible for the will to nill a perfect good. The disagreement is obvious if we consider what Scotus says immediately after the passage quoted above:

---

[25] John Duns Scotus, *Ordinatio* (Lyon, 1639), I, d. 1, q. 4, p. 162: "Ad tertium dico uno modo, quod licet non sit ibi aliquis defectus alicuius boni, vel aliqua malitia, et ideo non posset forte voluntas *nolle* illud, quia objectum actus nolendi est malum vel defectivum, potest tamen illud bonum perfectum *non velle*, quia in potestate voluntatis est non tantum sic vel sic velle, sed velle vel non velle, quia libertas eius est ad agendum vel non agendum; si enim potest alias potentias imperando movere ad agendum, non tantum sic et sic, sed ad determinate agendum vel non agendum, non videtur quod minor sit libertas sui respectu sui quantum ad actus determinationem; et hoc posset ostendi per illud August. I *Retractationum*, ubi supra contra primum articulum."

[26] John Duns Scotus, *op.cit.*, I, d. 1, q. 4, p. 157: "... in potestate voluntatis est avertere intellectum a consideratione finis, quo facto voluntas non volet finem, quia non potest habere circa ignotum."

> In another way, it is replied to the third preceding argument that it has not been proved that the will could not nill a good in which no appearance of badness or lack of goodness is found, just as it has not been proved that it could not will that in which no appearance of good is found, and this either beforehand in the thing, or in the apprehension, before it terminates the act of willing.[27]

For Buridan, the will can never nill a perfect good, although it can nill a greater good if and only if it is no longer considered as such by the intellect. For Scotus, on the other hand, it does not seem impossible to nill a perfect good as such and, a fortiori, to nill the greater of two goods. To summarize: Buridan claims that the will is free because it has a freedom of opposition that *does not* allow it to act directly against reason, and because it can not-will; Scotus claims that the will is free because it has a freedom of opposition that *does* allow it to act directly against reason, and because it can not-will. The possibility of not-willing belongs to what makes the will free. For Buridan, not-willing makes us free because unlike brute animals, we are not driven to react immediately to whatever appears before us. But this does not mean that the will is freer than the intellect, because any decision of the will ultimately depends upon the intellect. For Scotus, however, the will is freer than the intellect because it can order the intellect not to consider even a perfect good. These differences provide sufficient grounds to reject the idea that Buridan's theory of will is in any significant way Scotistic.

How, then, should we interpret Buridan's claim in his *Questions on Metaphysics*, viz., that "... the will does not need anything else to determine itself (*sine alio determinante*) to act in a way or another?" Does it mean – and this would indeed be voluntarist-sounding – that the will does not need anything other than itself for its complete determination? A little bit later, he says, "... the intellect is not sufficient to determine the will, but the will can do this on the basis of its own liberty (*intellectus non est sufficiens ad determinandum voluntatem, sed hoc habet voluntas de sua libertate*)."[28] Concerning the first remark, we should recall that Buridan clearly says that a previous judgment of the intellect is required for the will to act, so we could interpret this passage as meaning that the will does not need anything other

---

[27] John Duns Scotus, *op.cit.*, I, d. 1, q. 4, p. 162: "Aliter dicitur ad tertiam rationem praecedentem quod non est probatum quin voluntas possit bonum nolle in quo nulla invenitur ratio mali vel defectus boni, sicut non est probatum quin possit velle illud in quo non invenitur aliqua ratio boni, et hoc vel prius in re, vel in apprehensione, prius quam illud terminet actum volendi."

[28] This text has already been cited in note 19 above.

than a judgment of the intellect to determine itself. And, as the will is not really distinct from the intellect, it is in a sense true to say that the will does not need anything other than itself for its complete determination. This would be a way of affirming intellectualism in a voluntarist-sounding way.

The second claim, "… the intellect is not sufficient to determine the will, but the will can do this on the basis of its own liberty," could, I think, be interpreted in an intellectualist sense as well as in a voluntarist sense. Buridan is here discussing a case of contrary judgments about the same situation. He wants to show that what distinguishes us from brute animals is that a brute animal will necessarily and immediately choose according to the stronger judgment of sense if no impediment occurs, whereas a man has the opportunity to investigate further. Hence, the intellect might not be sufficient to determine the will if we understand that it is not sufficient to determine the will *immediately*, since *prima facie* judgments can be erroneous, especially when one and the same object is presented as being both good and bad. It is also possible to interpret the second part of the sentence, "… the will can do this on the basis of its own liberty," in an intellectualist sense, because it could be true to say that the will is able to determine itself because it is free, even if its freedom derives from the intellect's freedom. This again would be a way to affirm intellectualism in a voluntarist-sounding way.

Because it can be hard to accept that the above propositions could be interpreted in a strictly intellectualist sense, one could suggest that they should be rather interpreted along more moderate lines, following Thomas Aquinas, or that those affirmations are well and truly voluntarist in the sense that they claim that the will is autonomous vis-à-vis the intellect, and that Buridan changed his mind between writing his *Questions* on the *Metaphysics* and *Nicomachean Ethics*. Those interpretations are not impossible, but I do not think that they are plausible for the following reasons:

1. In general, Buridan is very consistent author, and I find it hard to imagine that he changed his mind so radically on such an important point.
2. The *Questions on Metaphysics* is not the place where we would expect to find Buridan presenting his theory of the will in all its details. We note, however, that the essentials of his theory are already present there, and that he speaks about not-willing in the same way that he does in his later *Questions on the Nicomachean*

Ethics. The fact that he already uses the notion of not-willing in the sense of an act of deferment, and also that he regards the will as not really distinct from the intellect, seems to show that Buridan's early position differs from the intellectualism defended by Aquinas.[29]

3. As I showed above, Buridan's affirmations in the *Questions on Metaphysics* can be interpreted along intellectualist lines, but in a voluntarist-sounding way.
4. It is easy to explain why Buridan would have used such a voluntarist-sounding way of speaking: most philosophers throughout the Middle Ages embraced one or another form of voluntarism and gave the will a certain kind of autonomy vis-à-vis the intellect. Buridan thus belongs to a minority group, with the theory he supports having been explicitly condemned at the very University where he was teaching. It is thus easy to understand why it was so important for him to try to save the appearances.
5. I would be inclined to think that it is because of his intellectualist tendencies in the *Questions on Metaphysics*, or, at least, because of the absence of a clear position (or of a clear voluntarist position) in that work, that Buridan felt compelled to express his allegiance to the Parisian Articles in the *Questions on the Nicomachean Ethics*. We should recall that at the end of Book IX, Question 4 of the *Questions on the Metaphysics*, Buridan says not only that it would be very difficult to prove that our will is absolutely indifferent, but also that, in a way, only faith can guarantee that we are different from brute animals.

Since Buridan's texts have thrown even modern interpreters into confusion about his exact position on this problem, it is not impossible that this confusion also prevailed among his own colleagues, some of whom perhaps reprimanded or attacked him because he was too ambiguous. This would explain why he so often expresses his adherence to the Parisian Articles in the *Questions on the Nicomachean Ethics*. And we should not blame him for that, as Saarinen seems to, first because it is quite understandable – Saarinen's comparison of Buridan to Albert the Great on this point strikes me as rather strange in view of the fact that Albert wrote well before and Buridan well after 1277 – and second because in expressing his adherence to

---

[29] Thomas Aquinas says that the freedom is in the will even if the cause of this freedom is the intellect. See Thomas Aquinas, *Summa Theologiae*, Ia–IIae, q. 10, a. 2; q. 13, aa. 1, 6.

the Articles, Buridan was not prevented from maintaining his own position, as we shall see.

This leads us to our second question: Was Buridan being sincere when he claimed that his conclusions are in accordance with the Parisian Articles? Or did he use "scotistic-sounding" or "voluntarist-sounding" terminology dishonestly, in order to deceive his opponents?

4. *The Agreement with the Parisian Articles*

The short answer is that, yes, Buridan was being sincere in saying that his conclusions were in agreement with Parisian Articles, but he did not necessarily understand those Articles in the same way the University Censors did. So let us examine those Articles one by one.

In Book III, Questions 1 and 3, and Book VII, Question 8 of the *Questions on the Nicomachean Ethics*, Buridan refers to Articles 159 and 160.[30]

Article 159 states: "Once all obstacles are removed, the appetite is necessarily moved by the object of its desire. This is an error where the intellectual appetite is concerned."[31] Buridan has no difficulty accepting this Article, first because he does not think that the desirable object is the only thing that determines an act of will, since a judgment of the intellect is always required; and second, because he distinguishes between natural and voluntary agents in terms of their liberty, the latter being not necessarily determined by a desirable object to will or to nill, whereas the former are so determined.[32]

Article 160 condemns the view "... that, once the will is in the state in which it is natural for it to be moved, and, when that which is naturally suited to move [it] remains readily disposed, it is impossible for the will to *not will*."[33] Buridan escapes from condemnation here because acts proper to the will are not necessary,[34] and also because

---

[30] John Buridan, *Quaestiones super decem libros Ethicorum Aristotelis* (Paris, 1513), III.1, fol. 44vb; III.3, fol. 43ra; VII.8, fol. 145ra. I here follow Hissette's classification of the Articles of Condemnation in Hissette, *Enquête*.

[31] "Cessantibus impedimentis, necessario movetur ab appetibili. Error est de intellectivo appetitus."

[32] John Buridan, *op.cit.*, III.3, fol. 43rb: "Ad secundum articulum, potest dici quod voluntas non necessario movetur ab appetibili quoad actum volendi aut nolendi."

[33] "Quod voluntate existente in tali dispositione, in qua nata est moveri, et manente sic disposito quod natum est movere, impossibile est voluntatem *non velle*."

[34] John Buridan, *op.cit.*, III.3, fol. 43^{ra-rb}: "But it seems to me that it must be

the will can always not-will – it can even not-will a perfect good, at least temporarily. We should also note that in Question 3, just before addressing himself to the aforementioned Articles, Buridan does not hesitate to say that he once heard a famous doctor of theology saying that he would not consider it improper to argue against Parisian Articles outside the diocese of Paris.[35]

In Book III, Question 5, he seems to refer again to Article 160, but the citation is not literal: "... once the will is in its final state, it cannot *will the opposite*: this view is in error."[36] The fact that Buridan uses the phrase "will the opposite (*velle oppositum*)" instead of "not will (*non velle*)" is very significant because it gives more strength to the Condemnation. We have seen that to be in agreement with Article 160, it was enough for Buridan to say that the will can not-will, because that Article does not specify whether "*non velle*" signifies only the absence of volition and nolition, or whether it also includes the positive concept of nilling (*nolle*). But here, "*velle oppositum*" seems to include "*nolle*," and Buridan should now prove that he agrees with the Article thus interpreted.

In reply to Article 160, Buridan is content to refer back to the first and third conclusions he has just argued for in the body of the Question.[37] The first conclusion states that the will can not-will that which is judged to be good (and it is clear that it can also will it), and that the will can not-nill that which is judged to be bad (and it is clear that it can also nill it).[38] It thus seems that the will can will something or will the opposite, i.e., not-will it, nill something or will

---

said as regards the first alleged Article [viz., 159] that the simple act of agreement or disagreement is not an act of the will strictly speaking, but willing or nilling, accepting and rejecting [certainly are]. For if it is asked of a continent man, 'Would you like to know such a married woman?', he will not reply, 'You bet!', but rather, he would say, 'I would, if it were not dishonest or sinful', indicating that he feels a certain agreement in that act on account of its pleasant appearance, even though he nills it (*Sed mihi videtur dicendum quantum ad primum articulum allegatum quod actus simplicis complacentiae vel displicentiae non est actus volendi aut nolendi proprie, sed nolle aut velle, acceptare et refutare: nam si petatur a continente viro 'vis tu cognoscere talem mulierem?', non respondebit 'volo', sed dicet 'vellem si non esset inhonestum vel peccatum', significans se propter delectationem apparentem habere in illo actu quamdam complacentiam nolle tamen ipsum*)."

[35] John Buridan, *op.cit.*, III.3, fol. 43ra: "Ad articulos allegatos, audivi semel ab uno doctore theologiae famoso valde dici quod non reputaret inconveniens extra parisiensem diocesim aliqua opinari contra episcopi parisiensis determinationem."

[36] John Buridan, *op.cit.*, III.5, fol. 44rb: "... existente voluntate in ultima dispositione, quod non possit *velle oppositum*: error."

[37] John Buridan, *op.cit.*, III.5, fol. 45ra: "Ad articulos parisienses potest responderi per primam conclusionem vel etiam per tertiam si quis consideret."

[38] Latin text is cited above, note 21.

the opposite, i.e., not-nill it. Buridan's theory thus appears to be in accordance with the Article, even understood in a very strong sense. The same is true of the third conclusion, which states that the will can nill that which is judged to be good in a way (and it is clear that it can also will it), and that the will can will that which is judged to be bad in a way (and it is clear that it can also nill it).[39] We should note here that Buridan carefully avoids referring to the second conclusion, which is the one the voluntarists would attack most. This conclusion says that the will cannot will that with respect to which no aspect of goodness is apparent to the intellect.

In Book III, Question 2, Buridan refers to Article 118, which condemns the view "… that the agent intellect is some separate substance superior to the possible intellect, and that it is separate from the body in substance, power, and operation, and is not the form of the human body."[40] Buridan does not hesitate to claim that he would consider this opinion probable if it had not been condemned. Then he suggests that the Article can be interpreted so that the Condemnation does not apply to his opinion of what agent intellect is.[41] It seems clear, then, that Buridan considers Parisian Articles to be open to interpretation.

In Book III, Question 4, Buridan refers to Article 157, which says, "… that when two goods have been proposed, the stronger one moves [the appetite] more strongly – This is in error, unless the degree in question is understood on the part of the moving good."[42] Here again, Buridan escapes the force of the Condemnation by saying that the will is not necessarily determined to will the greater good because it can always not-will.[43] But it remains clear from what he says

---

[39] Latin text is cited above, note 21.

[40] John Buridan, *op.cit.*, III.2, fol. 38vb: "Quod intellectus agens est quaedam substantia separata superior ad intellectum possibilem; et quod secundum substantiam, potentiam et operationem est separatus a corpore, nec est forma corporis hominis."

[41] *Ibid.*: "Et haec opinio videretur mihi probabilis nisi ei obstaret parisiensis articulus, ubi dicitur sic quod intellectus agens sit quaedam substantia separata superior ad intellectum possibilem et secundum substantiam potentiam et operationem sit separatus a corpore nec sit forma corporis humani est error. Et forte quod articulus potest exponi, scilicet quod positus fuit contra dicentes intellectum agentem esse humanae speciei appropriatum, et nullam formam substantialem humano corpori inhaerentem nisi corruptibilem et eductam de potentia materiae."

[42] John Buridan, *op.cit.*, III.4, fol. 43rb: "Quod duobus bonis propositis, quod fortius est, fortius movet. – Error, nisi intelligatur quantum est ex parte boni moventis."

[43] John Buridan, *op.cit.*, III.4, fol. 44rb: "Ad articulum parisiensem dicendum est quod voluntas non necessario movetur fortius a majori bono: quia potest a neutro moveri quoad actum volitionis non quoad complacentiam, sicut dictum fuit."

elsewhere that the will has to will the greater of two goods necessarily, unless that greater good is not considered any longer by the intellect.

In Book III, Question 5, Buridan refers to Article 166 which says: "That if there is rectitude in the reason, there is rectitude in the will – This is in error because it is contrary to Augustine's gloss on this verse of the Psalmist, 'My soul has thought to desire,' etc., and because according to this view, grace would not be necessary for the rectitude of the will, which is the error of Pelagius."[44] Buridan replies to this Article as we saw above, by referring back to his first and the third conclusions in the body of the Question,[45] carefully avoiding the second conclusion, which is the one voluntarists would attack most. But his reply to another argument later in the same Question cannot obscure his profound disagreement with this Article. Here he says that although the will could not-will what is honest, it cannot will what is dishonest when the reason is right.[46]

In Book III, Question 5, Buridan seems to refer to Article 169, which says that "... as long as passion and particular knowledge remain actual, the will cannot go against them," but the quotation he gives from the Article is not verbatim at all: "Once a universal major and particular minor are actually in place, the will cannot will the opposite – this is in error."[47] We find a similar formulation in Book VII, Question 7: "That, where there is knowledge of the universal and the particular, the will cannot will the opposite – this is in error."[48] He does not reply explicitly to this Article in Book III, Question 5, just as he gives no answer to the other two Articles mentioned there. But it is important to look more closely at this

---

[44] John Buridan, *op.cit.*, III.5, fol. 44rb: "Quod si ratio recta, et voluntas recta. – Error, quia contra glossam Augustini super illud psalmi: 'Concipivit anima mea desiderare' etc., et quia secundum hoc, ad rectitudinem voluntatis non esset necessaria gratia, sed scientiam solum, quod est error Pelagii." In both cases, Buridan just mentions the first sentence: "... si ratio est recta, voluntas est recta: error."

[45] See n. 37 above.

[46] John Buridan, *op.cit.*, III.5, fol. 45ra: "Ad aliam dicendum est quod ratio non est recta simpliciter et pertinens ad prudentiam nisi honestum omne praeferat delectabili cuicumque; immo non erit recta si aliquid inhonestum judicet delectabile vel utile, sicut visum est in undecima quaestione secundi libri; ideo secundum praedicta, voluntas licet possit non velle honestum, tamen inhonestum vel⟨le⟩ non potest recta existente ratione et cetera."

[47] John Buridan, *op.cit.*, III.5, fol. 44rb: "... existente majori universali in actu et minori particulari in actu quod voluntas non possit in oppositum: error." Cf. "Quod voluntas, manente passione et scientia particulari in actu, non potest agere contra eam."

[48] John Buridan, *op.cit.*, VII.7, fol. 143vb: "... stante scientia in universali et in particulari quod voluntas non possit velle oppositum, error."

Article because Buridan not only gives another interpretation of the Article, as he did for Article 118, he also modifies it so that it is difficult to recognize and understand what it means. The Parisian Article says that the will cannot act against a passion or an actual piece of particular knowledge, and this is condemned because it would lead to determinism. It is not absolutely clear to me what Buridan intends in his reformulation of the Article. He no longer speaks about the passions, even though it is evident that in the case he is discussing (adultery), passion is the reason why we have contrary judgments. The formulation we find in Book III ("Once a universal major and particular minor are actually in place, the will cannot will the opposite – this is in error") perhaps refers to the conflict between a greater good and a lesser one, the greater good being in general more universal than the lesser one. If this is what he means, then we have already seen why Buridan's position does not directly contradict Parisian Articles. The formulation we find in Book VII ("That, where there is knowledge of the universal and the particular, the will cannot will the opposite – this is in error") obviously refers to his discussion there of *akrasia*, in which he concludes that we can act against habitual knowledge, universal actual knowledge, and, although not directly, against particular actual knowledge. In any case, Buridan can find a way to avoid most of the implications of the Condemnation even if he cannot totally escape from it. As was the case with the "not will" in Article 160, it is enough for him to show that the will can act against particular actual knowledge, because in Buridan's reformulation, the Article does not specify whether the will acts directly or not. In Book VII, Question 7, Buridan likewise gives no explicit reply to this Article, once again referring to what he has said in that Question and in Book III.[49] But it is noteworthy that immediately before addressing himself to the Parisian Article, he once again reminds his audience that the will is free because it can defer its act of willing or nilling.[50]

Before concluding this discussion, it should be pointed out that there are other Parisian Articles relevant to Buridan's position, e.g., Articles 163: "That the will necessarily pursues what is firmly believed

---

[49] John Buridan, *op.cit.*, VII.7, fol. 144va: "De articulo autem parisiensi, dicendum est secundum exigentiam dictorum nunc et dictorum in tertio libro."

[50] *Ibid.*: "... dicendum est ex eo quod voluntas non dicitur libera quia possit velle quod judicatum est esse malum et hoc illo judicio stante, sed quia ipsa stante judicio de bonitate vel malitia potest differre actum volendi vel nolendi, sicut magis visum fuit in tertio libro."

by reason, and that it cannot abstain from what reason dictates. However, this necessitation is not compulsion, but the nature of the will;" 151: "That the soul wills nothing unless it is moved by another, on account of which this is false: 'the soul wills on its own' – this is in error, if 'moved by another' is understood in terms of the soul's being moved by something desirable or by an object, such that the desirable thing or the object is the entire reason for the motion of the will itself;" 164: "That a man's will is necessitated by his cognition, just like the appetite of a brute animal."[51] But it is clear that it would have been dangerous for Buridan to mention these Articles explicitly because he agrees with them, or at least with some of the theses included in them.

5. Conclusion

We can see that the focal point of Buridan's defense against the Parisian condemnations is the notion of not-willing. But it does not in the least affect his intellectualism. Indeed, when Buridan attempts to show that his views are not in violation of the Parisian Articles, he takes more and more liberties with them by giving his own interpretation of some of them, claiming that perhaps they could be held outside the Diocese of Paris, and even corrupting the citation of some of them.

Does this mean that Buridan used the notion of not-willing only to avoid running afoul of the condemnations? Here the answer is a resounding "no," for it is evident that not-willing has a very important role to play in practical matters, since the opportunity to deliberate is what makes us free and can make us always freer. And this, I think, is the main reason why Buridan insists so much on our ability to not-will and deliberate further. Buridan shares this view with Siger of Brabant,[52] who had been a target of the 1277 Condemnation. This explains why Buridan gave a "voluntarist-sounding" touch to

---

[51] Articles 163: "Quod voluntas necessario prosequitur quod firmiter creditum est a ratione, et quod non potest abstinere ab eo quod ratio dictat. Haec autem necessitatio non est coactio, sed natura voluntatis;" 151: "Quod anima nihil vult nisi mota ab alio. Unde illud est falsum: anima seipsa vult. – Error si intelligatur mota ab alio, scilicet ab appetibili vel objecto, ita quod appetibile vel objectum sit tota ratio motus ipsius voluntatis;" 164: "Quod voluntas hominis necessitatur per suam cognitionem, sicut appetitus bruti."

[52] Siger of Brabant, *Quaestiones in Metaphysicam* [Maurer], L.VII, q. 9, especially p. 326, ll. 15–19.

his theory. It is not for us to judge if this strategy was worthy of condemnation or dishonest; it is enough to know that Buridan was consistent and did not hesitate to affirm his dissent, even if it his position sometimes fell between the lines. In any case, if it was his intent to sow confusion, he certainly succeeded!

In his conclusion, Zupko speaks about "… the moderation and restraint with which Buridan expresses his own views."[53] I hope to have shown that this moderation is merely a façade, which in no way affects Buridan's strictly intellectualist position on free will.

---

[53] Zupko, "Freedom of Choice," p. 99.

Appendix: Schematic Outline of Buridan's Theory of Will

I. deliberation of the intellect upon the object

⇓

judgment of reason: the object is presented
to the will as good, as bad or as good and bad

II. first act of the will (necessary, spontaneous, not free)

at this stage the will is passive

⇓

if the object is presented as good	*if the object is presented as bad*	*if the object is presented as good and bad*
agreement	disagreement	both agreement and disagreement

III. second act of the will (contingent, deliberate, free)

at this stage the will is active

⇓

will	nill	not-will
⇓	⇓	⇓
acceptance	rejection	deferment
⇓	⇓	⇓
pursuit	avoidance	go back to I
⇓	⇓	
love	hate	
⇓	⇓	
desire	what is opposite to desire	

If I will X, I neither nill X nor defer
If I nill X, I neither will X nor defer
If I defer, I neither will X nor nill X
If I don't will X, I nill X or I defer
If I don't nill X, I will X or I defer
If I don't defer, I will X or I nill X

IDEO QUASI MENDICARE OPORTET INTELLECTUM HUMANUM: THE ROLE OF THEOLOGY IN JOHN BURIDAN'S NATURAL PHILOSOPHY

Edith Dudley Sylla

According to the currently accepted picture of medieval university thought, there was a unity between fourteenth-century philosophy and theology brought about by the use of common conceptions and methods of solving problems (for instance, the use of supposition theory, the language of ratios or proportions, maxima and minima, first and last instants, continuity and infinity, and so forth).[1] This unity is clear if we examine the widespread use of methods of the arts disciplines, especially logic but also mathematics, within fourteenth-century commentaries on Peter Lombard's *Sentences*. But at the University of Paris, although theologians might make use of the arts disciplines, Masters of Arts in this period were admonished to avoid unnecessary trespassing on the territory of theology. In a frequently quoted passage of his questions on Aristotle's *Physics*, John Buridan wrote:

> ... therefore some of my lords and masters in theology have blamed me for the fact that sometimes in my physical questions I mix some theological matters, since this does not pertain to those in the Arts Faculty. But I, with humility, respond that I very much wish not to be constrained only to this. All masters when they incept in arts vow that they will dispute no purely theological question, that is determining it, or about the incarnation. And they vow further that if it happens to them that they dispute or determine some question which touches both faith and physics, they will determine it for the faith, and that they will resolve the arguments [for the other side] accordingly as it seems to them they should be resolved. It is evident, however, that if any question touches faith and theology, this is one of them, namely whether it is possible that there be a vacuum. Therefore, if I want to dispute it, I must say what it appears to me should be said according to theology or else be perjured. And I must turn aside the arguments for the other side accordingly as it seems possible to me. And I cannot resolve them unless I pose them. Therefore I am compelled to do this.
>
> I say therefore that we can imagine a vacuum in two ways, as was said in another question, and it is possible that a vacuum exist in either

---

[1] See Murdoch, "From Social to Intellectual Factors."

way by divine power. And this is believed by me and not proved by natural reason, and so I do not intend to prove it, but only to state the way in which it seems possible to me.[2]

Now, if authors with theological credentials, such as Thomas Aquinas in the thirteenth century or Nicole Oresme in the fourteenth, after they had received their theological degrees, determined questions on the natural books of Aristotle into which they introduced theological matters, it might be less surprising, but Buridan was never a theologian, remaining far longer than most medieval scholars within the Arts Faculty. Moreover, Aristotle's own conception of knowledge divided knowledge into separate disciplines, each with its own principles and conclusions. In the Aristotelian conception, it was mistaken to confuse mathematics with physics, let alone theology with natural philosophy.[3] Buridan himself begins his questions on *On the Heavens* with the predictable question whether there is a discipline treating cosmology that is separate from the discipline of physics.[4] Why, then, did Buridan mix theology into his questions on the *Physics*? If we take him at his word, which seems plausible, it was at least in part because, when he incepted in the Arts Faculty, he vowed that if he happened to dispute a question which touched upon both faith and physics, he would determine it for the faith *and resolve the arguments for the other side accordingly as it seemed to him they should be resolved (eam*

---

[2] John Buridan, *Quaestiones super octo Physicorum libros Aristotelis* (Paris, 1509), fols. 73vb–74ra: "... ideo aliqui dominorum et magistrorum meorum in theologia improperaverunt michi de hoc quod aliquando in questionibus meis phisicalibus intermisceo aliqua theologica, cum hoc non pertineat ad artistas. Sed ego cum humilitate respondeo quod ego bene vellem non esse ad hoc astrictus. Sed unimos magistri cum incipiunt in artibus iurant quod nullam questionem pure theologicam disputabunt, ut pote determinate vel de incarnatione, et ultra iurant quod si contingat eos disputare vel determinare aliquam questionem que tangat fidem et philosophiam, eam pro fide determinabunt et rationes dissoluent prout eis videbuntur dissoluende. Constat autem si aliqua questio tangit fidem et theologiam, ista est una de illis, scilicet utrum possibile est esse vacuum. Ideo si eam volo disputare oportet me dicere quod de ea apparet michi dicendum secundum theologiam vel esse periurum et evadere rationes ad oppositum prout apparebit michi possibile. Et non possem solvere eas nisi moverem eas ergo sum ad hec facienda coactus. Dico ergo quod duplici modo possemus imaginari vacuum, sicut dictum est in alia questione. Et possibile est utroque modo vacuum esse per potentiam divinam. Et hoc est michi creditum et non ratione naturali probatum. Ideo nec istud intendo probare sed solum dicere modum secundum quem hoc apparet michi possibile." For another translation see Grant, *A Source Book*, pp. 50–51.

[3] Livesey, "William of Ockham" and "The Oxford Calculatores."

[4] "Utrum de mundo debet esse scientia distincta a scientia libri Physicorum." John Buridan, *Quaestiones super libris quattuor De caelo et mundo* [Moody], pp. 3–7.

pro fide determinabunt et rationes dissoluent prout eis videbuntur dissoluende).

It is the last clause of this vow that raises the most interesting issues in Buridan's case, namely that he will resolve the arguments for the side opposed to the faith accordingly as it seems to him they should be resolved. In this context, the arguments for the other side to be resolved were the arguments of Aristotle and Averroes. But on what premises or principles should such a resolution be based? Boethius of Dacia, representing the so-called integral Aristotelian or Averroist point of view, had earlier argued that a Master of Arts, determining a question in the Arts Faculty, should always base his arguments upon the principles of the discipline he professed, namely physics in this case. Many of Boethius of Dacia's theses had, however, been condemned in 1277. Should a Master of Arts disputing Aristotle's *Physics* resolve Aristotle's and Averroes's arguments using premises known purely by faith? Or might we expect Buridan to base his resolution of Aristotle's and Averroes's arguments on principles acceptable both to faith and to philosophy? Principles of logic or of metaphysics might fill such a role.

In an earlier paper on the physics of the Eucharist, I compared the ways in which Thomas Aquinas and William of Ockham used natural philosophy in determining questions about the Eucharist. I came to the conclusion that Aquinas and Ockham used physics in theology in different ways. Whereas Aquinas developed what I called a "sublimated physics" different from ordinary physics to account for what happens in and after transubstantiation, Ockham either transformed all of physics in light of the evidence of transubstantiation (accepting, for instance, that qualitative forms can exist without inhering in any subject) or else he said that what happened in the Eucharist was simply miraculous.[5]

In this paper I examine a selection of Buridan's questions on Aristotle's *Physics* with a complementary goal in mind. I am interested in how theology functions within Buridan's natural philosophy. On the basis of Buridan's discussion of questions on the eternity or creation of the world and on the possibility of a vacuum, I conclude that Buridan used theological as well as logical and metaphysical arguments where his vow at inception pushed him to resolve Aristotle's or Averroes's arguments. What resulted was often not natural philosophy but rather speculation about possible worlds. In his famous thesis

---

[5] Sylla, "Autonomous and Handmaiden Science."

about the effect of the Condemnations of 1277 on the rise of modern science, Pierre Duhem argued that the Condemnations pushed later natural philosophers to overturn Aristotelian physics, thereby moving them in the direction of Galileo or Newton. Other critics have, however, faulted fourteenth-century natural philosophers for being side-tracked by theological pressures into the consideration of imaginary cases that never occur in the real world (Galileo himself made this accusation). In this paper I want to examine Buridan's consideration of questions touching on both natural philosophy and faith to examine in more detail where his ideas fall within such a range of alternatives.

Boethius of Dacia, as I have said, refused to incorporate theological premises into his physics. Buridan, following his oath at inception, determined questions bearing on both natural philosophy and theology in accordance with faith, not physics. He did not, however, simply follow natural philosophical argument to its natural conclusion and then say, baldly, that he believed the opposite on the basis of faith. He followed the next clause of the oath at inception in resolving the natural philosophical arguments (Aristotle's and Averroes's arguments) as it seemed to him they should be resolved. This involved introducing additional hypotheses or conditions – in Buridan's terms it was necessary "... so to speak to beg the intellect" (*ideo quasi mendicare oportet intellectum humanum*). Thus, when the doctrines of faith were added to physics there was not a simple or inevitable conclusion to be drawn, but more work to be done, and various possible conclusions.

It is worth noting that Buridan was not adverse to modifying Aristotelian physics, whether theology was involved or not – he was not a rigid or dogmatic Aristotelian. In answering the question what causes the natural motion of elemental bodies, for instance, Buridan came to a non-Aristotelian solution. Having come to the conclusion that the natural motion of an element must be caused by a form inhering in it, Buridan asked whether it was the substantial form of earth, for instance, that moved it, or the accidental form of gravity. It is more likely the accidental form of gravity, he concluded, because the Eucharist, if dropped, falls just as it did before, but the previously existing substantial form is no longer there. Likewise, Buridan says, water vapor moves up by its lightness, even though its substantial form continues to be that of water of which the natural place is downwards.[6] Here Buridan has used ideas about the Eucharist as

---

[6] John Buridan, *Quaestiones super octo Physicorum libros Aristotelis* (Paris, 1509), VIII.5, fol. 114rb: "... si enim aliquis diceret quod movetur a sua forma substantiali,

evidence to help shape physical theory, not because he was compelled by his vow to do it – the cause of fall of a heavy body was not a theological question in any direct sense – but because he assumes that the doctrine of transubstantiation is true. Thus one way in which theology entered Buridan's physics was as a treasure trove of truths – about God, the Eucharist, angels, etc. – which could be used to test physical theories. For many in the twentieth century, science arbitrates what is believable in religion, but for Buridan the truths of theology might be sufficiently well-established to warrant their use as evidence in physics.

In what follows, I will consider two main cases. The first is the obvious case where theology and natural philosophy intersect in physics, namely the issue of the eternity vs. the creation of the world. And the second is the case that gave rise to the well-known passage quoted at the start, namely the question of the void or vacuum. In this second case, the link to theology is not some well-known theological doctrine such as the creation of the world, but rather the more general thesis, often linked to the Condemnations of 1277, that one should not deny to God the possibility of doing anything that is not a logical contradiction.

1. *Questions on Eternity and Creation*

The questions in Buridan's *Physics* that presuppose theology in the very statement of the question are most frequent in Book VIII.[7] There

---

apparet quod non vel tamen magis apparet moveri sic a sua gravitate vel levitate. Unde in sacramento altaris non manet aliqua forma substantialis in hostia que ante esset in ea et tamen quia manet gravitas caderet deorsum sicut ante. Et forte quia vapor elevatus ab aqua est adhuc substantialiter aqua et tamen quia factus est levis movetur sursum et non ad locum naturalem aque, ideo sic non movetur per formam substantialem aque sed levitatem." See Sylla, "Aristotelian Commentaries."

[7] The main questions involving theology are:

Book I, q. 15, Utrum necesse est omne quod fit fieri ex subiecto presupposito. [Alternate forms of the question: Utrum possibile sit aliquid ex nichilo fieri. Utrum aliquid potest fieri ex non ente vel ex nichilo.]

Book III, q. 15, Utrum est aliqua magnitudo infinita (and also other questions on the infinite involving God).

Book IV, q. 8, Utrum possibile est vacuum esse per aliquam potentiam.

Book VIII, q. 1, Utrum ad scientiam naturalem pertinet considerare de primo motore.

Book VIII, q. 2, Utrum Deus sic potuit facere motum de novo quod ante nullus fuisset motus neque mutatio.

Book VIII, q. 3, Utrum sit aliquis motus eternus.

Aristotle argued that creation *ex nihilo* is impossible and that the universe is eternal, but Buridan believed that it happened in fact. This is not a possibility, but a reality.

In one sense, medieval scholastics had little trouble explaining how Aristotle could reasonably argue against a beginning for the cosmos, while they believed it in fact had one. They said that creation is impossible naturally, but possible supernaturally. In the world as it exists, one body is always made out of another body, but God, with infinite power, can make a body out of nothing. This is what Buridan himself says in Book I, Question 15, when he asks whether it is necessary that everything that comes into existence come from something preexisting, or, alternatively, whether it is possible that something come into existence *ex nihilo*. His conclusions are, first, that something can come into existence *ex nihilo* – this he says he believes on faith and not because of any proof except the authority of the sacred scripture and the doctors of the Catholic faith. For thus God made and created angels and heaven and earth.[8] And his second conclusion is that it is necessary naturally that everything that comes into being comes from a preexisting subject. This second conclusion, too, he says he cannot demonstrate, but it can be shown by induction, there being no counter-instances. This is how Aristotle proves it, he says, and such conclusions should be deemed principles in natural science. Otherwise you could not prove that all fire is hot or that all rhubarb purges bile, or that all magnets attract iron. Such inductions, he says, are not demonstrations, because they do not follow formally and it is not possible to examine every instance. As is said in the *Posterior Analytics*, Book II, many indemonstrable principles are made manifest to us by sense, memory, and experience. Anyone who does not want to concede such propositions in natural and moral science is not worthy to take a large part in them.[9]

Book VIII, q. 6, Utrum primus motor sit immobilis.
Book VIII, q. 9, Utrum movens finitum potest movere per infinitum tempus.
Book VIII, q. 11, Utrum primus motor, scilicet Deus, sit infiniti vigoris.
Book VIII, q. 13, Utrum primus motor est indivisibilis et nullam habens magnitudinem.

[8] John Buridan, *op.cit.*, I.15, fol. 18[vb]: "Et tunc ego pono duas conclusiones. Prima est quod possibile est aliquid fieri sine subiecto presupposito ex quo vel in quo fiat et hanc conclusionem credo fide et non aliqua probatione nisi hoc auctoritate sacre scripture et doctorum fidei catholice. Sic enim Deus fecit et creavit angelum et celum et mundum. Et hanc conclusionem intendunt qui dicunt posse aliquid fieri ex nichilo vel etiam ex nichilo posse aliquid fieri."

[9] John Buridan, *op.cit.*, fols. 18[vb]–19[ra]: "Secunda conclusio est quod necesse est omne quod fit naturaliter fieri ex subiecto presupposito vel in subiecto presupposi-

Averroes, Buridan admits, labored long to demonstrate that nothing can be made *ex nihilo* – indeed he labored against the Catholic faith and against truth to demonstrate that it is impossible even by divine power for something to come into being *ex nihilo*. But, although it is true and to be firmly believed that God can create *ex nihilo* and that God has even created prime matter, nevertheless this cannot be demonstrated, according to Buridan, nor can it be shown by arguments arising from sense. Even the principal arguments at the beginning of Book I, Question 15, saying in various ways that God can create *ex nihilo* because he is omnipotent, are not demonstrative according to Buridan.[10]

Here, then, where faith and natural philosophy based on experience contradict each other and where creation *ex nihilo* is taken, on the basis of faith, as having happened in fact, Buridan is careful to argue that *neither* faith *nor* reason has proof or demonstration of its conclusion. The natural philosophical conclusion that things cannot come into being *ex nihilo* is known only by induction, no counter instances having been experienced. On the other hand, that the world was created *ex nihilo* is believed not because it can be demonstrated, but on the authority of sacred scripture and Catholic theologians.[11]

---

to… Sed tamen non puto quod hec conclusio sit demonstrabilis, sed est declarabilis per inductionem in qua non inventa est instantia. Sic probat enim eam Aristoteles et tale reputari debet principium in scientia naturali. Aliter enim tu non posses probare quod omnis ignis est calidus et omne reubarbarum est purgativum cholere, quod omnis magnes vel adamas est attractivus ferri. Et tales inductiones non sunt demonstrationes quia non concludunt gratia forme cum non sit possibile inducere in omnibus suppositis, sicut dicitur secundo posteriorum quod multa principia indemonstrabilia fiunt nobis manifesta sensu memoria vel experientia. Experientia ex multis significationibus et memoriis deducta non est aliud quam inductio in multis singularibus per quam intellectus non videns instantia nec ratione instandi cogitur ex eius naturali inclinatione ad veritatem concedere propositionem universalem et qui non vult tales declarationes concedere in scientia naturali et morali non est dignus habere in eis magnam partem …"

[10] John Buridan, *op.cit.*, fol. 19^{ra-rb}: "Verum est quod Commentator laboravit ad demonstrandum aliter dictam conclusionem, immo laboravit contra fidem nostram et contra veritatem ad demonstrandum quod impossibile est etiam per potentiam divinam aliquid fieri sine subiecto presupposito…quamvis sit verum et firmiter credendum quod Deus entia facere et creare ex nichilo potest, id est sine subiecto presupposito in quod agat, immo primum subiectum scilicet primam materiam ipse creavit, tamen hoc non credo esse michi demonstratum vel demonstrabile per rationes ex sensatis ortum habentes. Unde credo quod rationes que ad hec fiebant in principio questionis non sunt demonstrative. Ideo videndum est quomodo eas soluisset Aristoteles et Commentator."

[11] Elsewhere in the *Quaestiones super octo Physicorum libros Aristotelis* (Paris, 1509), Buridan sometimes argues on the counterfactual hypothesis that the world is eternal.

But the problems for creation *ex nihilo* do not end with such conclusions. Further difficulties arise from the Christian assertion that God is unchanging and outside of time, together with Aristotle's argument that if there were a beginning of the cosmos, it would have to have a cause. Aristotle had argued that there cannot be a first change, because, if there were, it must have had a cause. Why, then, would the cause first not act and then, at the moment of creation, act? To change from not acting to acting, the cause itself would have to change, but then this would be a change before the purported first change, a contradiction. A cause might first not act and then act, not because of a change in itself, but because of a change in the outside world. If I love good people and hate sinners, I can change from loving to hating a person if he begins to sin. But this requires change in the outside world, and there is no outside world before creation. And God himself, according to Christian doctrine, does not change. What if God had always intended to create the world ten thousand years before now? When the time came, without changing, God might create it. The problem with this solution is that, before creation, there is no time as we know it, time being, according to Aristotle, the measure of motion and so requiring a moving body and an intellect observing that motion.

It is Buridan's task, then, in accordance with the artists' oath at inception, to resolve Aristotle's and Averroes's arguments against the possibility of a first motion in time. It is not enough to say that he believes in accordance with Christian faith that the world had a beginning. He must also explain how God could have created the world some thousands of years before the present with no change in God preceding the moment of creation. In Book IV, Question 15 – whether rest is measured by time – Buridan considered a case in which before the creation of the world God first creates one angel and then creates two more angels while the first continues to exist. Suppose the angels themselves do not move. Then time could not be told from their motion. Can the passage of time be proven by the

---

Cf. John Buridan, *op.cit.*, fol. 115[va], concerning the question whether local motion is the first motion, where Buridan says, "I intend to talk in this question as if motion were eternal, as it is clear that Aristotle held, and that all procedes according to the natural course without miracle. I, indeed, [credo enim] believe that God, before all local motion and without local motion could have created or generated something and could have augmented it and altered it with diverse alterations. For it is in God's power, and the whole world and universal order come freely from God."

existence first of only one angel and then of three? After reaching several conclusions, Buridan goes on:

> But there is still a strong doubt, because even though there is no motion, nevertheless some things, which have no quantity other than duration, seem to be measurable with respect to their duration. And since the measure of a thing with respect to its duration is nothing else than time, it seems that time would exist although there were no motion. This is shown as follows. Suppose that before the world was created, and consequently before there is any motion, God created three angels, A, B, and C, and that he created A before, then he created B and C together, and afterwards he annihilated all three together. In this case, it would be necessary to concede that angel A lasts longer than angels B and C, even in a certain ratio, because it does not last any less than if there were a coexistent motion of the heavens. And nevertheless if there were a coexistent motion it would last longer in a certain ratio, for instance twice as long if the coexisting motion were double the motion coexisting with the other two ...
>
> But if this is conceded, other great anomalies follow, because the duration of angel A is nothing else than angel A, either intrinsically or extrinsically, unless it were said that the duration was God, because we have assumed that nothing existed then except God and that angel. And it cannot be said that the greater duration of one angel and the lesser duration of the other is God, because God is always indivisible and never greater or less ... This is difficult because sense or imagination cannot comprehend such a case, but only intellect, and even intellect may not be convinced by arguments taken from the senses that the said cases are possible. But this should simply be believed. Therefore one must so to speak beg the human intellect (*mendicare oportet intellectum humanum*).[12]

---

[12] John Buridan, *op.cit.*, fol. 82^{ra-rb}: "Sed adhuc est fortis dubitatio | quia quamvis nullus esset motus tamen adhuc res alique nullam habentes quantitatem nisi durationis viderentur esse mensurabiles quantum ad suam durationem et cum mensura rei quantum ad suam durationem non sit nisi tempus videtur quod tempus esset licet non esset motus et hoc sic declaratur ponamus quod antequam esset mundus creatus et per consequens antequam esset aliquis motus Deus creasset tres angelos A, B, C, and quod creavit prius A deinde creavit simul B et C et postea illos tres simul annichilavit. Isto casu posito oporteret concedere quod angelus A plus duravit quam angelus B vel C, etiam in certa proportione quia non minus duravit quam si motus celi fuisset coexistens eis et tamen si fuisset motus coexistens eis durasset plus in certa proportione ut forte plus in duplo si motus coexistens ei fuisset duplus ad motum coexistentem aliis. Ergo etiam [A] plus duravit in duplo sed angeli B et C equaliter duraverunt non unus plus vel minus quam alter. Cum ergo non sit equale vel duplum nisi sit certa mensura qua illa sint commensurabilia vel unum commensurabile per alterum ideo licet non esset aliquis motus tamen erat certa mensura quantum ad certam suam durationem.
Sed si hoc concedatur, sequuntur alie magne inconvenientie quia duratio angeli A non erat aliud quam angelus A nec intrinsece nec extrinsece nisi diceretur quod

As an ontological minimalist, Buridan argued that if nothing existed but God and angels, then the duration of angels could be nothing else but the angels themselves, so that a greater duration would have to imply something greater about the angel to which it applied. But the angels created later might be intrinsically greater in perfection or any other quality than the angel created first. In this way, Buridan is led to making a counterfactual hypothesis: when we say that the angel created first endures longer than the later created ones, what we mean is that if there were other moving bodies coexisting by which time could be measured, then the times of duration of the angels would have ratios to each other. It is in the midst of this discussion that Buridan argues that this case is beyond the capacity of the human imagination and even intellect and that it is necessary, so to speak, to beg the intellect in order to speculate about what might then be true. Continuing the preceding passage, Buridan says:

> ... therefore one must so to speak beg the human intellect. And on this it seems to me plausible to say that the duration of angel A or angel B or C is nothing else but angel A, B, or C. And therefore, since angel A would not be greater than angel B, the duration of A would not be greater than the duration of the latter, nor would the durations of B and C be equal, since those angels would not be equal. And nevertheless it appears to me that angel A was prior to angel B, not because it was in a prior time, but because it existed when B did not exist. And it can be said that that priority was temporal in a conditional sense, because if time had coexisted with those angels as time coexists with those things that are now coexisting with time, then angel A would have been prior in time to the succeeding time in which angel B or angel C existed. And thus similarly, in the aforesaid conditional sense and from an external denomination, we might say that angel A endured longer than angel B endured and even in a certain ratio, as twice or three times as long, because, that is, angel A would have coexisted with two or three times as long a time as the time with which angel B coexisted if there were

---

illa duratio erat Deus quia ponimus quod nichil tunc erat nisi Deus et ille angelus et non potest dici quod illa duratio maior unius angeli et duratio minor alterius sit Deus, cum ipse semper sit indivisibilis et nunquam maior vel minor. Si ergo duratio angeli A non erat aliud quam angelus A et sic de angelis B et C, tamen illi angeli nec erant equales nec unus maior altero cum omnes essent indivisibiles, vel si loquamur quod unus sit maior altero, id est perfectior, tamen sic erat possibile quod angelus A erat minor, id est minus perfectus et sic non apparet quod angelus A durasset plus vel quod eius duratio fuisset maior duratione angeli B et C. Istud est difficile quia sensus vel imaginatio non cadit super talia sed solum intellectus et adhuc apud intellectum forte non potest convinci per rationes et ex sensibus deductas quod dicti casus sint possibiles, sed hoc simpliciter credendum est, ideo quasi mendicare oportet intellectum humanum.

coexisting times. And the durations of angel B and angel C would be equal in the sense that an equal or the same time would coexist with the duration of angel B and with the duration of angel C if there were time coexisting with them. In no other way could we perceive or express how much earlier or how much longer angel A endured than angel B. And therefore we express in a different way how much your father was prior to you or how much longer he lasted than you. So too we cannot say that we perceive how great the motion we call local motion is except by the relation of the mobile to something else, which however, has no effect on that motion.[13]

When he comes to the question of the timing of creation in Question 2 of Book VIII, then, Buridan again has the problem how time could elapse if there were nothing moving. Buridan begins the question:

> It is asked second whether God could make motion *de novo* such that before there was neither motion nor mutation.
>
> It is argued that he could, because otherwise he would not be omnipotent.
>
> Aristotle and the Commentator seem to intend the opposite.
>
> This question has great and special difficulty on account of the immutability of God. But it is above all to be believed with pure and firm faith that God can and that God created the world and angels *de novo* and that there was no motion or mutation beforehand, nor any other thing besides God. Nevertheless Aristotle and the Commentator seem to have had the opposite opinion and, moreover, there is an extremely

---

[13] John Buridan, *op.cit.*, fol. 82^{rb-va}: "Et de hoc videntur michi probabile dicere quod non erat aliud duratio angeli A vel angeli B aut C quam ille angelus A vel B aut C. Et ideo cum angelus A non esset maior angelo B, duratio [A] non erat maior duratione illius nec durationes angelorum B et C fuerunt equales cum illi angeli non essent equales. Et tamen apparet michi quod angelus A erat prius angelo B non quia in priori tempore sed quia erat illo non existente. Et potest dici quod illa prioritas erat secundum tempus ad sensum conditionalem scilicet quia si illis angelis coextitisset tempus sicut coexistit his qui nunc sunt tempus coexistens angelo A fuisset prius tempore in succedendo tempus quo extitisset angelo B vel angelo C. Ita similiter ad sensum predictum conditionalem ex extrinseca denominatione diceremus angelum A plus durasse quam angelus B duravit et etiam in certa proportione ut in duplo vel triplo quia, scilicet, vel in duplo vel in triplo fuisset maius tempus coexistens angelo A quam coexistens angelo B si fuissent eis tempora coexistentia et equalis fuit duratio angeli B et angeli C tali modo quia equalis vel eadem fuisset | duratio temporis coexistentis angelo B et coexistentis angelo C si fuisset eis tempus coexistens. Et aliter non possumus percipere nec exprimere quanto prius fuit vel quanto plus duravit angelus A quam angelus B. Ideo etiam aliter exprimimus quanto pater tuus fuit prior te vel quanto plus duravit quam tu. Sicut etiam motum quem vocamus localem non possumus dicere nos percipere quantus esset nisi per habitudinem mobilis ad aliquid aliud quod tamen nichil operatur ad motum illum."

strong argument supposing that God is immobile, which is proved in Book VIII of this work and in Book XII of the *Metaphysics*, and which is conceded by faith as far as his divinity is concerned.[14]

Here again, then, the second part of Buridan's vow at inception comes into play. Having given four of Averroes' arguments against the possibility of an unchanging God first not creating the world and then creating it (to which I will return shortly), Buridan goes on:

> These things are difficult on account of the imagination which cannot easily conceive that, before all created things, there was not a time in which they did not exist. In the same way the imagination can hardly conceive that beyond this world, if it is assumed to be finite, there is not another space. On account of this it seems to us according to imagination that time and succession existed from eternity. If we understand that the motions which we call local motions can exist without external spaces with respect to which the mobiles have differing relations, and one can be faster than another in a certain ratio, but that nevertheless we cannot perceive those motions or express their velocities unless we imagine a space with respect to which they have differing relations, in the same way, although God can create several angels without motion or time, one before and the other after, nevertheless we cannot grasp this sort of priority and quantity of priority unless we imagine that time and continuous succession coexist with those angels created before and after, by the imagination of which we might express the quantity of priority and posteriority ...
>
> God, before infinite motion and infinite time could have made them earlier and could have made them later, and there is nothing that God could have made or could make that God does not know perfectly, just as perfectly as if they had been made. Thus infinite times and infinite motions are known by God as if they had existed. Therefore it is not to be wondered at if God knows perfectly how much longer angel A existed before angel B existed. For God knows how great the intermediate time between them would have been. Further, since from eternity God perfectly knows all the moments and instants which in infinite motion and time would have been if infinite had been time and motion [the word order here indicates a potential or syncategorematic infinite], and

---

[14] John Buridan, *op.cit.*, fols. 109vb–110ra: "Queritur secundo utrum deus potuit facere motum de novo quod ante nullus fuisset motus necque mutatio. Arguitur quod sic quia aliter non fuisset omnipotens. Oppositum videtur intendere Aristoteles et Commentator. Ista questio habet specialem et nimiam difficultatem propter Dei immutabilitatem. Sed omnino credendum est fide pura et firma quod sic et quod Deus de novo creavit mundum et angelos et quod nullus erat motus ante neque mutatio neque omnino alia res preter ipsum Deum. Sed tamen Aristoteles et Commentator visi sunt habuisse opinionem contrariam et adhuc fuit argumentum fortissimum supponendo quod Deus est immobilis quod probatur in octavo huius et in duodecimo Metaphisice et conceditur ex fide quantum ad ipsam divinitatem."

'IDEO QUASI MENDICARE OPORTET INTELLECTUM HUMANUM' 233

God knows how great the interval would have been between these sorts of moments and instants, it is no wonder if, from eternity, God's will could ordain that *then* the world would be created and *then* or *then* such an angel, and not earlier nor later, in such a way as we understand the words "then" and "then" to have supposition for determinate instants which would be in infinite time if infinite time existed. And I do not say that God understands time by means of the angel or the angel by means of time – indeed God understands everything by his most simple essence. And thus it is necessary for us to speak with circumlocution because we cannot understand things the way God does, but only one thing by means of another.[15]

When we think of the priority of possible angels one to the other before the creation of the world, we can only do it by thinking hypothetically or counterfactually of the relations that their durations would have to one another if there were motion and time passing,

---

[15] John Buridan, *op.cit.*, fol. 110rb–va: "Ista sunt difficilia propter imaginationem quem non bene potest capere quin ante omne factum fuerit tempus in quo ipsum non erat. Sic non potest bene capere quin ultra istum mundum, si ponatur [in]finitum, sit aliud spacium. Propter quod omnino nobis videtur secundum imaginationem quod ab eterno fuit tempus et successio quedam. Sed tamen oportet intelligere quod sint [*sic*, sicut?] motus quod motus vocamus locales possunt esse sine spaciis extrinsecis ad quem mobilia aliter et aliter se habeant et possent esse unus velocior et alter tardior etiam in certa proportione, tamen non possemus illos motus percipere nec eorum velocitates exprimere nisi imaginando spacium ad quod illa mobilia aliter et aliter se haberent, ita etiam quamvis Deus sine motu vel tempore posset creare plures angelos unum prius et alterum posterius, tamen huiusmodi prioritatem et quantitatem prioritatis non bene possumus capere vel exprimere nisi imaginando quod in illis sic prius et posterius factis coexistat tempus et successio continua per cuius imaginationem quantitatem experimeremus prioritatem et posterioritatem et quantitatem prioritatis in factione illorum. In tanto enim secundum durationem angelus A fuit prior angelo B quantum fuit tempus intermedium in[s]ter instancia creationum eorum si fuisset tempus. Deus ante infinitum motum et infinitum tempus potuisset fecisse a parte ante et posset facere a parte post et nichil unquam potuit vel potest facere quod ipse non cognoscat perfectissime et ita perfecte sicut si illa essent facta, ideo infinitum motum et infinitum tempus cognoscit ac si fuissent, et ideo non est mirandum si cognoscat perfecta quantum secundum durationem angelus A fuit prior angelo B. Ipse enim cognoscet quantum fuisset tempus intermedium inter creationes eorum et etiam cum ab eterno Deus perfecte cognoverit omnia momenta et instantia que in infinito motu vel tempore fuissent si infinita fuissent motus et tempus et cognoscit quanta inter huiusmodi momenta et instantia fuisset distantia, non est mirandum si ab eterno voluntas eius potuit ordinare quod tunc crearetur mundus et tunc vel tunc talis vel talis angelus et non prius nec posterius, ita quod intelligimus illas dictiones tunc et tunc supponere pro determinatis instantibus que in infinito tempore fuissent si infinitum tempus fuisset. Et non dico quod Deus intelligat tempus per angelum vel angelum per tempus. Immo omnia per suam essentiam simplicissimam intelligit. Et sic oportet nos loqui et circunloqui predicta quia non sic possumus res intelligere nisi unam per aliam et mediante alia."

as there are *not* in the case posited. But God, according to Buridan, knows all the instants that might exist in infinite time if there were motion going on in any possible world, and hence God distinguishes, in his simple essence, the moment of time in which God wants to and does create the world.

Thus Buridan resolves Aristotle's argument that before any purported first motion there would be another motion by which the agent causing the first motion changed from not causing the motion to causing it. He does it by making use of theological doctrines concerning God's knowledge of possible worlds. Without any change or motion, God knows all the possible instants that might exist within infinite possible time and creates the world according to a pre-existing plan – ten thousand years before the present, or twenty thousand years before the present, as the case may be. Although God never, so to speak, changes his mind and God always willed to create the world when God did create it, it was nevertheless a free choice, and God could have chosen to create it at a different time.[16] In the end, then, Buridan rejects Aristotle's and Averroes's arguments against the possibility of a first motion by asserting that no prior change is required for God to move from not creating the world to creating it. God has such absolute freedom of action that God can begin to act without any preceding internal change.[17] Thus for Buridan it is God's atemporal knowledge of possible worlds that explains how there can be a beginning in time without a change in the cause (God) before a change in the effect (the world). For humans, this can only be construed by begging or petitioning the intellect or by introducing an additional hypothesis, supposing, counterfactually, that there is some motion by which the passage of time could be measured. God does not need additional hypotheses: all is equally immediate to God.

Before coming to the conclusion that from eternity God wanted to create the world when he did create it, Buridan had listed four of Averroes's arguments against even the possibility of the beginning of the world and motion in time. First, if God had from eternity wanted to create the world, but did not do it until a certain time, there must have been something impeding God's action earlier and then this impediment must have been removed.[18] Second when it is

---

[16] John Buridan, *op.cit.*, fol. 110va.
[17] Cf. John Buridan, *op.cit.*, fol. 110va.
[18] John Buridan, *op.cit.*, fol. 110ra: "Contra istam solutionem obicit Commentator quia volens ad productionem alicuius non postponit volitum nisi propter aliquod impedimentum vel propter aliquam spositionem seu defectum alicuius dispositionis

said that God wanted to create the world "then," how is "then" to be understood?[19] If things are imagined in such a way, how is the creation of time to be understood?[20] Since God is omnipotent, there would always be a sufficient cause for the beginning of the world and so creation of the world should occur immediately and not be postponed.[21] Replying to Averroes's arguments after his own conclusion, Buridan says that even humans, because of free will, can begin to act without any preceding change. All the more so God, whose freedom far exceeds human freedom. The "then" in which God wanted to create the world is to be understood as the beginning of the past time that extends up to the present, for example the "then" will be twenty thousand years before the present. Or we could understand "then" to refer to the instants or times that could have coexisted with the events in question.[22] Thus Buridan resolved Averroes's arguments in this case most importantly by calling upon the special characteristics of God – God's absolute liberty or free will – that invalidate the arguments.

In other questions of Book VIII, Buridan again faces problems that what is firmly believed to be true by faith contradicts what Aristotle tried to prove, and he again solves the problems by referring to God's special characteristics. For instance, Aristotle tried to prove that motion was eternal in the past, but we know by faith that the world had a beginning in time. Not all of Aristotle's and Averroes's conclusions have to be rejected, Buridan claims. He begins his determination of Question 3 of Book VIII:

> On this question there are some conclusions which are probable to me, which Aristotle and the Commentator have conceded, and which also need not be negated according to our faith.[23]

---

volentis ad productionem illius voliti vel saltem quia expectat tempus et quocunque horum modorum hoc fuerit adhuc ante illius voliti productionem oportet esse aliam mutationem secundum quam auferetur impedimentum...ideo sic quod semper ante primum motum esset alia mutatio."

[19] *Ibid.*: "... quero quid tu vis intelligere per tunc et ante ..."

[20] John Buridan, *op.cit.*, fol. 110rb: "Item Commentator reiterat questionem de factura temporis vel illius tunc et videtur sequi processus in infinitum et sit ficta quedam locutio, scilicet quod Deus ab eterno voluit facere illud tunc et non ab eterno sed tunc et sic fecit tunc in tunc et ita pari ratione in infinitum."

[21] *Ibid.*: "Item positis causis omnino sufficientibus ad alicuius effectus productionem debet statim sequi illius productio, sed omnium effectuum a Deo immediate producibilium ad extra causa sufficiens est actio Dei immanens si non alia cum causa requiratur et ab eterno est actio Dei immanens quia non est aliud quam Deus ..."

[22] John Buridan, *op.cit.*, fol. 110^{va-vb}.

[23] John Buridan, *op.cit.*, fol. 111rb: "De ista questione sunt alique conclusiones

The first conclusion is that there never has been, nor is, nor will there be a perpetual or actually infinite motion. The second conclusion is that "Eternal or perpetual motion can be," understood in the potential or syncategorematic sense and referring only to the future.[24] In the end, however, having reviewed a number of Aristotle's arguments, Buridan says

> ... however, although Aristotle said this, we on the basis of pure faith and not having an evident demonstration arising from sense, ought to hold the opposite, namely that the motion before this day was neither eternal nor infinite, because before this day no greater motion existed, for God created the world anew, while there neither is nor was earlier time or motion. Positing and believing this on the basis for faith, the arguments for the other side are resolved ...[25]

As was the case in Question 2, so here Aristotle's arguments are resolved using the special characteristics of God and creation.[26] The prime mobile was not made by motion, but simply created, which does not differ from the mobile itself.[27] When it is asked why the creation did not occur earlier, Buridan replies that there is no reason except that the will of God wanted to create it at that moment.[28] There is a different relation between God and the world after creation than before simply because after creation the world exists, but this does not mean that there had to be a change before creation.[29] Buridan

---

michi probabiles quas concessissent Aristoteles et Commentator et quas etiam non oportet negare secundum fidem nostram."

[24] John Buridan, *op.cit.*, fol. 111^{rb-va}: "Prima est quod nullus est vel fuit vel erit motus perpetuus seu infinitus secundum durationem et nullum est vel fuit vel erit tempus perpetuum seu infinitum...hoc probabile est et verum... Secunda conclusio est quod eternus sive infinitus potest esse motus et sic de tempore saltem a parte post et non curo dicere nunc a parte ante propter hoc quod dicitur non esse potentiam ad preteritum...et secundum fidei veritatem non potest esse tantum quin possit esse maius et sic de motu ..."

[25] John Buridan, *op.cit.*, fol. 111vb: "... sed tamen quamvis Aristoteles ita dixerit nos fide pura et non per demonstrationem habentem ex sensibus ortum et evidentiam debemus tenere oppositum, scilicet quod nec eternus nec infinitus fuit motus finitus ante hanc diem, quia ante hanc diem nullus fuit maior. Deus enim creavit noviter mundum cum ante nullus esset vel fuisset mundus aut tempus aut motus. Et hoc ex fide posito et credito solvende sunt rationes ad oppositum ..."

[26] John Buridan, *op.cit.*, fols. 111vb–112ra.

[27] John Buridan, *op.cit.*, fol. 111vb: "... dicendum est quod factum fuit primum mobile non per motum sed per simplicem creationem que non differebat ab illo mobili facto et sic non erat proprie dicta mutatio ..."

[28] *Ibid.*: "... dicendum est quod in hoc nulla erat causa nisi voluntas divina que ab eterno voluit creare tunc et non ante ..."

[29] John Buridan, *op.cit.*, fol. 112ra: "... concedo quod non est possibile aliqua se habere adinvicem aliter et aliter sine aliqua mutatione vel aliqua novitate. Si ergo

replies similarly to ten arguments in all, defending the possibility that the world had a beginning.

To the question whether the prime mover, that is God, is of infinite power (Book VIII, Question 11), Buridan says that Aristotle and the Commentator say no because we see no effects by means of which, from sense, we could argue that there must be an infinite power.[30] Aristotle and the Commentator notwithstanding, Buridan believes by pure faith that God is of infinite power, but he doesn't know how to demonstrate it on the basis of principles evident on the basis of sense.[31]

In sum, comparing Buridan to Boethius of Dacia, it is clear that concerning the creation of the world, Buridan introduces far more theology into his questions on the *Physics* than Boethius of Dacia did. He does it where there are issues that involve both physics and theology. As the oath to be taken by inceptors commands, Buridan determines questions in accordance with faith, and he resolves the arguments for the other side accordingly as he believes they should be resolved. Whereas Boethius of Dacia had argued that a philosopher should conclude that the world is eternal because he argues on the basis of physical principles, Buridan accepts the creation of the world in physics as in theology, resolving Aristotle's and Averroes's arguments against the possibility of a new creation. For arguing in this way, Buridan might be criticized by upholders of the Aristotelian conception of the separate sciences (with its prohibition of *metabasis*), because he does insert into his questions on the *Physics* conclusions that do not follow from the principles of physics, such as the conclusion that God is of infinite power. But Buridan's theological superiors could not consistently fault him for his introduction here of theological matters: he is following the oath of inceptors in Arts to resolve arguments against faith as related to issues that involve both natural philosophy and theology. Most prominently, Buridan's resolutions make use of theological premises, such as the assumption of God's absolute freedom of choice. At the same time, however, Buridan uses additional premises of a philosophical origin, begging the human intellect, to conclude, for instance, that the ratios of duration before creation could be measured by the ratios of motions that would have occurred had there been motions coexisting, as there were not in fact.

---

dicamus quod Deus ad primum mobile se habeat aliqualiter quando fiebat vel quando factum erat qualiter non se habebat ad ipsum, hoc potest concedi ..."

[30] John Buridan, *op.cit.*, fol. 119vb.
[31] John Buridan, *op.cit.*, fol. 120ra.

2. *Questions on Vacua*

While questions of the eternity or the creation of the world obviously involved both faith and physics, questions on the vacuum did not so obviously involve theology. Yet Buridan believed they did, because of the assertion that it is possible for God to do or make anything that does not involve a self-contradiction. Buridan's determinations of questions concerning vacua, indeed, involve both theology and natural philosophy. Might Buridan have been more justly criticized by the theologians for his use of theology in questions concerning vacua? And, further, does Buridan's insertion of theology into questions on the *Physics* serve in the case of vacua to move physics in a fruitful direction, as Pierre Duhem claimed, or did it divert Buridan to the study of imaginary cases with no relevance to the real world?

The question that Buridan was debating when he made the statement cited at the start of this paper was whether it is possible that a vacuum exist by some power (*utrum possibile est vacuum esse per aliquam potentiam*) – this following the question whether it is possible that a vacuum exist (*utrum possibile est vacuum esse*), which he took to ask whether it is possible naturally. In the earlier question, Buridan had distinguished the senses of possible – whether something is possible naturally or supernaturally. In asking whether a vacuum is possible naturally, Buridan had argued that we may conclude that a vacuum is not possible naturally because in all the cases we have observed a vacuum has never occurred, proof by induction being sufficient in physics.[32]

In asking whether a vacuum is possible by some (that is, by a supernatural) power, the principal arguments for the opposing side were that it is not, because nothing is possible by any power which implies a contradiction or from which follows a contradiction, but from the existence of a vacuum contradictories follow. If everything were annihilated inside the sphere of the moon, it was argued, then the sides of the sphere would be next to each other because nothing would be between them, but they would not be next to each other

---

[32] John Buridan, *op.cit.*, fol. 73[va]: "Item omnis propositio universalis in scientia naturali debet concedi tanquam principium que potest probari per experimentalem inductionem, sic quod in pluris singularibus ipsius manifeste inveniatur ita esse et in nullo nunquam apparet instantia sicut enim bene dicit Aristoteles quod oportet multa principia esse accepta et scita sensu memoria et experientia. Immo aliquando non potuimus scire quod omnis ignis est calidus, sed per talem inductionem experimentalem apparet nobis quod nullus locus est vacuus ..."

because this is incompatible with the figure of a sphere.[33] Having answered the question in the affirmative, that a vacuum is possible by a supernatural power, Buridan asserted that this principal argument had been answered in Book III.

There, in Question 15 of Book III – whether there is some infinite magnitude–, Buridan concluded that if the contents of the sphere of the moon were annihilated, then the sphere would be void (*illud concavum esset vacuum*), the adjective "void" having supposition for the sphere itself, not for any supposed space inside it, because nothing would be inside it. As is apparent, in so answering Buridan used logic or supposition theory to show that if God annihilated everything inside the sphere of the moon there would be no self-contradiction if the propositions describing the situation were carefully formulated and interpreted.

These concerns were connected to the question concerning the possible existence of an infinite magnitude, because Aristotle had argued that if there were space outside the cosmos, it would be infinite, there being no rational grounds to deny its infinity if its existence were supposed at all. But, since an infinite space is self-contradictory, so Aristotle argued, therefore there is no space outside the cosmos. Buridan argued, to the contrary, that we must hold as a matter of faith that God could create space outside the cosmos. God could create a finite space outside the cosmos, he argues: creating a finite space does not compel the creation of an infinite space, contra Aristotle. It is true that Aristotle's argument against space outside the cosmos holds naturally, Buridan admits. And theology itself has little or nothing to say about space outside the cosmos – save perhaps for the passage in Genesis about the waters above the firmament. So why, then, does faith demand that one accept that God could create space outside the cosmos? Because one should not deny to God the possible creation of anything that does not involve a logical contradiction, Buridan replies.

Beginning his determination of the question, Buridan asserts that it must be believed by faith that God could create other spheres and worlds outside this one in any ratio of finite to finite.[34] It is

---

[33] John Buridan, *op.cit.*, fol. 73[vb].
[34] John Buridan, *op.cit.*, fol. 57[va]: "Primo igitur credendum est fide quod deus posset ultra istum mundum formare et creare alias speras et alios mundos et omnino alias magnitudines finitas quantascunque vellet, ita quod omni finita creata posset creare maiorem in duplo in decuplo in centuplo et sic de omni alia proportione finite ad finitum."

not necessary to conclude that God could create an actually infinite magnitude – if he created something actually infinite, it would be equal to God in power, which is not to be believed.[35] Having stated what seems reasonable to him, Buridan hastens to add that he leaves all this to the determination of the theologians,[36] but then he immediately forges on to additional tentative conclusions:

> Although it cannot be demonstrated that outside the world there is no space and magnitude, because God could make space and magnitude there, nevertheless I am of the opinion that there is no space or magnitude there, nor another world. And Aristotle adduces natural arguments for this in Book I of *On the Heavens*, which are to be dealt with there, and so I only offer the persuasion here that it is not plausible that God makes another world or worlds there, because if God had wanted to make more worldly creatures than he made, it was not necessary to make other worlds, because he could have made this world twice or a hundred times larger. And if God did not make there another world or worlds, there is no apparent reason why God would have made space there, because it would serve no purpose beyond this world and would seem to be useless ... if there were such a space there, it would have no profit either for saving the appearances or for avoiding difficulties. Such a thing should not be posited to exist unless it follows from the words of sacred scripture.[37]

Besides, Buridan argues, it should be believed on faith – and he believes – that God could annihilate everything inside the sphere of the moon. But the same arguments are used against space outside the cosmos as are used of the space that would result if God annihilated everything inside the sphere of the moon. If the arguments are not valid for the situation inside the sphere of the moon, neither are they valid outside the cosmos.

---

[35] *Ibid.*: "... sed forte non oportet credere quod Deus posset creare magnitudinem actu infinitam, quia illa creata non posset creare maiorem. Repugnat enim quod actu infinito sit aliud maius. Et tamen inconveniens est quod Deus possit facere creaturam potentie sue proportionatam, sic quod non possit maiorem et perfectiorem facere. Et michi videtur simile de magnitudine sicut et perfectione. Quamvis enim omni creatura formata et formabili Deus posset formare sive creare perfectiorem, tamen non posset creare aliquam infinite perfectionis. Illa enim esset eque perfecta vel non minus perfecta quam Deus. Nam si esset minus perfecta ipsa non esset infinite perfectionis, sed posset creari perfectior et ita videtur mihi de magnitudine et velocitate et intensione caliditatis et huiusmodi ..."
[36] John Buridan, *op.cit.*, fol. 57[va–vb]: "... tamen de isto, et de pluribus, immo de omnibus que dicam in ista questione ego dimitto determinationem dominis theologis et acquiescere volo determinationi eorum."
[37] John Buridan, *op.cit.*, fol. 57[vb].

How then can Buridan answer the critics who say that to suppose God annihilates everything inside the sphere of the moon leads to contradictions? If this happened, Buridan says, there would be nothing inside the sphere of the moon, nothing for which any noun could have supposition. What would exist would be the empty sphere of the moon and what was outside of that. Inside, there might either be nothing or, supernaturally, a disembodied extension. Take the case in which there is no extension. If one asks what would happen if a stone were moved within this emptiness, Buridan answers that it is hard to answer because we have difficulty not imagining that there is space within the emptiness, just as it appears that the sun is not larger than a horse and that the sun is much smaller than the earth. In such a case the imagination has to be corrected by the intellect.

So, he says, since we are not talking of what is naturally possible, but of what is miraculously possible, it is possible that within the place of a grain of millet or within its magnitude, God could create a very big body, even one that is greater than the world – it being understood that the big body is not there circumscriptively. This is to be believed just as it is believed that Christ is within the small quantity of the Eucharist, and that Christ is just as large there as he was at the last supper or as he is now in paradise, and with the same shape. Within a small space, say of ten feet, God could move a stone or a large body very fast in a straight line for a whole year and yet that body would not exit from the space, nor even get farther from one side or closer to the other.[38] Having thus reasoned

---

[38] John Buridan, *op.cit.*, fols. 57vb–58ra: "Sed tunc queritur quid esset de distantia laterum polorum orbis lune ad invicem et quid esset de motu lapidis in approprinquando vel elongando ad polum. Dico quod in hoc est difficile satisfacere ymaginationi quia semper apparet imaginationi quod ibi esset spacium, sicut semper sensui apparet quod sol non sit maior equo et quod sit valde minor terra, tamen in talibus intellectus debet corrigere illas apparentias sensus et ymaginationes. Dico ergo quia non loquimur de talibus naturaliter possibilibus sed miraculose possibilibus quod in uno parvo spacio et loco ut in loco grani milii vel sub eius magnitudine Deus posset formare valde magnum corpus, scilicet maius quam sit mundus et verum est quod illud corpus non esset in illo parvo corpore vel loco circumscriptive et commensurabiliter. Hoc credendum est quia | sub parva quantitate hostie et in eius parvo loco est corpus Christi ita magnum sicut erat in cena et sicut est in paradiso et ita figuratum, immo in qualibet parte quantitativa hostie quantuncumque parva est totum corpus Christi magnum et optime figuratum. Sed hec magnitudo corporis Christi non se habet in hostia modo commensurabili ad magnitudinem hostie. Et non minus posset Deus facere maius corpus in loco et cum magnitudine grani milii propter quod etiam concludo quod intra convacuum orbis lune posset Deus facere maius corpus in centuplo quam sit mundus non mutata magnitudine et figura orbis lune. Sed non esset ibi circumscriptive nec modo mensurabili ad magnitudiem orbis. Immo si esset

about what is possible miraculously, Buridan again submitted the matter to the final word of the holy church and the doctors of sacred scripture.[39]

Yet again, he forged ahead. If there were nothing in the world but three stones in a row, he says, and if God annihilated the middle stone, then the remaining two stones would neither touch nor be distant from one another. Where the middle stone had once been, God could create a stone a hundred times as large and make the two end stones a hundred times further apart than at the start without moving them. All this is true but hard to understand, he concludes, because it cannot happen naturally and because it is beyond our imagination.[40]

---

ibi corpus per modum circunscriptivum et mensurabilem ad magnitudinem orbis, illud non posset esse maius quam nunc est, quia oporteret eius diametrum esse tertiam partem vel circiter linee circularis protracte in concavo orbis lune. Postea etiam dico quod in parvo spacio ut in spacio decem pedum Deus posset movere unum lapidem vel unum corpus valde magnum per unum totum annum continue valde velociter motu recto et tamen ille lapis non exiret ab illo spacio quiescente immo nec aliqui cono illius spacii approximaretur nec ab aliquo cono elongaretur. Hoc declaratur quia si Deus moveat manum de capite ipsius ad pedes, erit longus motus quantum corpus Christe est longum, et tamen nullo cono hostie fiet illa manus proximior quam ante vel remotior, quia non movetur manus secundum comparationem vel situm ad mangitudinem hostie, sed secundum situm et comparationem ad magnitudinem corporis Christi."

[39] John Buridan, *op.cit.*, fol. 58ra: "... et non assero hec omnia sed in asserendo hec vel aliqua eorum aut non asserendo, submitto me totaliter decreto et ordinationi sancte ecclesie et doctorum scripture sacre."

[40] John Buridan, *op.cit.*, fol. 58^{ra-rb}: "... videtur michi quod si essent tres lapides pedales consequenter se habentes secundum rectitudinem et tangentes se et nichil esset plus in mundo et Deus annichilaret lapidem medium non approximando vel coniungendo adinvicem lapides extremos, illi lapides remanentes nec tangerent se invicem nec essent longe ab invicem nec prope. Sicut etiam si Deus annichilaret magnitudinem lapidis substantia eius remanente lapis adhuc haberet partem aliam a parte sed non haberet partem extra partem situaliter, lapides ergo illi non distarent quia nichil esset medium per quod distarent et tamen necesse est quod per spacium medium sit et mensuretur distantia distantium nec tamen oporteret eos esse contiguos quamvis nichil esset inter eos, sicut nec poli orbis lune tangerent se quamvis nichil esset inter eos secundum rectitudinem. Et ita concluditur quod plus requiritur ad hoc quod duo corpora tangant se quam quod inter ea nichil sit, et hoc forte quod plus requiritur est quod se habeant adinvicem communsurabiliter et secundum determinatos situs partium alterius. Unde nec proprie magnitudo corporis Christi tangit mangitudinem hostie et illi lapides non sic se haberent adinvicem. Sed tunc queritur utrum inter illos lapides absque motu eorum posset Deus iterum creare lapidem maiorem quam erat lapis annichilatus qui erat intermedius. Et ego credo quod sic, scilicet lapis in centuplo maior et faciens magis distare illos lapides in centuplo quam distarent per lapidem annichilatum antequam annichilaretur. Nec mirum quia sine motu corporis B et sine distantia ipsius A, scilicet, posset Deus facere

Thus, in supporting Aristotle's argument that there is no infinite magnitude, answering Question 15 of Book III in the negative, Buridan nevertheless allowed that God could annihilate everything inside the sphere of the moon and that God could, though it is unlikely that God did, create a finite space outside the cosmos. In supposing that the sphere of the moon could be evacuated, Buridan agrees with Aristotle that, if this were to occur, there would be no extension there. Space in the sense of an empty three-dimensional extension is, he believed, naturally impossible. But it is not supernaturally impossible. Just as in the Eucharist there is extension of the appearance of bread with no thing that is extended, so God could, supernaturally, create a disembodied extension, even one without any coextensive quality. If there were scriptural grounds to assume that God did so, as there are scriptural grounds to assume that Christ entered a room through a closed door and hence that two bodies could be in the same place, Buridan would have no trouble accepting this hypothesis.

In summary, adding the theological truth that God is omnipotent to Aristotle's discussion of the possibility of a vacuum, Buridan upheld Aristotle as far as the natural world is concerned, allowed that God could possibly create an empty three dimensional space, but decided that God probably has not done so. In the course of his deliberation, Buridan inserted speculations about what might be the case miraculously if God did annihilate everything within the sphere of the moon. Drawing upon the theological doctrine that Christ is in place definitively and not circumscriptively within the Eucharist – so that this mode of being in place is possible – Buridan speculated that God could insert large bodies and even worlds into evacuated places, given that these new bodies or worlds could be in place definitively. Buridan thus imagined a supernaturally possible world quite different from our own.

Or was it so different? What is the nature of the real world? And what is the basis of our knowledge of it? Should we assume that physics in its current form is true? At the end of Question 15 of Book III, whether there is some infinite magnitude, Buridan said:

> … the preceding things I do not say, nor have I said assertively, but only disputatively, putting forward doubts so that I might be taught by others the truth concerning these things.[41]

---

quod ipsum esset in diversis locis ab invicem distantibus, sed talia non possunt fieri naturaliter, nec imaginatio cadit super ea, ideo difficiliora ad intelligendum sunt."

[41] John Buridan, *op.cit.*, fol. 58rb: "Protestor ut prius quod predicta non dico nec

There are many possible relations between natural philosophy and theology as disciplines making knowledge claims. Following Pierre Duhem, it has been supposed that the Condemnation of 1277 promoted arguments *secundum imaginationem* about what might happen if *de potentia Dei absoluta* God brought about a vacuum. These *secundum imaginationem* arguments are supposed to have helped lead to modern ideas of inertia, as a body supposed to be set in motion in a vacuum might be supposed to continue on indefinitely, with no outside force acting on it. Duhem's historical narrative about the importance of the Condemnations of 1277, I might point out, assumes that Newtonian physics represented progress over Aristotelian physics, so that it was a good thing to promote the discovery of inertia.

In the passages I have just considered, on the other hand, Buridan does not use theological reasoning in a direction leading in any obvious way to Newton. Buridan's initial assumption, like Aristotle's, is that in pure emptiness there would be no extension, no three dimensions. When one begs the human intellect, however, quite different conceptions may come to mind. In his discussions of vacua, as in his discussion of the creation of angels before the creation of the world and motion, we see Buridan bypassing the intellectual route leading to Newtonian absolute time and space to pursue issues that arose later when Newtonian physics was questioned, leading to the development of relativity.

I conclude then, that in Buridan's questions on the *Physics*, theological doctrines pushed Buridan to beg the human intellect, going beyond the accepted truths of natural philosophy to conditional or counterfactual hypotheses and to speculations about possible worlds. I suppose that, for Buridan, there was greater parity between the claims of theology and of physics than is common today. If one assumed, as Buridan seems to have done, that the knowledge claims of physics and of theology both deserve respect, that they cannot ultimately contradict each other, but that they are distinct disciplines or sciences with distinct principles and methods, then the areas of overlap between theology and physics are subject to negotiation "begging the human intellect."

Buridan does not exclude theology from physics, along the lines of Boethius of Dacia, nor does he overwhelm physics with theology, along the lines of today's Creationists. Rather, in a moderate way, Buridan introduces theological truths into the body of Aris-

---

dixi assertive sed disputative movendo dubitationes ut de his docear ab aliis veritatem, etc."

totelian physics and then shows, plausibly, that to draw inferences from physics plus theology, it is necessary to add other hypotheses – to beg the human intellect. In the cases considered here, in saying that to perceive or measure local motion outside the cosmos we have to suppose a reference frame, or that to speak of ratios of duration before creation we have to suppose that something coexisted that could serve as a clock, Buridan was grappling with issues worthy of consideration by physicists many centuries later. When we reserve judgment about the value or blame to be attached to mixing theological matters into natural philosophy, we can see that, in Buridan's case, it had consequences worthy of our interest and attention.

ARISTOTELIAN METAPHYSICS AND EUCHARISTIC THEOLOGY: JOHN BURIDAN AND MARSILIUS OF INGHEN ON THE ONTOLOGICAL STATUS OF ACCIDENTAL BEING[*]

Paul J. J. M. Bakker

1. *Introduction*

In his *Questions on Aristotle's Metaphysics,* John Buridan investigates repeatedly and from several different angles the problem of the ontological status of accidental being. In most cases, he recognizes the existence of a fundamental incompatibility between the position of Aristotle and the doctrine of faith. In this article, I shall examine the way in which Buridan tries to resolve this apparent tension between metaphysics and the faith, and the way in which it may have affected his own position with respect to the problem of accidental being. An interesting contrast to Buridan's theory is provided by Marsilius of Inghen, who is commonly reckoned among the exponents of the school of Buridan. In his commentary on the *Metaphysics,* he too takes up the question of accidental being and its ramifications for the Christian faith.

The methodological background of Buridan's approach to faith is provided by a statute that was issued on April 1, 1272 at the Arts Faculty of the University of Paris. On the penalty of being excluded from the Faculty community, this statute prescribed the following three methodological principles. First, that no master or bachelor of arts shall presume to determine or even to dispute any purely theological question, since this would imply a transgression of the limits assigned to him. Second, that whenever a master or bachelor of arts happens to dispute a question that appears to touch on both faith and philosophy, he shall not determine it contrary to the faith. Third, that whenever a master or bachelor of arts happens to read or to dispute any difficult passage or question that seems to undermine the faith, he shall either refute the arguments or the text insofar

---

[*] The writing of this paper was made possible through financial support from the Netherlands Organization for Scientific Research (NWO), grant 200-22-295.

as they are against the faith, or concede that they are absolutely false.[1] This statute from the 1370s remained in force throughout the fourteenth and fifteenth centuries by means of a vow, which every candidate had to profess at the moment of his inception in the Arts Faculty. For this reason, it is of major importance for the study of the relation between philosophy and theology in the later Middle Ages.[2] Now, in a well known and frequently quoted passage from his

---

[1] Denifle e.a. (eds.), *Chartularium Universitatis Parisiensis*, vol. I: n. 441, p. 499: "... statuimus et ordinamus quod nullus magister vel bachellarius nostre facultatis aliquam questionem pure theologicam, utpote de Trinitate et Incarnatione sicque de consimilibus omnibus, determinare seu etiam disputare presumat, tanquam sibi determinatos limites transgrediens, ... Statuimus insuper et ordinamus quod si questionem aliquam, que fidem videatur attingere simulque philosophiam, alicubi disputaverit Parisius, si illam contra fidem determinaverit, ex tunc ab eadem nostra societate tanquam hereticus perpetuo sit privatus, ... Superaddentes iterum quod si magister vel bachellarius aliquis nostre facultatis passus aliquos difficiles vel aliquas questiones legat vel disputet, que fidem videantur dissolvere, aliquatenus videatur; rationes autem seu textum, si que contra fidem, dissolvat vel etiam falsas simpliciter et erroneas totaliter esse concedat ..." An English translation of the entire statute is to be found in Grant, *A Source Book*, pp. 44–45. The traditional interpretation of this statute, which was first developped by Mandonnet, *Siger de Brabant et l'averroïsme latin*, vol. I: pp. 198–202, and then diffused by Van Steenberghen, *Maître Siger de Brabant*, pp. 83–84, has recently been revised by Putallaz e.a., *Profession: philosophe. Siger de Brabant*, pp. 128–134, on the one hand, and by Bianchi, *Censure et liberté intellectuelle*, pp. 165–201, on the other.

[2] Denifle e.a. (eds.), *Chartularium Universitatis Parisiensis*, vol. I: n. 501, p. 587: "Item nullam questionem pure theologicam disputabitis quamdiu rexeritis in artibus, utpote de Trinitate et Incarnatione. Item si contingat vos determinare aliquam questionem, que tangat fidem et philosophiam, eam pro fide determinabitis, et rationes contra fidem dissolvetis secundum quod vobis dissolvende videbuntur." At the beginning of the fifteenth century, Jean Gerson explicitly refers to this vow: Jean Gerson, *Trilogium astrologiae theologizatae* [Glorieux], p. 107: "Falluntur in hoc plures astronomorum et simplicium christianorum vel aliorum non imbutorum litteris sacris, qui protinus ut legunt vel audiunt aliqua de libris talium, credunt et assentiunt, cum tamen reperiantur illic errores contrarii fidei. Nec mirum, cum etiam in libris Aristotelis et Avicennae et Averrois et aliorum qui quotidie leguntur simile inveniatur. Sed (sec *ed.*) cautissima religiositate provisum est in praeclara artium facultate almae Parisiensis Universitatis, tam per articulos Parisienses quam per statutum jurejurando firmatum per quemlibet qui licentiatur in artibus, ubi sententialiter sic habetur: 'jurabitis quod dum contingetvos determinare aliquam quaestionem de philosophia, illam semper pro parte fidei determinabitis et responsiones Philosophi in oppositum factas dissolvetis'." At the beginning of the sixteenth century, the vow seems to have become obsolete, as is shown by the following passage from Ludovicus Coronel: "Quarto, ego minus sufficiens et indignus non memini, quando fui ad gradum in artibus promotus, tale iuramentum fecisse, nec intellexisse aliquem sociorum meorum ita iurasse." (Luis Coronel, *Physice perscrutationes* (Lyon, 1530), fol. xcii[vb]) Gerson's passage was pointed out by Pluta, "Einige Bemerkungen," p. 110, n. 7. Coronel's quotation was translated into English by Grant, *A Source Book*, pp. 51–52.

Questions on Aristotle's Physics, Buridan explicitly appeals to this vow in order to justify his having mixed certain theological considerations into his commentary. According to him, given the case of a question touching both faith and physics, any master of arts must at least give the position of theology in order to be able to determine the question in accordance with faith, as the statute requires.[3] Against this background, then, I will first outline the way in which Buridan resolves the tension between philosophy and theology raised by the problem of accidental being. Second, I will investigate the way in which Buridan's solution was evaluated by Marsilius of Inghen.

2. *Buridan's Theory of Accidental Being*

The problem of the ontological status of accidental being falls into the category of questions touching both theology and philosophy. On the philosophical side, the problem immediately relates to the Aristotelian doctrine of the categories. This doctrine, as far as it concerns us at present, may be summarized in the following dictum: *accidentis esse est inesse,* that is, the being of an accident is nothing but its inherence in a substance.[4] On the theological side, the problem of the ontological status of accidental being relates to the Eucharist,

---

[3] John Buridan, *Quaestiones super octo Physicorum libros Aristotelis* (Paris, 1509), Book IV, q. 8, fols. 73vb–74ra: "Et ideo aliqui dominorum et magistrorum meorum in theologia improperauerunt michi de hoc quod aliquando in questionibus meis phisicalibus intermisceo aliqua theologica, cum hoc non pertineat ad artistas. Sed ego cum humilitate respondeo quod ego bene vellem non esse ad hoc astrictus, sed omnes magistri, cum incipiunt in artibus, iurant quod nullam questionem pure theologicam disputabunt vt pote de Trinitate (determinate *ed.*) vel de Incarnatione; et vltra iurant quod, si contingat eos disputare vel determinare aliquam questionem que tangat fidem et philosophiam, eam pro fide determinabunt et rationes dissoluent prout eis videbuntur dissoluende. Constat autem, si aliqua questio tangit fidem et theologiam, ista est vna de illis, scilicet vtrum possibile est esse vacuum. Ideo, si eam volo disputare, oportet me dicere quod de ea apparet michi dicendum secundum theologiam vel esse periurum et euadere rationes ad oppositum prout apparebit michi possibile, et non possem soluere eas nisi mouerem eas; ergo sum ad hec facienda coactus." An English translation of Buridan's text is to be found in Grant, *A Source Book,* pp. 50–51. See also Edith Sylla's contribution to the present volume. A general outline of the relation between theology and philosophy at the Arts Faculty of the University of Paris in the fourteenth century is given by Zupko in "Sacred Doctrine, Secular Practice."

[4] For the Aristotelian and Boethian sources of this dictum, see Thomas Aquinas, *Expositio libri Posteriorum,* 11, n. 40. An equivalent, but slightly different dictum is to be found in Hamesse, *Les Auctoritates Aristotelis,* 1, n. 160, p. 128: "Accidentia non sunt entia, sed quid entis."

in particular to the so-called doctrine of transubstantiation. According to this doctrine, the substance of the bread is changed into the substance of the body of Christ while the accidents of the bread remain unchanged.[5] Now, according to the general consensus of the theologians following Peter Lombard, this doctrine implies that the accidents of the consecrated bread subsist "without a subject" (*sine subiecto*), that is, without inhering in a substance.[6] Both the philosophical and theological dimensions of the problem of the ontological status of accidental being were explicitly taken into account by John Buridan in his commentary on the *Metaphysics*, written in Paris between 1346 and 1355.[7] Two questions, taken from books IV and V respectively, are of particular importance.[8]

(a) The first asks whether the term "being" signifies substances and accidents by means of a single nature or concept (*utrum hoc nomen "ens" significet substantias et accidentia secundum unam rationem sive secundum unum conceptum*). Right at the outset, Buridan announc-

---

[5] For the origins of this doctrine, see Hödl, "Der Transsubstantiationsbegriff," Jorissen, *Die Entfaltung der Transsubstantiationslehre* and Goering, "The Invention of Transubstantiation." From the beginning of the fourteenth century onwards, the decree *Firmiter credimus* of the fourth Lateran Council (1215) was commonly understood as being an official ratification of this doctrine: See Bakker, *La raison et le miracle*, vol. I: pp. 14, 232, 248, 267, and 281. For the most crucial passage of the decree, see García y García (ed.), *Constitutiones*, p. 42.

[6] See Peter Lombard, *Sententiae*, p. 304: "Si autem quaeritur de accidentibus quae remanent, scilicet de speciebus et sapore et pondere, in quo subiecto fundentur, potius mihi videtur fatendum exsistere sine subiecto, quam esse in subiecto; quia ibi non est substantia nisi corporis et sanguinis dominici, quae non afficitur illis accidentibus. Non enim corpus Christi talem in se habet formam, sed qualis in iudicio apparebit. Remanent ergo illa accidentia per se subsistentia, ad mysterii ritum, ad gustus fideique suffragium, quibus corpus Christi, habens formam et naturam suam, tegitur." For an historical survey of the problem of the Eucharistic accidents, see Jansen, "Eucharistiques (accidents)," Sylla, "Autonomous and Handmaiden Science," Imbach, "Metaphysik, Theologie und Politik," Imbach, "Philosophie und Eucharistie," Imbach, "Le traité de l'eucharistie de Thomas d'Aquin" and Imbach, "Pourquoi Thierry de Freiberg." See also Bakker, *La raison et le miracle*, I: pp. 293–430.

[7] For the dating of John Buridan, *In Metaphysicen* (Paris, 1518), see Michael, *Johannes Buridan*, II: pp. 809–810.

[8] Both questions (IV.6 and V.8) have been studied within a more general framework by De Rijk, "On Buridan's View of Accidental Being," De Rijk, *Jean Buridan* and De Rijk, "Foi chrétienne et savoir humain," pp. 406–408. All quotations from Buridan's questions on the *Metaphysics* are based on the following two manuscripts: Carpentras, Bibliothèque Municipale, 292 and Paris, Bibliothèque nationale de France, lat. 14.716. I am very much obliged to Professor L. M. de Rijk (University of Maastricht) for having allowed me to use his transcription of both manuscripts. In the incunabular edition of Buridan's commentary (John Buridan, *op.cit.* (Paris, 1518)), the concerning questions are to be found on fols. 16va–17vb and 31rb–32ra respectively.

es that, with regard to this question, Aristotle's theory differs radically from the position of the faith.[9] In order to specify the differences, he first gives a general outline of the Aristotelian conception of the ontological status of accidental being. This outline is built up from three notions: "the white" (*album*), "being-white" (*esse album*), and "whiteness" (*albedo*). According to Buridan, Aristotle's view consists of saying that the white and being-white are two different things, but that whiteness is the same thing as being-white. From this equation of whiteness and being-white, Aristotle draws the following three conclusions. First, he maintains that accidents can in no way be separated from their subjects. For it includes a contradiction to say that there is being-white without there being some (white) thing. Likewise, it includes a contradiction to claim that there is whiteness in absence of some (white) thing.[10] Second, Aristotle holds that neither the concept of being-white nor the concept of whiteness are absolute concepts, but that both are connotative concepts. In other words, both the concept of being-white and the concept of whiteness connote something other than being white or whiteness.[11] Third,

---

[9] John Buridan, *In Metaphysicen*, Book IV, q. 6: MS Carpentras, Bibliothèque Municipale, fol. 60vb and MS Paris, Bibliothèque nationale de France, fol. 130rb: "Ista questio est iudicio meo valde difficilis propter hoc (ut puto) quod Aristotiles valde aliter opinatus est de accidentibus quam fides nostra ponit." One of the eight arguments given in support of an affirmative answer to the question alludes to the existence of the Eucharistic accidents (MS Carpentras, Bibliothèque Municipale, fol. 60rb and MS Paris, Bibliothèque nationale de France, fol. 130ra): "Item, ab illo modo essendi qui est *per se stare* sumitur unus conceptus communis omnibus per se subsistentibus; sed accidentia possunt per se subsistere, scilicet separata a substantia, saltem per potentiam divinam; ergo possibile est quod ipsis substantiis et accidentibus conveniat unus conceptus communis; et ille non videtur esse nisi conceptus a quo sumitur hoc nomen 'ens'." According to Buridan, words signify real existing things through the mediation of concepts. For this point, see Biard, *Logique et théorie du signe au XIVe siècle*, pp. 168–172.

[10] John Buridan, *op.cit.*, IV.6: MS Carpentras, Bibliothèque Municipale, fol. 60^{va-vb} and MS Paris, Bibliothèque nationale de France, fol. 130^{rb-va}: "Credo enim quod ipse (i.e., Aristotiles) opinabatur, si homo est albus, quod ad hoc non concurrit nisi homo et albedo. Modo ultra: quamvis album sit idem quod homo, tamen Aristotiles credidit quod esse album non sit homo, quia dicetur in septimo huius quod in dictis secundum accidens non est idem ipsum et esse ipsum, ut album et esse album. Et ideo, cum esse album non sit homo, Aristotiles credidit quod esse album est idem quod albedo, et esse magnum idem quod magnitudo, et esse figuratum idem quod figura, et sic de aliis. Et ideo universaliter Aristotiles credidit quod accidentia nulla virtute sunt separabilia a subiectis suis, quia videtur implicare contradictionem quod sit esse album nisi aliquid sit quod est album. Et ita pari ratione: cum esse album sit idem quod albedo, videtur implicare contradictionem quod esset albedo nisi aliquid esset quod esset album."

[11] John Buridan, *op.cit.*, IV.6: MS Carpentras, Bibliothèque Municipale, fol. 60vb

Aristotle asserts that both being-white and whiteness are not beings absolutely speaking (*simpliciter*), but merely in a relative way (*secundum quid*). For being-white is not a being on its own, but only when joined to some (white) thing. Similarly, whiteness is a being only when linked to some (white) thing.[12] By virtue of these three conclusions, then, Aristotle answers in the negative the question under discussion. In his view, accidents and substances are not beings in a univocal way, but only in an analogous way, that is, accidents are beings "by attribution" to substance. The term "being," then, signifies substances by means of an absolute concept, without connoting any other thing, and accidents by means of a connotative concept. In other words, the concept of an accident necessarily includes the concept of a substance.[13]

In contrast to the Aristotelian conception of the ontological status of accidental being, Buridan offers a different theory by taking into account the point of view of the faith. Referring explicitly to

---

and MS Paris, Bibliothèque nationale de France, fol. 130va: "Postea Aristotiles credidit quod esse hominem vel lapidem est esse aliquid, sed esse album non est esse aliquid, immo est esse aliquale, et sic esse tricubitum non est esse aliquid, sed est esse aliquantum, nec esse in domo est esse aliquid, sed est esse alicubi, et sic de aliis predicamentis. Sicut ergo esse album non dicitur secundum conceptum simplicem, scilicet absolutum a connotatione, ita credidit Aristotiles quod hoc nomen 'albedo' non diceretur secundum conceptum absolutum a connotatione, immo secundum conceptum connotativum."

[12] John Buridan, *op.cit.*, IV.6: MS Carpentras, Bibliothèque Municipale, fol. 60vb and MS Paris, Bibliothèque nationale de France, fol. 130va: "Et ultra Aristotiles credidit quod esse album, simpliciter loquendo, non esset aliquid, sed esset aliquid secundum quid, scilicet cum additione, quia esse album est bene aliquid esse album et esse magnum est aliquid esse magnum, et sic de aliis. Et ideo sequitur ultra quod albedo, simpliciter loquendo, non est aliquid, sed cum additione: albedo est aliquid esse album." For the expression "cum additione," see Aristotle, *Metaphysics*, VII, c. 5, 1031a1–3 (Hamesse, *Les Auctoritates Aristotelis*, 1, n. 164, p. 129).

[13] John Buridan, *op.cit.*, IV.6: MS Carpentras, Bibliothèque Municipale, fol. 60vb and MS Paris, Bibliothèque nationale de France, fol. 130va: "Sed tunc videre possumus opinionem Aristotilis de substantiis et accidentibus et de conceptu entis, accipiendo istum terminum 'ens' nominaliter, ita quod ista sunt nomina synonima: 'ens' et 'aliquid,' tunc diceret Aristotiles quod hoc nomen 'ens,' vel hoc nomen 'aliquid,' non dicitur secundum eandem rationem de terminis substantialibus et de terminis accidentalibus, quia secundum rationem simplicem sine connotatione aliena dicitur de terminis substantialibus. Homo enim, simpliciter loquendo, est aliquid, et asinus est aliquid, sed albedo vel nigredo non est aliquid (ut dicebatur), ita quod hoc nomen 'ens' vel 'aliquid' non dicitur de albedine vel magnitudine secundum rationem simplicem, immo secundum rationem connotativam. ... Ideo patet quod accidentia non dicuntur simpliciter entia, immo entia secundum quid, scilicet cum additione et per attributionem ad substantiam, quia conceptus accidentis explicatur per conceptum substantie cum additione."

the subsistence of the Eucharistic accidents *sine subiecto*, he takes his point of departure in an affirmation of Gods power to separate accidents from their substances. From this, he deduces that whiteness, in order to exist on its own (*per se*), must be a real being, and hence that it possesses the status of a being not only while existing separately from a substance, but also while inhering in a substance. He concludes, then, that the concept of whiteness is an absolute concept and not a connotative concept. Therefore, the term "being" signifies substances and accidents in a univocal way, by means of an absolute concept.[14] The main difference between Aristotle's and Buridan's conception of the ontological status of accidental being consists therefore in the fact that, in Buridan's view, whiteness is not the same thing as being-white. For, assuming that whiteness comes to exist separately from a substance by way of divine intervention, it can still be signified by the term "whiteness," but not by the expression "being-white," which necessarily requires the presence of some (white) thing.[15] Thus, Buridan proposes a new definition of the no-

---

[14] John Buridan, *op.cit.*, IV.6: MS Carpentras, Bibliothèque Municipale, fol. 61ra–rb and MS Paris, Bibliothèque nationale de France, fol. 130vb: "Nunc dicendum est ad questionem motam, tenendo que debemus tenere ex fide. Dico ergo quod nos tenemus quod per potentiam Dei accidentia possunt separari a substantiis et separatim subsistere [et] sine substantia sibi subiecta. Unde dicimus quod sic de facto subsistunt in sacramento altaris. Si ergo ponamus quod albedo sic per se subsistat absque hoc quod alicui subiecto inhereat, tunc manifestum est quod illa albedo vere est ens et vere est aliquid. Et etiam ex hoc manifestum est quod conceptus a quo sumitur hoc nomen 'albedo' est ita simplex sine aliqua connotatione sicut aliquis terminus substantialis. Et si de isto termino 'albedo' predicetur hoc nomen 'ens' vel hoc nomen 'aliquid,' non oportet quod de eo predicetur secundum aliquam attributionem ad substantiam subiectam vel ad aliquem terminum substantialem, quia sine substantia subiecta ipsa est ens et aliquid, et non minus ipsa est ens vel aliquid quando inheret quam si subiectum esset ablatum. Ipsa enim est illud idem quando inheret quod ipsa esset si subiectum esset ablatum. Ideo hoc nomen 'ens' vel hoc nomen 'aliquid' eque simpliciter et secundum conceptum eque simplicem dicitur de albedine sicut diceretur de lapide vel de asino. ... Et istis visis, videtur quod oportet concedere quod hoc nomen 'ens' vel hoc nomen 'aliquid' dicuntur vere univoce secundum unum conceptum communem simpliciter absolutum a connotatione de terminis significantibus substantias et de terminis significantibus talia accidentia."

[15] John Buridan, *op.cit.*, IV.6: MS Carpentras, Bibliothèque Municipale, fol. 61ra and MS Paris, Bibliothèque nationale de France, fol. 130vb: "Postea etiam sequitur ex hoc quod albedo non est idem quod esse album, quia in casu posito quod albedo sic est separata, verum est dicere quod hoc est albedo, et tamen hoc non est esse album, quia non est esse album nisi aliquid sit album, et tamen illa albedine nichil est album." In Buridan's view, the ontological difference between "white (*album*)," "being-white (*esse album*)," and "whiteness (*albedo*)" is reflected at the level of language. In order to establish the differences between the terms "white," "being-white," and "whiteness," he uses the technical notions of *signification, supposition*, and *con-*

tions *substance* and *accident*, including in his definition the distinction between the order of nature and the order of miracles. According to him, *substance* ought to be defined as a thing that exists on its own in the order of nature without inhering in any other thing. Likewise, *accident* ought to be defined as a thing which does not exist on its own in the order of nature but which, in a miraculous way, may come to exist on its own. Thanks to this new definition, Buridan is able to maintain that whiteness subsisting separately from its subject continues to be an accident. For Aristotle, on the other hand, separately existing whiteness inevitably changes into a substance, by reason of its *per se* existence.[16]

---

notation. The difference between the term "white" and the expression "being-white" is not situated at the level of signification (both terms signify whiteness), but at the level of supposition and connotation: See John Buridan, *op.cit.*, VII.3: MS Carpentras, Bibliothèque Municipale, fol. 86[ra–rb] and MS Paris, Bibliothèque nationale de France, fol. 154[va–vb]: "Et puto quod sit de intentione Aristotilis quod iste terminus 'album' supponit solum pro subiecto cui inheret albedo, et appellat vel connotat albedinem sibi inherentem. Sed iste terminus 'esse album,' quia res est alba formaliter per albedinem, supponit vel pro illa albedine secundum quam formaliter est esse album vel supponit pro congregato ex illa albedine et subiecto cui inheret. Et tunc statim manifestum est quod non est idem album et esse album, quia non est idem subiectum et forma sibi inherens, nec etiam est idem subiectum et congregatum ex forma et subiecto. ... Non alia significat 'esse album' quam 'album,' sed diversimode. Significant ea que significant, quia 'album' significat albedinem, non supponendo pro ea, sed pro congregato ex ea et aliquo alio, sed 'esse album' significat albedinem, supponendo pro ea vel aggregato ex ea et subiecto." In spite of this difference, both the term "white" and the expression "being-white" connote whiteness as a disposition adjacent to the subject (*dispositio adiacens*): See John Buridan, *op.cit.*, VII.4: MS Carpentras, Bibliothèque Municipale, fol. 86va and MS Paris, Bibliothèque nationale de France, fol. 155ra: "Ideo credo esse dicendum quod pro tanto differunt album et esse album, quia hoc nomen 'album' sic diversimode plura significat quod pro uno illorum supponit, et non pro altero, sed illud alterum connotat tamquam rem vel dispositionem adiacentem illi rei pro qua supponit. Tunc enim 'esse ipsum' non supponit pro ipso, sed pro congregato, verbi gratia, 'esse album' non supponit pro albo, sed pro congregato ex albo et albedine sibi adiacente, scilicet per quam dicitur album." The term "whiteness," on the other hand, does not connote any disposition adjacent to whiteness (MS Carpentras, Bibliothèque Municipale, fol. 86va and MS Paris, Bibliothèque nationale de France, fol. 155ra): "Nos, dicentes albedinem esse separabilem, diceremus quod hoc nomen 'albedo' non connotat dispositionem aliquam adiacentem albedini secundum quam albedo formaliter dicatur albedo." For details of Buridan's semantics of accidental terms, see De Rijk, "On Buridan's View of Accidental Being," pp. 45–47 (concerning "*album*" and "*esse album*") and pp. 47–48 (concerning "*albedo*") and De Rijk, *Jean Buridan*, pp. 28–35. For the notion of a *dispositio adiacens*, see n. 17 below.

[16] John Buridan, *op.cit.*, IV.6: MS Carpentras, Bibliothèque Municipale, fol. 61rb and MS Paris, Bibliothèque nationale de France, fol. 131ra: "Sed tunc restant duo dubia. Si accidentia ponantur sic per se subsistere, quomodo ponetur differentia inter

(b) A similar opposition between Aristotelian philosophy and the doctrine of faith arises within the context of the question of whether causal relations are dispositions added to causes and effects (*utrum causalitates sive dependentiae rerum ad invicem sunt dispositiones additae illis causis et illis causatis*).[17] For present purposes, the importance of this question lies in the fact that it immediately concerns the ontological status of inherence. At this point, the question of the separability of accidents is again at the center of the disagreement between Aristotle and the faith. According to Aristotle, no accident is able to exist separately from a substance. For this reason, the Philosopher thought that an accident inheres by itself in its subject and that, conversely, a subject is by itself the subject of an accident. Therefore, in order to account for the inherence of an accident in a subject, Aristotle considered it superfluous to posit some disposition added to either the accident or the subject.[18] Buridan, on the other hand, claims that the inherence of an accident in a subject possesses the status of an added disposition. In order to defend his position, he in-

---

substantias et accidentia? ... Ad primum istorum ego dico quod Aristotiles dixisset quod albedo esset vere substantia si sic posset per se subsistere et esset hoc aliquid, quia, si queratur 'quid est hoc?', non potest responderi nisi quod hoc sit albedo. Sed nos possumus aliter dicere, scilicet quod omne illud est substantia quod naturaliter per se subsistit, ita quod non inheret alteri; et omne illud etiam est substantia quod est pars talis nature per se subsistentis. Et omne illud est accidens quod sic non subsistit per se naturaliter, nec est pars per se subsistentis, non obstante quod subsistit per se miraculose. Et sic albedo, quamvis per se subsisteret, non diceretur substantia, quia non sic subsistit naturaliter, sed miraculose."

[17] It is most likely that the notion of an *added disposition* (*dispositio addita*) relates to the notion of an *adjacent disposition* (*dispositio adiacens*), which Buridan uses in developing his semantics of accidental terms (see n. 15 above). Yet it should be stressed that, to my knowledge, Buridan does not give any formal definition of these two types of disposition. Moreover, in his question on added dispositions, the notion of an *adjacent disposition* does not occur. For a provisional interpretation of the two notions, see De Rijk, "Foi chrétienne et savoir humain," p. 407. On the notion of an *added disposition* in particular, see Normore, "Buridan's Ontology," pp. 198–199 and Zupko, "Sacred Doctrine, Secular Practice," pp. 588–590.

[18] John Buridan, *op.cit.*, V.8: MS Carpentras, Bibliothèque Municipale, fol. 74va and MS Paris, Bibliothèque nationale de France, fol. 144ra: "Ista questio non erat Aristoteli difficilis. ... Erat enim eius opinio quod albedo lapidis non est separabilis a lapide sive a subiecto suo, nisi per eius corruptionem, et ita etiam ipse ponit de omnibus accidentibus, et ita etiam ponit de omnibus formis materialibus, quod non sunt separabiles ab ipsa materia nisi per corruptionem earum. ... Non enim credidit quod per potentiam divinam posset fieri talis separatio. Et ideo in octavo huius et in multis aliis locis ipse omnino negat quod albedo inhereat suo subiecto per aliquam dispositionem sibi additam, immo per se ipsam formaliter inheret, et subiectum per se ipsum sibi subicitur. Nec ibi oportet esse aliud quid quam subiectum illud et illam albedinem ad hoc quod illud subiectum sit album."

geniously makes use of the argumentation of Aristotle himself, who had recognized that, if it were true that accidents could exist separately from their subjects, one would have to admit the existence of an added disposition by virtue of which accidents inhere in their subjects. Referring to the miraculous mode of existence of the Eucharistic accidents, then, Buridan is able to argue along Aristotelian lines in order to reach a non-Aristotelian conclusion. For Buridan knows from the doctrine of faith that accidents are indeed separable from their subjects. As a consequence, following Aristotle's reasoning (*secundum Aristotelem*), one has to acknowledge the existence of an added disposition in order to explain the inherence of accidents in their subjects.[19] As Buridan openly admits, he would never have acknowledged the existence of such a disposition if the faith had not taught him the separability of accidents.[20]

In summary, in the two questions I have just considered, Buridan diverges from Aristotle's metaphysics of accidental being by reason of the doctrine of faith. In particular, referring to the miraculous mode of subsistence of the Eucharistic accidents, he offers a new definition of the notions *substance* and *accident* in order to justify, at the general level of metaphysics, the separability of accidents. In his view, the doctrine of the faith according to which accidents are able to exist separately from their substances reveals the insufficiency of Aristotelian metaphysics. Without knowing the position of the

---

[19] John Buridan, *op.cit.*, V.8: MS Carpentras, Bibliothèque Municipale, fol. 74^{va-vb} and MS Paris, Bibliothèque nationale de France, fol. 144ra: "Et tamen ipse (*sc.* Aristotiles) ultra bene dicit quod, si accidens esset separabile a subiecto, ut albedo lapidis a lapide et scientia tua ab anima tua, quod tunc preter illum lapidem et illam albedinem exigeretur alia dispositio ad hoc quod lapis esset albus, et preter animam et scientiam exigeretur alia dispositio ad hoc quod subiectum, scilicet anima, esset sciens. ... Modo ultra secundum fidem catholicam nos concedimus huiusmodi separationem, unde ponimus accidentia in benedicto sacramento altaris, que fuerunt accidentia panis, remanere, et tamen nulli subiecto inherere. Igitur secundum Aristotilem nos ultra debemus concedere quod requiritur dispositio addita." See also the following passage (MS Carpentras, Bibliothèque Municipale, fol. 75ra and MS Paris, Bibliothèque nationale de France, fol. 144^{rb-va}): "Ad aliam de albedine et lapide dico quod oportet quod sit dispositio addita ad hoc quod albedo inhereat lapidi, vel etiam ad hoc quod dependeat a lapide, propter hoc quod possibile est quod ipsa maneat non inherens et non dependens, scilicet a lapide. ... Credo quod dicendum est quod illa dispositio requisita ad hoc quod homo sit albus, preter hominem et albedinem, est inherentia albedinis ad hominem."

[20] John Buridan, *op.cit.*, V.8: MS Carpentras, Bibliothèque Municipale, fol. 75ra and MS Paris, Bibliothèque nationale de France, fol. 144rb: "Sic ergo probabile est quod in predictis sint tales dispositiones addite. Quas tamen nullo modo concederem nisi esset illa separabilitas quam ex fide tenemus."

faith at this point, he would undoubtedly have adhered to Aristotle's doctrine.[21] On the other hand, it should be stressed that Buridan refrains from giving any explanation of the mode of existence of the Eucharistic accidents. As a philosopher, that is, as subordinate to the statute of 1272, he must determine a question touching on both philosophy and theology in accordance with faith, without trespassing on the boundary between these two disciplines.[22]

3. Marsilius of Inghen's Theory of Accidental Being

Before moving on to examine the *Questions on Aristotle's Metaphysics* of Marsilius of Inghen, composed at Heidelberg between 1386 and 1396, it must be emphasized that a measure equivalent to the Parisian statute of 1272 was in force at the university of Heidelberg. From the perspective of academic discipline, then, it seems that Marsilius worked in an environment similar to Buridan's.[23] Yet Marsilius's case is not altogether identical to Buridan's. Buridan never studied theology. He spent his entire academic career in the Arts Faculty. Marsilius, on the other hand, pursued his theological studies while preparing his commentary on the *Metaphysics*. For this reason, his commentary provides a very interesting source for the study of the relation between philosophy and theology.[24] Now in Marsilius's commentary, as

---

[21] John Buridan, *op.cit.*, VII.1: MS Carpentras, Bibliothèque Municipale, fol. 85ra and MS Paris, Bibliothèque nationale de France, fol. 153va: "Et sine dubio hec omnia concederem, si crederem, sicut credebat Aristotiles, scilicet quod albedo vel caliditas, et sic de aliis talibus accidentibus, non possent separari per aliquam virtutem a substantiis suis nec separatim conservari." For an appreciation of Buridan's procedure, see also De Rijk, *Jean Buridan*, pp. 44-47 and De Rijk, "Foi chrétienne et savoir humain," p. 408.

[22] De Rijk, *Jean Buridan*, pp. 37-38, proposes (on behalf of Buridan, it seems, but without referring to any specific passage) an interpretation of transubstantiation. To my knowledge, Buridan completely avoids any positive statement about what actually happens at the moment God annihilates the substance of the bread while conserving its accidents. This is indeed not the philosopher's concern, but the theologian's.

[23] See Winkelmann, *Urkundenbuch*, I, p. 41: "Item quod si contingat aliquem eorum post sui incepcionem seu birreti magistralis recepcionem aliquam determinare questionem que concernit fidem et phisicam, eam pro fide determinabit et raciones contra fidem laborantes pro posse dissolvere studeat et dissolvat."

[24] For Marsilius' *Questions on Aristotle's Metaphysics*, see Reina, "*Comprehensio veritatis*." For the dating of this work, see Hoenen, *Marsilius of Inghen*, p. 16. All quotations from Marsilius's commentary are based on the following two manuscripts: Kraków, Biblioteka Jagiellońska, Cod. 709 and Wien, Österreichische Nationalbibliothek, CVP 5297. For these manuscripts, see Markowski, "Les Questiones de Marsile d'Inghen sur la Métaphysique d'Aristote," pp. 12-15, 17-20 and 31.

well as in Buridan's, the problem of the separability of accidents and, hence, the problem of the relation between Aristotelian metaphysics and Eucharistic theology occupies an important place. Yet Marsilius does not argue along the same lines as Buridan. Indeed, in his question on whether the term "being" signifies substances and accidents univocally (*utrum "ens" univoce significet substantias et accidentia*), Marsilius introduces Buridan as an advocate of the *communis opinio* of the theologians, though expressing some doubt as to whether Buridan really continued to adhere to this opinion until the end of his life. In contrast to Buridan's opinion, he himself proposes to uphold a "more metaphysical" doctrine. The main part of his question, then, can be interpreted as the confrontation between a theological and a metaphysical theory of the ontological status of accidental being.[25]

In order to present the *communis opinio*, Marsilius summarizes in a series of eight propositions the corresponding question by Buridan. In fact, the first four propositions paraphrase the Aristotelian theory of the ontological status of accidental being. (a) According to Aristotle, the being of an accident is nothing but its inherence. This means that it is impossible for an accident to exist on its own. (b) On the basis of this first proposition, the philosophers argue that accidents are absolutely inseparable from their subjects. For only substances possess the status of a "*quid*" or an "*aliquid;*" accidents merely have the status of a "*quale*" or an "*aliquale*." (c) For this reason, there is no essential agreement (*essentialis convenientia*) between an accident and a substance. (d) It follows that no concept can signify substances and accidents univocally.[26] Only the last four propositions, however,

---

[25] Marsilius of Inghen, *op.cit.*, Book IV, q. 5: MS Kraków, Biblioteka Jagiellońska, fol. 26va and MS Wien, Österreichische Nationalbibliothek, fol. 27vb: "Ista questio difficilis est, quia non est facile aliquam partem eius efficaciter demonstrare. Et ideo probabiliter procedendo, primo recitanda est opinio communis quam multi theologi tenent; et est bone memorie magistri Iohannis Biridani in questionibus suis super isto quarto, quamuis non credam quod in fine uite fuerit illius opinionis. Secundo respondebitur aliter ad questionem et ut michi apparet magis methaphisicaliter."

[26] Marsilius of Inghen, *op.cit.*, IV.5: MS Kraków, Biblioteka Jagiellońska, fol. 26va and MS Wien, Österreichische Nationalbibliothek, fols. 27vb–28ra: "Ad primum articulum est primo notandum quod opinio communis uidetur posse reduci ad octo conclusiones. Quarum prima est ista: quod secundum Philosophum accidentis esse est purum inesse et nullo modo posse per se existere. ... Ex qua inferunt secundam quod ideo philosophi dicunt accidens non esse simpliciter, nec esse aliquid, sed pocius esse in alio uel aliquid alterius. Probant ex precedenti quia: esse quid uel aliquid seu esse simpliciter non uidetur competere nisi hiis que per se esse possunt. Qualiter accidens secundum eos esse non potest per precedentem conclusionem; unde in particulari dicunt Aristotelem non concessisse albedinem esse aliquid, sed esse aliquale, eo quod nunquam per se est nec esse potest, ut credidit, sed semper

give Buridan's own theory of the ontological status of accidental being. This theory is, so to speak, grounded on the doctrine of faith. (e) As is demonstrated by the subsistence of the Eucharistic accidents, it is indeed possible for accidents to exist without inhering in a subject. (f) Consequently, one has to assume a certain essential agreement between an accident and a substance. (g) On account of this agreement, the term "being" signifies substances and accidents univocally, at least those accidents which, by divine power, are able to exist separately from their subjects. (h) It follows that being possesses the status of a most general genus with respect to substance and accident. Conversely, substance and accident are merely subordinated genera of being.[27]

---

dat esse quale suo subiecto. ... Ex quo inferunt terciam conclusionem quod iuxta mentem Philosophi nulla est essencialis conueniencia accidencium cum substancia. ... Quarta conclusio, quod iuxta hanc uiam Philosophi nullus potest esse conceptus essencialis simplex communis substancie et accidentibus. Probant ex precedenti quia: nulla est essencialis conueniencia inter ea."

[27] Marsilius of Inghen, *op.cit.*, IV.5: MS Kraków, Biblioteka Jagiellońska, fol. 26^{va-vb} and MS Wien, Österreichische Nationalbibliothek, fol. 28ra: "Quinta conclusio: secundum rei ueritatem accidencia possunt per se et sine subiecto existere. Patet ex fide, in sacramento altaris. Sexta conclusio: quod in posse per se existere conueniunt substancia et accidens. Patet ex precedenti. Ex quo infertur correlarie quod aliqua est essencialis conueniencia substancie et accidentis. Septima conclusio: quod ens est uniuocum ad omnes substancias et omnia accidencia, saltem illa que possunt seorsum existere per potenciam diuinam. Nam ex quo inter ea est essencialis conueniencia, ab illa potest sumi conceptus communis utrique. Octauo inferunt quod ens est genus generalissimum ad hunc terminum 'substancia' et alium terminum qui absolute significat omnia accidencia, saltem potencia per se existere, si esset ei nomen impositum, et quod hic terminus 'substancia' est genus subalternum contentum sub hoc genere 'ens'." With regard to the last proposition, Marsilius tells us that the advocates of the *communis opinio* nevertheless concede the Aristotelian doctrine according to which the number of categories is fixed at ten (MS Kraków, Biblioteka Jagiellońska, fol. 26vb and MS Wien, Österreichische Nationalbibliothek, fol. 28ra): "Ad racionem adductam dicunt quod nichilominus manent decem predicamenta, quia licet ens sit genus ad omnes terminos absolutos, tamen manent nouem modi generales predicandi distincti nouem predicamentorum accidencium, quia non per res, sed per modos predicandi uel significandi circa primas substancias distinguuntur predicamenta. Verum est secundum eos quod isti termini 'color,' 'odor' et alii formas accidentales significantes absolute non sunt de predicamento qualitatis, sed entis. Nec oportet, ut dicunt, concretum et abstractum semper poni in eodem predicamento, cum secundum ueritatem sint genera distincta." The last proposition and Marsilius's specification are derived from Buridan's answer to the question, "*Utrum erit unum solum genus generalissimum an adhuc decem erunt generalissima*" (John Buridan, *In Metaphysicen*, VI.4 (*secundum dubium*)). In his determination, Buridan provides an extensive analysis of the semantics of concrete and abstract terms. For Buridan's claim that being has the status of a genus with respect to substance and accident, see also Ebbesen, "Proof and its Limits," p. 107 (and n. 39). On the other hand, it

The "more metaphysical" doctrine defended by Marsilius himself comes remarkably close to the Aristotelian metaphysics of accidental being. It is developed from two presuppositions. The first claims that, in the order of nature (*in lumine naturali*), it is just as impossible for a substantial form to exist without matter as it is for an accident to exist without a subject. The second presupposition affirms that, this notwithstanding, both the term "substance" and the term "being" signify the composite of matter and substantial form univocally.[28] On the basis of these two presuppositions, Marsilius formulates his first proposition: it is not because of the fact that accidents are unable to exist on their own that Aristotle denied that accidents are beings in the proper sense of the word (*simpliciter*). For, by the same token, the Philosopher should have admitted that substantial forms are not beings in the proper sense of the word, which is inconsistent with the second presupposition.[29] In order to formulate his second proposition, Marsilius proceeds to make four preliminary remarks, the latter two of which are the most pertinent. Taken together, these two remarks affirm that substances have a different mode of dependence on the first cause than accidents. The mode of dependence proper to a substance gives it the status of a "*quid*" existing on its own. On the other hand, the mode of dependence proper to an accident gives it the status of a disposition of a substance, which, leaving aside the pos-

---

should be noted that in one of his earliest writings, his *Questions on Porphyry's Isagoge* (q. 14), Buridan rejects the notion that the number of categories is fixed at ten. See Zupko, "Sacred Doctrine, Secular Practice," p. 658.

[28] Marsilius of Inghen, *op.cit.*, IV.5: MS Kraków, Biblioteka Jagiellońska, fol. 26vb and MS Wien, Österreichische Nationalbibliothek, fol. 28[ra–rb]: "Ad secundum est prima supposicio quod forme substanciales in lumine naturali non plus esse possunt sine subiecto quam accidencia. Patet, quia Aristoteles equaliter inpossibile reputaret materiam stare sine forma, uel formam substancialem sine materia, sicut accidencia esse sine subiecto. Apparet ex 2° *De anima*, quia actus actiuorum et forme substanciales semper sunt in paciente disposito. Eciam illud apparet primo *De generacione* et determinatiue ex processu Philosophi 7° et 8[uo] *huius*. Secunda supposicio: quod hoc non obstante, iste terminus 'substancia' uel 'ens' uniuoce dicitur de substancia composita, forma substanciali et materia. Patet per Philosophum 7° et 8° *huius* et in principio 2[i] *De anima*, ubi substanciam tanquam uniuocum diuidit in materiam, formam et compositum."

[29] Marsilius of Inghen, *op.cit.*, IV.5: MS Kraków, Biblioteka Jagiellońska, fol. 26vb and MS Wien, Österreichische Nationalbibliothek, fol. 28rb: "Ex hiis duabus supposicionibus infertur prima conclusio huius articuli: quod Philosophus non negauit accidencia esse encia simpliciter ex eo quia reputauit ea non posse per se existere, sed solum esse in alio. Patet, quia eadem racione negasset formam substancialem esse ens simpliciter; cuius contrarium dicit secunda supposicio. Et consequencia patet per supposicionem primam."

sibility of a miracle, can in no manner exist on its own. Between these two modes of dependence there is no essential agreement, since the former belongs to a being on its own whereas the latter belongs to a being which, in the order of nature, merely possesses the status of a disposition or affection.[30] Proceeding from these two preliminary remarks, then, Marsilius formulates his second proposition, according to which there can be no absolute concept signifying substances and accidents univocally, since, for a concept to be able to signify two things univocally there must be some essential agreement between the two things, which is not the case for substances and accidents.[31] This second proposition ultimately forms the basis of Marsilius's answer to the question under discussion: the term "being" does not signify substances and accidents univocally, but equivocally.[32]

---

[30] Marsilius of Inghen, *op.cit.*, IV.5: MS Kraków, Biblioteka Jagiellońska, fol. 27ra and MS Wien, Österreichische Nationalbibliothek, fol. 28^{rb-va}: "Tercio notandum quia duo uidentur modi dependencie rerum a prima causa, extendendo nomen 'rei' ad substancias et accidencia. Qui primus modus est ut sit quid secundum se existens uel saltem huiusmodi rei constitutiuum ut illud sit quid secundum se; et hic modus omnibus substanciis conuenit. Nam etsi forma uel materia non sit quid secundum se existens, nichilominus componunt et constituunt hoc aliquid ut partes essenciales eius. Secundus modus est emanacio per modum afficientis huiusmodi quid siue ens per se existens uel partem eius. Et huiusmodi modus est omnium accidencium; nam quodlibet eorum secundum suam naturam propriam inititur subiecto, quod per ipsum afficitur siue disponitur. Nec quouismodo potest hoc esse per se sine miraculo, nec aliquid per se existens cum alio constituere, cum accidens nunquam componat hoc aliquid cum subiecto suo. Et ut clarior sit sermo, primus modus uocetur 'dependencia rerum per se existencium,' secundus 'dependencia rerum afficiencium uel disponencium substancias.' Quarto notandum quod uidere meo Philosophus uoluit hos duos modos in nulla essenciali conueniencia conuenire, cum primus sit encium per se, uel simpliciter uel in componendo, secundus sit encium que secundum suam naturam sine miraculo nullo modo habent esse per se, nec simpliciter nec compositiue, sed solum habent esse affectiue." The first two *notabilia* are rather confirmations of the first proposition.

[31] Marsilius of Inghen, *op.cit.*, IV.5: MS Kraków, Biblioteka Jagiellońska, fol. 27^{ra-rb} and MS Wien, Österreichische Nationalbibliothek, fol. 28va: "Quibus premissis, sit secunda conclusio hec: quod nullus est conceptus uniuocus et absolutus substancie et accidentis. Probatur sic: conceptus uniuocus communis absolutus fundatur super aliqua essenciali similitudine rerum; sed nulla est talis inter substanciam et accidens; igitur."

[32] Marsilius of Inghen, *op.cit.*, IV.5: MS Kraków, Biblioteka Jagiellońska, fol. 27rb and MS Wien, Österreichische Nationalbibliothek, fol. 28va: "Correlarium responsale ad quesitum: iste terminus 'ens' non significat uniuoce substanciam et accidencia. Sequitur ex conclusione, cum sit terminus absolutus." Following this "corollary," Marsilius formulates two further propositions in the same vein (MS Kraków, Biblioteka Jagiellońska, fol. 27rb and MS Wien, Österreichische Nationalbibliothek, fol. 28vb): "Tercia conclusio: nec 'esse' dicitur uniuoce de substancia et accidente. Patet ex tercio et quarto notabilibus, cum illi duo modi dependenciarum rerum substancialium et

In short, Marsilius agrees with Aristotle in saying that an accident is not a being in the proper sense of the word, but he denies the Aristotelian explanation in terms of the inseparability of an accident and its subject. For, if the Aristotelian explanation were correct, one would have to admit for the same reason that a substantial form is not a being in the proper sense of the word, which, in Marsilius's view, is false. According to him, the only reason why an accident is not a being in the same way as a substance consists in the fact that an accident is a "*quale*," a substance a "*quid*." Now the mere fact of being a "*quale*" and not a "*quid*" does not imply the inseparability of the accident and its subject. But the existence of an accident separately from its subject belongs entirely to the order of miracles and can in no way be accounted for by natural or philosophical argument.[33]

---

accidentalium nullam habeant conuenienciam nec in esse nec in ⟨aliquo⟩ alio, cum unus totus innitatur modo existendi per se et alius totus modo existendi in alio, cum quo non componit nec componere potest. Quarta conclusio: huiusmodi 'ens' et 'esse' conueniunt equiuoce substancie et accidenti. Patet, quia conueniunt eis; et non uniuoce, ut dicunt precedentes conclusiones; igitur."

[33] Marsilius of Inghen, *op.cit.*, IV.5: MS Kraków, Biblioteka Jagiellońska, fol. 27rb and MS Wien, Österreichische Nationalbibliothek, fol. 28va: "Et ad racionem alterius opinionis dicitur quod, licet accidens per primam potenciam possit separari, et separatum conseruari in esse et agere, a suo subiecto, nichilominus hoc totum est preter nature cursum solitum et miraculosum, nam naturalis tendencia accidentis est ad subiectum, sicut forme ad materiam. Et ergo ex tali conseruacione miraculosa non fit accidens quid simpliciter." It should be noted here that as far as the notion of an *added disposition* is concerned, Marsilius advocates the view of Aristotle. See Marsilius of Inghen, *op.cit.*, V.9: MS Kraków, Biblioteka Jagiellońska, fol. 57vb and MS Wien, Österreichische Nationalbibliothek, fol. 66ra: "Ad secundum (*sc.* articulum) est conclusio responsalis hec: quod inhesio forme in materia est forma inherens; et pari modo dependencia effectus a sua causa est effectus dependens. Hec est de mente Philosophi, octauo huius, ut recitatum fuit in primo articulo. Et probatur racione sic: inhesio forme substancialis in materia est forma substancialis inherens; igitur pari racione ubique debet ita poni." Before arriving at this conclusion, Marsilius rejects an opinion similar to Buridan's. This opinion holds, on the one hand, that, from Aristotle's perspective, it is superfluous to accept any added disposition in order to explain the nature of inherence (MS Kraków, Biblioteka Jagiellońska, fol. 57rb and MS Wien, Österreichische Nationalbibliothek, fol. 65va): "Secundo notandum quod Aristoteles loquens in lumine naturali de hiis inhesionibus uel dependenciis faciliter se expediret. Reputaret enim quod forma inherens, siue substancialis siue accidentalis, nullo modo esse ualeret sine subiecto cui inheret. Et propter hoc diceret causam esse suam inhesionem et non requirit aliquam coniunctionem superadditam qua forma substancialis uel accidentalis uniatur cum suo subiecto." On the other hand, the very same opinion claims that, from the perspective of faith, inherence does indeed possess the status of an added disposition (*Ibid.*): "Tercio notandum quod secus est, secundum hanc opinionem, supponendo ex ueritate fidei katholice quod Deus quarumlibet duarum rerum absolutarum quamlibet potest ab alia separare et separatim conseruare, sicut albedinem in sacramento altaris benedicto conseruat sine

4. Appraisal

On the basis of the foregoing comparison between Buridan and Marsilius, one has to conclude that the former moves away from a fundamental tenet of Aristotelian metaphysics in order to be able to explain philosophically an important claim of Eucharistic theology, viz., the separability of accidents. To this end he introduces, at the very level of the definition of *substance* and *accident*, the distinction between the order of nature and the order of miracles. Marsilius, on the other hand, separates the order of nature from the order of miracles far more radically than Buridan in order to safeguard the Aristotelian metaphysics of accidental being. Even though Marsilius acknowledges God's capacity to make accidents subsist without their subjects, he refrains from explaining this possibility within the context of his metaphysics. From a philosopher's perspective, it is wholly impossible for an accident, as a disposition of a substance, to subsist on its own. From the perspective of faith, on the other hand, one has to concede the possibility for God to make an accident subsist separately from a substance. In Marsilius's eyes, these two perspectives remain fundamentally at odds with one another.

From the viewpoint of an historian of medieval philosophy, the foregoing inquiry leads to a somewhat surprising outcome. For the position of the theologian Marsilius comes remarkably close to that of the philosopher John of Jandun, who recognizes the claim of the faith while at the same time refusing to give any philosophical explanation of it. On the other hand, the position of the philosopher Buridan seems comparable to that of the theologian Duns Scotus, insofar both authors seek to make the theological doctrine of the separability of accidents philosophically comprehensible. Moreover, Scotus and Buridan both try to achieve this by defending the doctrine of the univocity of being.[34] In terms of the Parisian statute of 1272, Buridan actually proceeds according to the methodological principles prescribed by the statute: he determines on the side of

---

subiecto. ... Quibus premissis, ponit conclusionem responsalem quod inhesio forme in materia est res superaddita forme inherenti; similiter et dependencia effectus absoluti a causa secunda est res superaddita rei dependenti."

[34] For Scotus and Jandun, see Bakker, *La raison et le miracle*, I: pp. 367–381 and 409. I will publish shortly an article on the relation between the doctrine of the univocity of being and the problem of the Eucharistic accidents from the second half of the thirteenth century to the end of the fifteenth century.

the faith a question touching on both theology and philosophy. Marsilius proceeds differently: he "neutralizes" the opposition between the Aristotelian metaphysics and the Eucharistic theology through a radical separation of both disciplines.[35]

---

[35] In his *Sentences*-commentary, too, Marsilius advocates a strict separation of philosophy and theology. For this point, see Hoenen, "Marsilius von Inghen."

PHILOSOPHICAL THEOLOGY IN JOHN BURIDAN[*]

Rolf Schönberger

The title of my paper is somewhat indeterminate, but it points to a complex of issues that suggests at least the following three questions:

1. What are the constituent elements of the philosophical theology advanced in Buridan's writings?
2. What role does theology play in Buridan's philosophy as a whole?
3. How is this theology developed?

If one makes the effort to compile all of the passages in Buridan's writings where he raises the issue of God, a more or less traditional picture emerges. I have sketched this picture in some detail in my book on Buridan's theory of relations, and will not repeat it here.[1] It cannot be said that the concept of God developed by Buridan is as specific to his philosophy as we find in other great masters of the Middle Ages, e.g., the *ipsum esse subsistens* of Thomas Aquinas, the *ens infinitum* of Duns Scotus, the *unum* of Meister Eckhart, etc. To note this, however, is not necessarily to state a criticism of Buridan's concept. As Buridan himself claims, none of the relevant attributes taken in isolation is sufficient to establish the existence of something specific. This concept is first established through the union of these attributes:

> ... but from the collection of those attributes and several others, e.g., that [God] is both agent and final cause of the entire world, we can conclude the greatest perfection of God's substance, and most of all when it has been concluded that there is nothing added to God through which those diverse attributes agree with him.[2]

This conventional – or, shall we say, "ordinary" – view is actually quite extraordinary in the context of the fourteenth century. For institutional reasons, Buridan had to refrain from addressing himself publicly to theological debates. When he did engage in such debates,

---

[*] Translated by Mark D. Fisher.
[1] Schönberger, *Relation als Vergleich*, p. 296 sqq.
[2] John Buridan, *In Metaphysicen* (Paris, 1518), Book XII, q. 8, fol. 71va: "... sed ex congregatione istorum praedicatorum et plurium aliorum ut quod est causa totius mundi et agens et finalis, nos possumus concludere maximam perfectionem substantiae eius, maxime quando fuerit conclusum quod nihil est ipsi additum per quod ipsi ista diversa praedicata convenient."

however, he did so in full awareness of possible reproach for having transgressed boundaries, and in such a way as immediately to forestall anyone who would object to his account on this basis. Regarding the question of the possibility of a vacuum, Buridan defends himself by saying that one cannot push aside all theological considerations without doing damage to the discussion.[3] Respect for disciplinary boundaries is not practicable in all questions. But Buridan had neither the incentive for, nor a tendency towards, engaging in those infamous speculations about the possible contents of the divine will. In this case, at least, Buridan certainly did not need to be restrained.

It is important to note the traditional character of Buridan's interpretation of the divine attributes, especially in comparison with his contemporaries. For instance, Buridan claims that when applied to the subject "God," predicates are understood denominatively relative to the sense they have in other cases of predication. This is the doctrine of analogy, but it is significant that Buridan neither nostalgically returns to a pre-Scotistic doctrine, nor naively tries to resurrect the older tradition. Rather, he adheres to the doctrine of analogy in full awareness of its rejection by philosophers such as Ockham, and even makes use of the same, or nearly the same, examples in his explanation of this doctrine.

I would like to address the second question regarding the role philosophical theology plays in Buridan's thought as a whole only very briefly here. My position, for which I have sought to provide an argument in my book, is as follows: the concept of divine omnipotence does not indicate a break with systematic philosophy, nor even with the special authority of philosophy more generally. Since the concept is logically equivalent to the principle of non-contradiction, and Buridan expressly subscribes to the latter,[4] fundamental degra-

---

[3] John Buridan, *Quaestiones super octo Physicorum libros Aristotelis* (Paris, 1509), Book IV, q. 8, fols. 73^v-74^r. This, too, brought reproach upon Buridan. Cf. Grant, "The Condemnation of 1277," pp. 232-235; Grant, "Science and the Medieval University," p. 85; Pluta, "Ewigkeit der Welt," pp. 854 sq. For Buridan's relation to the Oath of 1272, cf. Miethke, "Bildungsstand und Freiheitsforderung," pp. 239 sq. For Buridan's relation to academic freedom, cf. Courtenay, "Inquiry and Inquisition," pp. 179-181. For Buridan's relation to the university oath in general, cf. McLaughlin, *Intellectual Freedom*, pp. 23 sq.

[4] John Buridan, *In Metaphysicen* (Paris, 1518), XII.14, fol. 77ra: "... deus est omnipotens non quia posset facere impossibilia fieri, sed quia potest facere omnia possibilia fieri; ideo non est contra omnipotentiam eius, si non potest facere alium deum;" John Buridan, *Quaestiones super octo Physicorum libros Aristotelis* (Paris, 1509), I.15, fol. 19rb: "... agens non dicitur omnipotens ex eo quod possit facere impossibilia fieri vel etiam ex eo quod possit possibilia fieri aliter quam sint possibilia fieri; Deus

dations of human rationality are just as fundamentally ruled out. And non-contradiction is clearly one of the principles of human rationality.

But Buridan's philosophy can address itself only to the category of possibility with this principle. "Possibility" has one and the same meaning in the field of theology: supernatural possibility. Nevertheless, Buridan provides a philosophical determination of the logical status of this possibility in terms of the aforementioned principle of non-contradiction. The question of how such things as miracles, creation, and so on, are possible is always answered by Buridan in the same way, and, accordingly, never in a specific way. The predicate "supernaturally" (*supernaturaliter*) is a pure negation, without the possibility, but also without the necessity, of being critically examined. Perhaps the expression "synthesis," which people have often used to describe the great philosophies of the thirteenth century, is too superficial to capture their actual intellectual development. It would simply be false to apply this term to Buridan's philosophy, however.[5]

Philosophical reflection cannot establish for itself a productive relation to those things of whose truth we are convinced on the basis of faith. There is no role for the doctrine of analogy to play there – nor can it have any role, as only theology could decide the question of whether a particular analogy is appropriate and acceptable. Therefore, Buridan's philosophy must appear strangely sterile with regard to the field circumscribed by the concepts of theology, religion, faith, etc. In contrast, it appears quite fertile in its recognition of these boundaries, as occurs, for instance, in Buridan's transformation of the concept of substance.[6] It seems to me, however, that one should keep in mind that Buridan's philosophy has neither the aim nor the methodological potential to penetrate philosophically the totality of human thought and action insofar as that pertains to God. I need only mention such things as the structuring of history, the symbolism of ritual actions, the vast field of spirituality (*De amicitia*

---

non potest facere alium Deum sibi aequalem. Sed ipse est omnipotens quia potest facere omnia possibilia fieri ex modo quo sunt possibilia fieri; immo omne quod fit ipse facit et omne quod fiet ipse faciet; sed hoc erit ex subiectis praepositis et concurrentibus aliis agentibus particularibus;" John Buridan, *op.cit.*, I.22, fol. 26ra: "... ipse est illa potentia qua ipse est omnipotens et ex eo dicitur omnipotens quia omnia factabilia ipse potest facere et omnia creabilia creare et omne annihilabilia annihilare."

[5] Cf. Schönberger, "Eigenrecht und Relativität."
[6] Schönberger, *Relation als Vergleich*, pp. 329 sqq.

spirituale) and asceticism (including the examination of conscience and everything proper to a normative account of self-knowledge), or even the formation of one's life in the consciousness of the immediate presence of God that is the foundation of mysticism. Buridan does speak of God's omnipresence,[7] but the most critical aspect of this presence, which appears in human consciousness as the power of representation, is understandably not a theme of Buridan's philosophy.

1. *The Category of Motion in Medieval Philosophical Theology*

The most interesting question for a historical understanding of Buridan's philosophical theology seems to be the third of our initial questions: how does Buridan develop his theory? The phrase "philosophical theology" – which was to my knowledge first used by the young Thomas Aquinas[8] – does not indicate a separate discipline, but only an element of certain treatises. In medieval commentaries, the titles of disciplines often coincide with the titles of Aristotle's books. The "discipline" in question here, philosophical theology however, is part of at least two of these, i.e., the *Physics* and *Metaphysics*. This corresponds to the cosmological, or rather, to the natural-philosophical foundation of metaphysics in Aristotle, which is why its meta-character is sometimes disputed.[9] It is only in the modern

---

[7] John Buridan, *In Metaphysicen* (Paris, 1518), XII.14, fol. 77ra: "... princeps mundi est immaterialis et incorporeus et sic non determinat sibi aliquem situm; ideo potest sufficienter esse praesens toti mundo ad regendum totum;" John Buridan, *op.cit.*, VII.15, fol. 50vb; John Buridan, *Expositio et quaestiones in Aristotelis De anima* [Patar], Book II, q. 7, p. 556: "Deus, qui est summe simplex adest toti mundo et cuilibet parti mundi, per indistantiam et immediatam assistentiam, licet non per informationem vel inhaerentiam;" John Buridan, *Expositio et quaestiones in Aristotelis De anima* [Patar], II.6; John Buridan, *Quaestiones super octo Physicorum libros Aristotelis* (Paris, 1509), VII.4, fol. 112vb: "... ineffabilis modus unionis dei ad caelum et ad omnia alia, scilicet per indistantiam et praesentialem assistentiam; a nulla enim re distat, nulli rei abest, sed omnibus adest sine distantia vel aliquo medio;" John Buridan, *op.cit.*, II.5, fol. 32vb.

[8] Thomas Aquinas, *Super Boetium De Trinitate*, q. 5 a. 4, p. 154 and 199: "Theologia vero philosophica."

[9] Hegel claims that Aristotelian philosophy "... would actually be the metaphysics of nature for the contemporary physicist" (Hegel, *Vorlesungen* [Glockner], XVIII, p. 337). Similarly, Heidegger remarks that "Aristotelian 'physics' is the concealed and, accordingly, never sufficiently considered land register of western philosophy ... it makes very little sense to say that the *Physics* precedes the *Metaphysics*, since the *Metaphysics* is just as physical as the *Physics* is metaphysical." (Heidegger, *Vom*

period that philosophy has attempted to distance itself completely from this attitude, whether for the benefit of practical reason (and its priority) or of history (as a category of totality).

Already in the Middle Ages, however, the question was raised about which of these two disciplines, i.e., physics or metaphysics, could generate the more adequate conception of God. Averroes had maintained against Avicenna that evidence for the existence of God must be provided in (Aristotelian) physics. Avicenna argued to the contrary that the concept of being (*ens*), since it is the most general and fundamental, was the only appropriate basis for conceptualizing God. Since metaphysics takes being as being (*ens ut ens*) as its subject, the word "God" must intellectually permeate the discipline of metaphysics as well. This line is followed by Thomas Aquinas,[10] and, even more strictly, by Duns Scotus.[11] Buridan does so as well,[12] dismissing outright Averroes's criticisms of Avicenna.[13] The argument that Buridan advances here is well known: God must be thought of as the universal cause; but if he were determined merely to be the cause of motion, he would not be the cause of immobile things as well. Buridan reclaims expressly for the philosophers what is advanced in other authors as a break – sometimes quiet, sometimes strident – with the cosmologically-grounded metaphysics of Aristotle:

> The natural philosopher and the metaphysician consider causes differently, since the natural philosopher considers them only insofar as the things they cause depend upon them for generating or corrupting, either with respect to a mode [of substance] or in changing [a substance] completely, actively as well as passively. But the metaphysician

Wesen und Begriff der Physis, IX, p. 242). Cf. Heidegger, *Einführung in die Metaphysik*, p. 14: "The *Physics* determines from the beginning on the essence and history of metaphysics ..." Cf. Heidegger, *Der Satz vom Grund*, p. 107.

[10] Thomas Aquinas, *Summa contra gentiles*, II cap. 16, nr. 935: "... esse autem est universalis quam moveri."

[11] John Duns Scotus, *Ordinatio*, II d. 1 q. 3 n. 115: "... efficiens in plus est quam movens;" Cf. John Duns Scotus, *op.cit.*, esp. n. 153, p. 77; John Duns Scotus, *Quaestiones super libros Metaphysicorum Aristotelis* [Andrews e.a.], I q. 1 n. 83: "... quattor causae, inquantum quaelibet in suo genere dat esse circumscribendo rationem motus et mutationis, pertinet ad Metaphysicum; materia et forma, inquantum sunt partes essentiae; efficiens, inquantum dat esse circumscribendo motum, licet enim non ageret nisi movendo, tamen ratio dantis esse, prior est ratione moventis ..."

[12] John Buridan, *Quaestiones super octo Physicorum libros Aristotelis* (Paris, 1509), VIII.1.

[13] John Buridan, *op.cit.*, VIII.1, fol. 109rb: "... credo quod in praedictis Commentator inepte reprehendit Avicennam et quod Avicenna melius locutus est in isto proposito."

considers causes insofar as the things they cause depend upon them in being. For this reason, Aristotle would not say that God is the cause of the intelligences with respect to moving or changing them, because he thought that they are immutable. But all the same, intelligences and all other things depend for their being upon God himself, who is of no concern to the natural scientist.[14]

It is no easy task, however, for medieval philosophy to distinguish strictly between these disciplines. Of course, one could call upon the Aristotelian distinction between being as being (*ens ut ens*) and being as mobile (*ens ut mobile*), but even then there remains a great deal of thematic overlap, e.g., in the doctrine of causes or of the unmoved mover. The doctrine of causes in particular was invoked by authors as different as Duns Scotus and Meister Eckhart in order to clearly separate metaphysics from physics.[15] But this could be successful only to the extent that the concept of being (*ens*) is successful in establishing a full-fledged metaphysics. To cite two authors once again, Thomas understands metaphysics as the penetration of that which constitutes being *qua* being; but since all knowledge consists in the knowledge of reasons, this leads to the question of why there exists anything at all. For Duns Scotus, in contrast, being (*ens*) is the formal aspect that makes it possible to determine the infinite; but with "infinite" (*infinitum*) we make a modal determination that cannot be applied to any being in nature.

The nominalist concept of the *res absoluta* does not allow for a sharp distinction between physics and metaphysics. Each reality is considered, with respect to its independence, against the standard of the *res absoluta* (teleology, causation, relation, motion, time, etc.). This does not mean, of course, that Ockham and other defenders of nominalism in its various forms smoothed over the distinction between God and creation, as some had feared, based on their univocal interpretation of the concept of being. This difference is merely construed in a different manner, via the predicates that have been ascribed to God as first cause (*causa prima*). But if being *qua* being is

---

[14] John Buridan, *op.cit.*, VIII.1, fol. 109^{rb-va}: "Physicus et metaphysicus differenter considerant de causis, quia de his considerat physicus solum inquantum ex eis dependent causata in generando vel corrumpendo vel in modo vel omnino in transmutando tam active quam passive; metaphysicus autem de his considerat secundum quod ex eis dependent causata in essendo; unde Aristoteles non diceret quod deus esset causa intelligentiarum in movendo vel transmutando eas, quia reputavit eas intransmutabiles. Sed tamen intelligentiae et omnia alia in esse suo dependent ab ipso Deo, de quo non intromittit se naturalis."

[15] Cf. Schönberger, "*Causa causalitatis*."

said to be an empty notion – and this is the common denominator of all forms of nominalism – then the most natural alternative is to return to the concept of causality as developed by Aristotle. Buridan does not simply revise the scholastic distinction, unknown to Aristotle, between the cause of motion and the cause of being; he actually develops a concept of God in the category of motion. God is the first cause (*causa prima*) and therefore the unmoved (*immobilis*). Most of the questions in Buridan's commentaries on the *Physics* and the *Metaphysics* concern this attribute.

At first glance, it seems that Buridan turns in a new direction after the concept of being (*ens*) appeared to signal a dead end, as we saw above. This does not mean that his new attempt to ground the concept of God in that of motion is not equally problematic. Buridan now has to deal with precisely the same problems that led thirteenth-century metaphysicians to prefer the alternative. A few difficulties of this alternative are as follows:

1. The concept of motion applies to only a certain region of reality. Thus, no idea of God can be formed wherein this foundation of the world (God) has a universal significance. It is not certain that Aristotle ascribed such a universal significance to his θεός, but the Christian understanding of God stands or falls with this universality.[16]
2. There is no agreement about how large a domain is actually covered by the category of motion. Does it apply to all things in general that undergo change? It certainly applies to nature, but does it also apply to "motions" of thought and impulses of the will?
3. Even the restricted sense of motion seems to lead to the ideas, which Aristotle developed with great consequence, that the world is eternal and that the first mover is unmoved and in no way a creator – not even the efficient cause (*causa efficiens*) of that which is moved, but only the "that-for-the-sake-of-which" (οὗ ἕνεκα).

---

[16] Following Avicenna, Buridan sometimes speaks of a *universale secundum causalitatem*. But he also frequently uses the expression *deus et intelligentiae*: John Buridan, *Expositio et quaestiones in Aristotelis De caelo* [Patar], Book I, tract. 4, cap. 2, p. 62; John Buridan, *Expositio et quaestiones in Aristotelis De anima* [Patar], Book I, tract. 1, p. 6; Book II, tract. 1, cap. 2, p. 51, and Book II, tract. 1, cap. 3, p. 58; John Buridan, *Quaestiones super decem libros Ethicorum Aristotelis* (Paris, 1513), Book III, q. 1, fol. 31[rb]; III.2, fol. 38[vb]); John Buridan, *In Metaphysicen* (Paris, 1518), XII.11, fol. 74[ra]; John Buridan, *Quaestiones super octo Physicorum libros Aristotelis* (Paris, 1509), VIII.7, fol. 116[ra], etc.

4. The very principle on the basis of which Aristotle believed he could establish the unmoved mover (πρῶτον κινοῦν ἀκίνητον) is somewhat problematic – though not from the standpoint of later philosophers, but with respect to the conceptual possibilities in Buridan's time. That is, the principle on the basis of which Aristotle thought "heaven and nature depend"[17] is, "everything that is moved is moved by another."[18]

The influence exerted by this last problem in the later Middle Ages can be hinted at only briefly here. Aristotle came up with the concept of the unmoved mover (πρῶτον κινοῦν ἀκίνητον) as an alternative to the theory of self-motion, in connection with which Plato had sought to establish the World Soul as the principle of motion in nature. The relevant passages in the Platonic dialogues were unknown in the Middle Ages,[19] yet this idea remained discernible and even developed new strength. It was transmitted through Macrobius,[20] but especially through Augustine and Anselm of Canterbury, both of whom had defined freedom of the will in terms of self-motion.[21] Accordingly, Augustine and Anselm were called upon in the thirteenth century to support efforts to criticize the Aristotelian principle of motion and limit its application to inorganic nature. The neo-Augustinians, Duns Scotus, the nominalists, and even Buridan considered the will to be free to the extent that it "moves itself according to volition."[22]

---

[17] Aristotle, *Metaphysics*, XII. 7, 1072b14.

[18] Aristotle, *Physics*, VII.1, 241b34; cf. Weisheipl, "The Principle," and Verbeke, "L'argument du livre."

[19] Plato, *Phaedrus*, 245; Plato, *Laws*, 990 sqq.

[20] Macrobius, *Commentarii in Somnium Scipionis* [Willis], II. 14–16, pp. 1357–1517).

[21] In a short text with the heading *Utrum per se anima moveatur*, Augustine takes a thought from Cicero (*Tusculanarum disputationum libri V* [Pohlenz e.a.], I, 55) and refines it: "Moveri per se animam sentit, qui sentit in se esse voluntatem; nam si volumus, non alius de nobis vult. Et iste motus animae spontaneus est" (Augustine, *De diversis quaestionibus octoginta tribus* [Mutzenbecher], 83, q. 8). This passage is not obscure; on the contrary, it is frequently cited by such authors as Peter John Olivi, *Quaestiones in secundum librum Sententiarum* [Jansen], II q. 58, p. 363; John Duns Scotus, *Lectura*, II d. 25 n. 15, p. 232; and Anselm, *De concordia praescientiae et praedestinationis et gratiae Dei cum libero arbitrio* [Schmitt], p. 283 sq.: "... voluntas quidem instrumentem movet omnia alia instrumenta quibus sponte utimur, et quae sunt in nobis – ut manus, lingua, visus –, et quae sunt extra nos – ut stilus et securis –, et facit omnes voluntarios motus; ipsa vero se suis affectionibus movet. Unde dici potest instrumentum se ipsum movens. Dico voluntatem instrumentem omnes voluntarios motus facere; sed si diligenter consideramus, ille verius dicit ut facere omne quod facit natura aut voluntas, qui facit naturam et instrumentum volendi cum affectionibus suis, sine quibus idem instrumentem nihil facit."

[22] John Buridan, *Quaestiones super decem libros Ethicorum Aristotelis* (Paris, 1513),

Thus, Aristotelian metaphysics is constrained indirectly, but also fundamentally.

With this, it seems that we are sufficiently clear about the nature of the problem for Buridan. He is faced with the following dilemma: either God is reduced to a mere power of nature by a highly specified concept of motion, or the world must be considered a beginningless cosmos. Neither alternative can be interpreted as an act of creation, nor does either leave room for actual divine interventions and initiatives. In the second part of this study, I will try to show how Buridan's philosophical theology, though unfailingly based on the category of motion,[23] comes to terms with this dilemma.

2. *The Problem of Self-Motion*

The first question is whether Buridan is willing, or able, to preserve the Aristotelian principle of motion. He addresses this issue in the opening questions of his commentary on Book VII of Aristotle's *Physics*. His answer is on the one hand rather restrictive of the principle, though it is not entirely clear how far this restriction is supposed to extend. In his first conclusion, he claims that: "the same thing can act upon itself either in itself, or accidentally, or mediately, or immediately"[24] (*idem potest agere in seipsum vel per se vel per accidens vel mediate vel immediate*). In this case, Buridan makes use of precisely those arguments that were brought forward at the beginning of the Question to establish the importance of the problems associated with it, i.e., what makes it worthy of inquiry. Here, as at the end of the Question, these arguments are expressly confirmed by Buridan despite the fact that they are precisely the arguments that were advanced since the 1360's against the universal validity of the principle of motion. They call upon phenomena, or interpretations of phenomena, which would be relegated to an utterly counterintuitive passivity if they were subsumed under the principle of motion. They are at least in part the same phenomena that Aristotle himself

---

III.1 sq.; John Buridan, *op.cit.*, X.2–3; etc.

[23] For this concept, see Maier, *Metaphysische Hintergründe*, pp. 1–57, here pp. 53–56, "Motus est actus entis in potentiae." Also Schönberger, *Relation als Vergleich*, pp. 442 sq. (im Rahmen der Unterscheidung von actus und potentia); Biard, "Le statut du mouvement."

[24] John Buridan, *Quaestiones super octo Physicorum libros Aristotelis* (Paris, 1509), VII.1, fol. 133vb.

discussed critically in the last book of the *Physics*: (1) the motions of living things not taken to have purely external sources; (2) the motion of the elements (i.e., heavy things fall downwards "on their own").[25] In both cases, Aristotle made an indispensable external factor responsible for the initiation of motion. According to Aristotle, such a complex case of initiation of motion – i.e., one requiring both an inner and an outer cause – is possible in these cases because they concern complex realities, i.e., objects of nature. As such, they allow for internal causal relations in which one part acts on another.

This account does not hold, however, for two further cases of activity, namely, the motions of the mind and the will. As mental powers, they cannot be thought of as having parts. The first example, as is well known, is taken from Aristotle's doctrine of the soul. Aristotle claims that the agent intellect (*intellectus agens*) is "essentially active," an activity interpreted by one representative tradition in terms of the activation of the possible intellect (*intellectus possibilis*). In his Questions on *De anima*, Buridan poses not only the usual questions about the immortality and individuality of the intellect, but also a question concerning the necessity of positing an agent intellect.[26] He speaks of two intellects, which distinguish themselves through their activity. First, the divine intellect must be considered as purely active. This intellect is identical with God, but considered according to that aspect according to which God, as universal cause, cooperates in the formation of acts of cognition (*ad intellectionem formandum*). Concerning the human form of the *intellectus agens*, Buridan thinks that the theory of Averroes is in substantial agreement with that of Aristotle,[27] though it is in fact false. He dismisses the theory of Avempace in like manner. The third alternative is that of the Christian faith. Since Buridan upheld the doctrine of the species, he had to recognize that the intellect has a passive side. Yet this is also precisely the reason why an *intellectus agens* is necessary. The image belonging to a judgment (whether affirmative or negative) is not proper to the species, but can only result from the formative power of an intellect.[28]

---

[25] Cf. John Buridan, *op.cit.*, VIII.5, fols. 113va-114vb. For the history of this discussion, see Maier, *An der Grenze*, here pp. 143–180, "Die Ursache der Fallbewegung;" and Maier, "Les commentaires."

[26] John Buridan, *Expositio et quaestiones in Aristotelis De anima* [Patar], III.5, pp. 426–431.

[27] John Buridan, *op.cit.*, III.5, p. 428: "... satis consonat dictis Aristotelis."

[28] John Buridan, *op.cit.*, III.5, p. 430: "... intelligendo possumus componere et dividere, et ad hoc faciendum oportet ponere aliquod determinans ad sic componendum vel dividendum."

And since neither the phantasm nor God can be considered specific causes, only the intellect is left as the explanation of this image.

The most important question for the problem under consideration is how the active and passive intellects are distinguished. It is precisely this characterization as active and passive that appears to have provided the most compelling reason for the Aristotelians to assign these two functions to distinct capacities. The Aristotelian principle of motion had expressly ruled out the possibility of a self-initiated activity. Buridan, however, rejects this, claiming that it is one and the same intellect that is passive with respect to the species and active with respect to the act of knowing: "it is the same intellect in us which acts and is acted upon with respect to understanding"[29] (*idem est intellectus in nobis qui agit et patitur respectu intellectionis*).

Accordingly, there are not distinct capacities, but distinct functions of the same capacity. But does this not contradict the Aristotelian teaching, according to which activity is of a higher order than passivity?[30] Since a thing cannot be nobler than itself, a distinction in these capacities of the soul must follow from a distinction in their perfection. However, Buridan does not give much weight to these objections: perhaps Aristotle does speak of the divine intellect, which would be more perfect than the human intellect, but perhaps he wishes only to maintain the general distinction of rank between activity and passivity, and in this case we can certainly agree with him.

According to Buridan, there is also no need to introduce a double potency on the basis of the aforementioned distinction. He considers two analogous cases. Heavy things fall downwards not through external influence, but on their own.[31] The second analogous ex-

---

[29] John Buridan, *op.cit.*, III.5, p. 430. Ockham also rejects the idea that with the distinction between activity and passivity, one must accept an additional ontological commitment (William Ockham, *Quodlibeta septem* [Wey], 14, p. 89): "... dico quod idem potest esse activum et passivum respectu eiusdem, nec ista repugnant."

[30] Aristotle, *De Anima*, III. 5, 430a18–19. This claim is advanced by the scholastics in an almost formulaic fashion: Thomas Aquinas, *Summa contra gentiles*, I, cap. 65, nr. 537, and II, cap. 45, nr. 1222; Thomas Aquinas, *De substantiis separatis*, cap. 12 (nrs. 69, 79–80); Henry of Ghent, *Quodlibet I* [Macken], q. 14, p. 85; John Duns Scotus, *Lectura*, II d. 9 q. 1–2 n. 14, p. 18; and finally by Buridan himself: John Buridan, *In Metaphysicen* (Paris, 1518), XII, 3, fol. 66rb; John Buridan, *Quaestiones super octo Physicorum libros Aristotelis* (Paris, 1509), VII, 1, fol. 103va. In addition, this formula has an Augustinian parallel, according to Augustine, *De Genesi ad litteram libri duodecim* [Zycha], XII, 16, p. 401 sq.): omne agens est praestantius passo. This is also often cited, frequently in parallel to the Aristotelian text: Henry of Ghent, *Quodlibet I* [Macken], 14, p. 85; John Duns Scotus, *Lectura*, II d. 9 q. 1–2 n. 14, p. 293.

[31] John Buridan, *Expositio et quaestiones in Aristotelis De anima* [Patar], III, 5, p. 430: "... mediante illa gravitate movet seipsum deorsum, cim ipsum est extra locum suum

ample is taken from the doctrine of the agent sense (*sensus agens*).[32] Here, too, Buridan does not hesitate to adopt an active meaning, and so again he need not fear the multiplication of entities beyond necessity.[33] As far as I can tell, Buridan does not ask about the consistency of these two theories within Aristotelian thought. He is much more concerned to show that the will can be considered free only if its potential for activity is realized, not through an external influence, but through self-motion.

The relation this has to Buridan's philosophical theology should be apparent. If he constrains or completely rejects the principle of motion, the whole idea of the first unmoved mover is undermined. He is no longer the mover of *everything*. Of course, the objection that this is far too cosmological a conception of God, which does not wholly agree with what Christians mean when they use the word "God," is misplaced here. This objection would not have been decisive for Buridan because all that is meant by "unmoved mover" is a single unmistakable determination. Think of the remark Thomas appends to the conclusion of his proof for the existence of God from motion (*ex motu*): "And this everyone understands to be God" (*Et hoc omnes intelligunt deum*), or the way Dante famously ends the *Divine Comedy*, referring to "The love that moves the sun and the other stars" (*L'Amour che mouve il sole e l'altre stelle*).[34]

I cannot discuss here Buridan's theory of the intellect and will in a sufficiently thorough manner. We need not, however, consider each aspect of this theory, but only ask to what extent it forces Buridan to modify the principle of motion. He claims throughout – and this is most surprising to me given his connections with the tradition

---

et non impeditum." This example was also used throughout discussions of the self-motion of the will, as a few instances will show: Thomas Aquinas, *De veritate*, 22, 6; Thomas Aquinas, *Summa Theologiae*, I, q. 83, a. 1; q. 59, a. 2; Dietrich of Freiberg, *De elementis corporum naturalium* [Pagnoni-Sturlese], pp. 56–93, there p. 76 sq.; Henry of Ghent, *Quodlibet IX* [Macken], 5, p. 115 sq.; Siger of Brabant, *Impossibilia* [Bazán], V ad 4, pp. 89–92; John Duns Scotus, *Ordinatio*, II, d. 2, pars 2, q. 6: *Utrum angelus possit movere se*; Godfrey of Fontaines, *Les Quodlibet cinq, six et sept* [De Wulf e.a.], VII, q. 6, p. 341; and still later: Peter Aureoli, *Commentarium in secundum librum Sententiarum* (Roma, 1596), II d. 25 a. 1, fols. 264aD sqq. For its use in Meister Eckhart, cf. Schönberger, "Das gleichzeitige Auftreten," p. 428 sq.

[32] Cf. Pattin, *Pour l'histoire du sens agent*.

[33] John Buridan, *Expositio et quaestiones in Aristotelis De anima* [Patar], II, 10, pp. 308–315.

[34] Dante, *La Divina Commedia* [Petrocchi], Par. XXXIII, p. 145. Furthermore, Dante says here that this love, despite the deficiencies of fantasia, *moves* desire and the will.

surrounding the idea of spontaneity – that the principle "everything that is moved is moved by another" (*omne quod movetur ab alio movetur*) should not be entirely rejected, but in a certain sense remains intact. Accordingly, even the motion of the intellect or the will requires an external cooperating factor. The intellect relates itself to its object. Likewise, the will is not able to relate itself to something unknown. Its object, or what it wills, must be mediated by the understanding.

Whereas the first Question concerns the explanation of the principle, the second asks whether the foundation Aristotle gives for this principle is a good one. Buridan is forced to answer the latter question in the negative because Aristotle makes use of two unacceptable assumptions. First, Aristotle's version of this principle deals only with change of location.[35] Local motion, that is, change in an object with respect to its determination in a place, can occur only as regards divisible things (a football is moved by a foot). To be sure, Aristotle distinguishes various senses of the concept of motion, the primary (though certainly not the only!) sense being motion from place to place. Nevertheless, for Buridan the principle of motion applies exclusively to local motion, and hence to things in three-dimensional space alone.

The second restriction is directly related to the first. It concerns the Aristotelian claim that motion ceases if part of the thing in motion comes to rest. Again, the principle obviously applies only when we speak of complex things. For psychological forms of motion, this means that the structure of *mental* activity must be determined in such a way that although it is essentially initiated spontaneously, it never comes about without recourse to some external object that is likewise mediated through mental activity.[36] The crucial point for

---

[35] This is more precisely determined in John Buridan, *Quaestiones super octo Physicorum libros Aristotelis* (Paris, 1509), VII, 2, fol. 113rb: "… se habentia circumscriptive seu commensurative ad loca vel spacia in quibus vel super quae aut circa quae dicuntur moveri."

[36] Nothing can positively force me to will something, but I can be hindered from realizing my volition, or from completely determining it myself, such as when I am being tortured – in which case I am no longer doing what I will. In addition, Ockham expressly exempted effects of the will when he taught that God can bring about any of the effects of secondary causes in the absence of these causes (William Ockham, *Quodlibeta septem* [Wey], III, 14, p. 254): "… omnis actus alius a voluntate potest fieri a solo Deo. Some examples of this doctrine;" Thomas Aquinas, *Summa contra gentiles*, IV, 65, nr. 4017: "… divina autem virtus potest producere effectus quarumcumque sine ipsis causis secundis: sicut potuit formare hominem sine semine, et sanare febrem sine operatione naturae. Quod accidit propter infinitatem virtutis suae, et quia omnibus causis secundis largitur virtutem agendi. Unde et effectus causarum

Buridan's discussion of this principle is that the relation to external things involved in mental activity follows not from being immediately subsumed by the principle of motion, but from its specific analysis. Therefore, the universal significance of the first unmoved mover (*primum movens immotum*) is preserved, at least *de facto*.

Buridan must then discover what kind of motion proceeds from this "mover." As stated at the beginning, it seems that only those options found in Aristotle are considered: (1) a motion eternally identical with itself – otherwise, the predicate "unmoved" would not be legitimately applied to the mover; (2) only the final determination of those things moved on its behalf.[37] Buridan involves himself here in a discussion in which we find not only a meeting of distinct intellectual traditions – ancient cosmology confronting the biblical belief in creation – but also a case in which Arabic interpreters of Aristotle (especially Averroes) have defended the autonomy of the Aristotelian doctrines.

This is not the first time Buridan is confronted with the task of reconciling truths of faith with what is taught in Aristotelian physics, which is based on evident principles advanced with considerable consistency and remarkable insight. The problem becomes: how can an immutable being create a world that has a beginning in time, or, more precisely, that has a point of time first in the sense that it has no predecessor? Aristotle expressly denied this possibility. In order to think about this, one must assume that the cosmos was governed by entirely different conditions at its inception than is the case now. If this is true, however, then we could not know anything about these conditions, or about such a beginning, which would constitute an astonishing, even irrational, break with the principles we invoke for understanding and explaining the world as it exists for us now. It follows from the aforementioned principle that there is no motion in nature that is not preceded by some other change. The creation of the world, however, implies a change – a universe

---

secundarum conservare potest in esse sine causis secundis;" *Quodl.* IX, q. 3, a. 5, pp. 98 sq.; Thomas Aquinas, *De rationibus fidei*, cap. 8, p. B 68; Henry of Ghent, *Summa (Quaestiones ordinariae)* [Wilson] a. 35, q. 6 ad 4, p. 69: "... immo immediate potest facere supernaturaliter quae agit mediantibus causis naturalibus et contrarium erroneum est." Additional examples from Henry of Ghent, Giles of Rome, Peter John Olivi, Duns Scotus, Ockham, Adam Wodeham, William Crathorn, Gregory of Rimini, and Buridan can be found in my book on Buridan, *Relation als Vergleich*, p. 331 sq. n. 135.

[37] In German there are two different forms of the perfect: "ich wurde bewegt" and "ich wurde bewogen."

exists, which was not there in the preceding moment. How can this be entertained without recourse to time? Furthermore, this change would also appear to affect the Creator himself, since he now stands in a relation to his creation. So, we must inquire further whether it is possible to fix such a first moment at all. The fixing of a moment seems to presuppose a temporal flux, in whose flow the moment in question must be localized; or, more precisely, it first becomes a moment in time through this localization. But such a process seems to rule out an assumed first moment of time, just as the discussion of change seems inevitably to presuppose a first moment because it has a temporal connotation, which already assumes time. But if the concept of change is inapplicable, then we must speak about the eternity of the world, and with it the infinity of time. However, this appears to contradict once again the doctrine of the creation of the world: In the beginning God created the heavens and the earth (*In principio creavit Deus caelum et mundum*).

Yet this discussion already differs from that of Aristotle, for in it there is mention of the will of God. The concept of the will allows for the possibility of a spontaneous origin, but this still constitutes a change. So it must also be assumed that the will to create has endured from eternity. If this is taken together with the claim that the world has a beginning in time, then it follows that the will to create and the existence of creation are not identical. An old answer to this dilemma was to claim that the ability to decide when to realize one's intentions belongs directly to the capacity to determine the will (e.g., I have the intention today to meet someone tomorrow). Averroes had previously objected that we can only make sense of such delayed acts of realization with respect to finite beings who can already relate themselves to time, and thus to changes that are already occurring. God knows no tomorrow.

In this connection, Buridan refers to a fundamental requirement of this discussion, viz., that our power of representation can articulate itself only temporally. For purposes of this investigation, we must speak about something "pre"-temporal (*einem "Vor" der Zeit*) because the word "before" (*vorher*) already has temporal connotations. All of these temporal issues – including the sequence (!) of the acts of creation – are due to the ineliminably temporal character of our thinking, which cannot simply be placed on the same level as reality.[38]

---

[38] John Buridan, *Quaestiones super octo Physicorum libros Aristotelis* (Paris, 1509), VII, 2, fol. 110rb-va.

Buridan makes the following proposal for the solution of this dilemma: the fixing of a moment of time does not necessarily presuppose the idea of a "flux" of time. A first moment need not be conditioned by some predecessor; it is sufficient if it has succeeding moments. The first "now" of creation can be determined by a certain interval from the present.[39]

The change in question, which is linked to the origin of the world, is as little a natural process for Buridan as it was for thirteenth-century theologians. This is not an irrational (because it is arbitrary) renunciation of the principles for explaining the world where its origin is at issue. Rather, it is necessary on pain of circularity that we not try to think of the constitution of things that act, move themselves, generate other things, and influence those that already exist, by using the same concepts that apply to what has already been constituted. From this it must follow that the state of affairs at the creation of the world cannot be represented at all. Since representation is based on our world of experience, it would completely outstrip its power to apply it to the establishment of the universe. The only thing we can assert on the basis of reason, according to Buridan, is the following: the sentence "The world exists" can be asserted with a claim to truth, but the temporal index according to which it began to be true must be determined with reference to the present, as has already been suggested.

Buridan also takes part in the decades-long debate about whether it is through his permanence, his power, or both, that God's infinity should be determined.[40] To which logic does the predicate "infinite" belong? Can this infinity be demonstrated without recourse to assurances from faith? Unfortunately, there is no room to treat these questions here.

I hope to have shown that although he holds fairly traditional views if only their content is considered, Buridan is involved in providing new foundations for them, or else he views the possibility of

---

[39] The belief in a temporal beginning of the world up to this point did not contain any idea of abstract quantity. Accordingly this beginning did not lie in some unrepresentable pre-time (*unvordenkliche Vorzeit*), but rather in a determinable past time; the world, according to Augustine, was "... not yet six thousand years" old: Augustine, *De civitate Dei*, XII, 11, p. 365; Meister Eckhart, *Predigten* [Quint e.a.], 30, vol. 2, p. 96: "Allez, daz got geschouf vor sehs tûsent jâren, dô er die werlt machete, ...;" Meister Eckhart, *op.cit.*, 10, vol. 1, 166. Also, consider that Judaism has begun its calendar since the tenth century with the creation of the world, which it fixes at 3716 B.C.

[40] John Buridan, *Quaestiones super octo Physicorum libros Aristotelis* (Paris, 1509), VIII, 9, fols. 116vb-118ra.

providing them with foundations as rather limited. To a certain degree, he follows Avicenna in maintaining that the most adequate conception of God comes not from physics, but metaphysics. But clearly, the actual elucidation of this concept – regardless of its programmatic direction – is profoundly influenced by considerations of natural philosophy. This is especially true in the case of his new conception of self-motion, only part of which I was able to address (think of the famous impetus theory of projectile motion,[41] or the lesser-known discussion of the motion of the elements).[42] This new understanding of self-motion undermined the theoretical stability of the concept of the "first unmoved mover." In light of this, I think that another claim, which does not merely concern the internal coherence of Buridan's philosophy, becomes plausible: Buridan's understanding of philosophical theology does not permit analogies between the contents of faith and those of philosophy – and this for obvious reasons. There can only be occasional points where it is proclaimed that these are not in conflict. Buridan's innovative natural philosophy also tends to make its connection to metaphysics somewhat unstable. While his originality in the field of natural philosophy was a strong incentive for others to carry his project further, his ongoing commitment to a metaphysics of the "first unmoved mover" was far less attractive. Despite all this, the fact that other particularly fruitful traditions such as Scotism opted for a strict division of physics and metaphysics only strengthens the impression that must have been evoked in Buridan's readers. In any case, we should not downplay the intensity of Buridan's effort to reconcile the idea of self-motion (spontaneity) with what Aristotle thought we could understand of the ground of reality, viz., that it is a first unmoved mover.

---

[41] John Buridan, *op.cit.*, VIII, 12, fols. 120rb-121rb.
[42] Cf. n. 24.

BIBLIOGRAPHY

1. Manuscripts

JOHN BURIDAN, *Quaestiones in Metaphysicam, secundum ultimam lecturam.* Carpentras, Bibliothèque Municipale, 292, fols. 45ra–118rb
—, *Quaestiones in Metaphysicam, secundum ultimam lecturam.* Paris, Bibliothèque nationale de France, lat. 14.716, fols. 117ra–191vb
—, *Quaestiones super octo Physicorum libros Aristotelis, secundum ultimam lecturam.* Città del Vaticano, Biblioteca Apostolica Vaticana, Cod. Vat. lat. 2163.
—, *Quaestiones super octo Physicorum libros Aristotelis, tertia lectura.* Città del Vaticano, Biblioteca Apostolica Vaticana, Cod. Chigi E VI 199, fols. 1ra–99vb
—, *Quaestiones super octo Physicorum libros Aristotelis, secundum ultimam lecturam.* Frankfurt am Main, Stadt- und Universitätsbibliothek, Cod. Praed. 52, fols. 1ra–138rb
—, *Quaestiones super octo Physicorum libros Aristotelis, secundum ultimam lecturam.* København, Kongelige Bibliotek, Cod. Ny kgl. Saml. 1801 fol.
MARSILIUS OF INGHEN, *Quaestiones in Metaphysicam.* Kraków, Biblioteka Jagiellońska, Cod. 709, fols. 1ra–182vb
—, *Quaestiones in Metaphysicam.* Wien, Österreichische Nationalbibliothek, CVP 5297

2. Pre-Modern Published Sources

ALBERTUS MAGNUS, *De anima.* Ed. C. Stroick. Münster, 1968. (*Opera omnia*, 7/1)
ANSELM, *De concordia praescientiae et praedestinationis et gratiae Dei cum libero arbitrio.* Ed. F. S. Schmitt. Roma, 1940. Repr. Stuttgart & Bad Cannstatt, 1968. (*Opera omnia*, 2)
ARISTOTELES LATINUS, *Physica. Translatio vetus.* Ed. F. Bossier & J. Brams. Leiden & New York, 1990. (*Aristoteles Latinus*, VII, 1, fasc. 2)
ARISTOTLE, *The Complete Works of Aristotle. The Revised Oxford Translation.* Ed. J. Barnes. 2 vols. Princeton, 1984. (*Bollingen Series*, LXXI:2)
—, *Posterior Analytics.* Translated with notes. Ed. J. Barnes. 2nd edition. Oxford, 1994.
—, *On the Soul, Parva Naturalia, On Breath.* Translated by W. S. Hett. London, 1975. (*Loeb Classical Library*, 288)
AUGUSTINE, *De civitate Dei.* Ed. B. Dombart e.a. Turnhout, 1955. (*Corpus Christianorum Series Latina*, 47–48)
—, *De diversis quaestionibus octoginta tribus.* Ed. A. Mutzenbecher. Turnhout, 1975. (*Corpus Christianorum Series Latina*, 44 A)

—, *De Genesi ad litteram libri duodecim.* Ed. J. Zycha. Praha, Vienna & Leipzig, 1894. (*Corpus Scriptorum Ecclesiasticorum Latinorum,* 28)
AVERROES, *Aristotelis opera cum Averrois commentariis.* Venezia, 1562-1574. Repr. Frankfurt am Main, 1962.
—, *Tahafut al-Tahafut (The Incoherence of the Incoherence).* Translated by S. van den Burgh. 2 vols. London, 1954.
AVICENNA, *Liber De anima seu Sextus De naturalibus, IV-V.* Ed. S. van Riet. Louvain & Leiden, 1968. (*Avicenna Latinus,* 1)
CICERO, *Tusculanarum disputationum libri V.* Ed. M. Pohlenz & O. Heine. 2 vols. Stuttgart, 1957.
DANTE ALIGHIERI, *La Divina Commedia.* Ed. G. Petrocchi. 4 vols. Verona, 1966-1967.
—, *Monarchia.* Translated with a commentary by R. Kay. Toronto, 1998. (*Studies and Texts,* 131)
DENIFLE, H. & A. CHATELAIN (EDS.), *Chartularium Universitatis Parisiensis.* 4 vols. Paris, 1889-1897.
DIETRICH OF FREIBERG, *De elementis corporum naturalium.* Ed. M. R. Pagnoni-Sturlese. Hamburg, 1985. (*Opera omnia,* 4)
Die Fragmente der Vorsokratiker. Ed. H. Diels & W. Kranz. 7th edition. Vol.1. Berlin, 1954.
GEORGIUS LOKERT SCOTUS (ED.), *Quaestiones et decisiones physicales insignium virorum.* Paris, 1516 &1518. Repr. Frankfurt am Main, 1969.
GODFREY OF FONTAINES, *Les Quodlibets cinq, six et sept de Godefroid de Fontaines.* Ed. M. de Wulf & J. Hoffmans. Louvain, 1914. (*Les Philosophes Belges, Textes et études,* 3)
HAMESSE, J., *Les Auctoritates Aristotelis. Un florilège médiéval. Étude historique et édition critique.* Louvain & Paris, 1974. (*Philosophes médiévaux,* 17)
HENRY OF GHENT, *Quodlibet I.* Ed. R. Macken. Louvain & Leiden, 1979. (*Opera omnia,* 5)
—, *Quodlibet IX.* Ed. R. Macken. Louvain, 1983. (*Opera omnia,* 13)
—, *Summa (Quaestiones ordinariae).* Art. XXXV-XL. Ed. G. A. Wilson. Louvain, 1994. (*Opera omnia,* 28)
—, *Summae quaestionum ordinariarum.* Paris, 1520. Repr. Louvain & Paderborn, 1953. (*Franciscan Institute Publications, Text Series,* 5)
JEAN GERSON, *Trilogium astrologiae theologizatae.* Ed. P. Glorieux. Paris, 1973. (*Oeuvres complètes,* 10)
JOHN BURIDAN, *De dependentiis, diuersitatibus, et conuenientiis.* In J. M. M. H. Thijssen, "Buridan on the Ontological Status of Causal Relations. A first Presentation of the Polemic 'Questio de dependentiis, diversitatibus et convenientiis'." In Zimmermann e.a. (eds.), *Mensch und Natur im Mittelalter,* pp. 234-255.
—, *Summulae de dialectica.* Translated by G. Klima, with introduction and notes. New Haven, Conn. (forthcoming in the *Yale Library of Medieval Philosophy Series*)
—, *Expositio et quaestiones in Aristotelis De caelo.* Ed. B. Patar. Louvain-la Neuve, Louvain & Paris, 1996. (*Philosophes médiévaux,* 33)
—, *Summulae. De fallaciis.* Ed. H. Hubien. [unpublished manuscript]

—, *In Metaphysicen Aristotelis questiones.* Paris, 1518. Repr. Frankfurt am Main, 1964.
—, *De memoria et reminiscentia.* Ed. Georgius Lokert Scotus. In Georgius Lokert Scotus (ed.), *Quaestiones et decisiones physicales insignium virorum* (Paris, 1516 & 1518), fols. 41ra–42vb.
—, *Expositio et quaestiones in Aristotelis De anima.* In B. Patar, *Le Traité de l'âme de Jean Buridan [De prima lectura]. Édition, étude critique et doctrinale.* Louvain-la-Neuve & Longueil, 1991. (*Philosophes médiévaux*, 29)
—, *Quaestiones in Aristotelis De anima liber secundus, de tertia lectura.* In P. G. Sobol, *John Buridan on the Soul and Sensation. An Edition of Book II of his Commentary on Aristotle's Book on the Soul with an Introduction and a Translation of Question 18 on Sensible Species.* Ph.D. diss. Indiana University, 1984.
—, *Quaestiones in Aristotelis De anima, liber tertius.* In J. Zupko, *John Buridan's Philosophy of Mind. An Edition and Translation of Book III of his "Questions on Aristotle's De anima" (Third Redaction), with Commentary and Critical and Interpretative Essays.* 2 vols. Ph.D. diss. Cornell University, 1989.
—, *Quaestiones in Posteriorum Analyticorum libros.* Ed. H. Hubien. [unpublished manuscript]
—, *Quaestiones in Praedicamenta.* Ed. J. Schneider. München, 1983. (*Bayerische Akademie der Wissenschaften. Veröffentlichungen der Kommission für die Herausgabe ungedruckter Texte aus der mittelalterlichen Geisteswelt*, 11)
—, *Quaestiones in Priorum Analyticorum libros.* Ed. H. Hubien. [unpublished manuscript]
—, *Quaestiones super decem libros Ethicorum Aristotelis.* Paris, 1513. Repr. Frankfurt am Main, 1968.
—, *Quaestiones super libris quattuor De caelo et mundo.* Ed. E. A. Moody. Cambridge, Mass., 1942. (*The Mediaeval Academy of America*, 6) Repr. Cambridge, Mass., 1970. (*The Mediaeval Academy of America*, 40)
—, *Quaestiones super octo Physicorum libros Aristotelis, secundum ultimam lecturam.* Paris, 1509. Repr. Frankfurt am Main, 1968.
—, *Questiones longe super librum Perihermeneias.* Ed. R. van der Lecq. Nijmegen, 1983. (*Artistarium*, 4)
—, *De somno et vigilia.* Ed. Georgius Lokert Scotus. In Georgius Lokert Scotus (ed.), *Quaestiones et decisiones physicales insignium virorum* (Paris, 1516 & 1518), fols. 41ra–49ra.
—, *Sophismata.* Ed. T. K. Scott. Stuttgart & Bad Cannstatt, 1977. (*Grammatica Speculativa*, 1)
—, *Summulae. De demonstrationibus.* Ed. L. M. de Rijk & H. Hubien [unpublished manuscipt]
—, *Summulae. De suppositionibus.* Ed. R. van der Lecq. Nijmegen, 1998. (*Artistarium*, 10-4)
—, *Tractatus de consequentiis.* Ed. H. Hubien. Louvain & Paris, 1976. (*Philosophes Médiévaux*, 16)
—, *Tractatus de differentia universalis ad individuum.* Ed. S. Szyller. *Przeglad Tomistyczny*, 3 (1987), 135–178.

—, *Tractatus de infinito. Quaestiones super libros Physicorum secundum ultimam lecturam, Liber III, quaestiones 14–19*. Ed. J. M. M. H. Thijssen. Nijmegen, 1991. (*Artistarium Supplementa*, VI)

JOHN DUNS SCOTUS, *Lectura*. Città del Vaticano, 1960-. (*Opera omnia*, 16-)

—, *Ordinatio*. Città del Vaticano, 1963. (*Opera omnia*, 6)

—, *Ordinatio*. Lyon, 1639. Repr. Hildesheim, 1968. (*Opera omnia*, 5–10)

—, *Quaestiones super libros Metaphysicorum Aristotelis*. Ed. R. Andrews, G. Etzkorn, G. Gál, R. Green, F. Kelley, G. Marcil, T. Noone & R. Wood. 2 vols. St. Bonaventure, New York, 1997. (*Opera philosophica*, 3 & 4)

—, *Quaestiones super universalia Porphyrii*. Paris, 1891. (*Opera omnia*, 1)

KAHN, C. H., *The Art and Thought of Heraclitus. An Edition of the Fragments with Translation and Commentary*. Cambridge, 1979.

KING, P., "Jean Buridan on Mental Language." In Normore (ed.), *An Anthology* (forthcoming).

—, *Jean Buridan's Logic: The Treatise on Supposition, the Treatise on Consequences*. Dordrecht, 1985. (*Synthese Historical Library*, 27)

LUIS CORONEL, *Physice perscrutationes*. Lyon, 1530.

MACROBIUS, *Commentarii in Somnium Scipionis*. Ed. J. Willis. Leipzig, 1963. (*Bibliotheca scriptorum graecorum et romanorum teubneriana*)

MEISTER ECKHART, *Predigten*. Ed. J. Quint & G. Steer. Stuttgart, 1958-. (*Die deutschen Werke*, 1-)

NORMORE, C. (ED.), *An Anthology on Mental Language*. (forthcoming)

BLAISE PASCAL, *Pensées*. Ed. L. Lafuma. Paris, 1962.

PETER AUREOLI, *Commentarium in secundum librum Sententiarum*. Roma, 1596.

PETER JOHN OLIVI, *Quaestiones in secundum librum Sententiarum*. Ed. B. Jansen. 3 vols. Quaracchi, 1922–1926. (*Bibliotheca franciscana scholastica medii aevi*, 4–6)

PETER LOMBARD, *Sententiae in IV libris distinctae, II: liber III et IV*. Grottaferrata, 1981. (*Spicilegium Bonaventurianum*, 5)

PLATO, *The collected dialogues of Plato, including the letters*. Ed. E. Hamilton & H. Cairns. Princeton, 1961. (*Bollingen Series*, LXXI)

—, *Timaeus a Calcidio translatus commentarioque instructus*. Ed. J. H. Waszink. London & Leiden, 1962. (*Plato latinus*, 4)

PLUTARCH, *Plutarch's Lives. With an English translation*. Ed. B. Perrin. London & Cambridge, Mass., 1959. (*The Loeb Classical Library*)

RIJK, L.M. DE (ED.), *Nicholas of Autrecourt. His Correspondence with Master Giles and Bernard of Arezzo*. Leiden, New York & Köln, 1994. (*Studien und Texte zur Geistesgeschichte des Mittelalters*, 42)

ROGER BACON, *De multiplicatione specierum*. In D. C. Lindberg, *Roger Bacon's Philosophy of Nature. A Critical Edition, with English Translation, Introduction, and Notes, of "De multiplicatione specierum and De speculis comburentibus"*. Oxford, 1983.

SENECA, *Ad Lucilium*. Ed. L. D. Reynolds. 2 vols. Oxford, 1965.

SIGER OF BRABANT, *Quaestiones in Metaphysicam*. Ed. A. Maurer. Louvain-la-Neuve, 1983. (*Philosophes médiévaux*, 25)

—, *Impossibilia*. In B.-C. Bazán (ed.), *Siger de Brabant. Ecrits de logique, de morale et de physique*. Louvain & Paris, 1974. (*Philosophes médiévaux*, 14)

THOMAS AQUINAS, *Aristotelis Librum De anima*. Ed. A. M. Pirotta, O.P. Turin, 1959.
—, *Summa contra gentiles*. 3 vols. Turin, 1961–1967.
—, *De ente et essentia*. Roma, 1976. (*Opera omnia*, 43)
—, *Expositio libri Boetii de ebdomadibus*. Roma & Paris, 1992. (*Opera omnia*, 50)
—, *Expositio libri Posteriorum. Editio altera retractata*. Roma & Paris, 1989. (*Opera omnia*, 1/2)
—, *De rationibus fidei*. Roma, 1969. (*Opera omnia*, 40).
—, *Scriptum super Sententiis*. Ed. P. Mandonnet & M. P. Moos. 5 vols. Paris, 1929–1947.
—, *De substantiis separatis*. Turin, 1954. (*Opera Philosophica*) —, *Summa contra gentiles*. 3 vols. Roma, 1918–1930. (*Opera omnia*, 13–15)
—, *Summa contra gentiles*. Translated and edited by K. Albert & P. Engelhardt. In *Summe gegen die Heiden*. 4 vols. Darmstadt, 1974–1996. (*Texte zur Forschung*, 15–19)
—, *Summa Theologiae*. Roma, 1888–1906. (*Opera omnia*, 4–12)
—, *Super Boetium De Trinitate*. Roma & Paris, 1992. (*Opera omnia*, 50)
—, *De veritate*. Roma, 1975. (*Opera omnia*, 22/1)
THOMAS CAJETAN, *Super Librum De ente et essentia. Commentary on Being and Essence. Translated from the Latin with an Introduction*. Ed. L. H. Kendzierski & F. C. Wade. Milwaukee, 1964. (*Mediaeval philosophical texts in translation*, 14)
FEDERICI VESCOVINI, G., *Les "Quaestiones De anima" di Bagio Pelacani da Parma*. Firenze, 1974. (*Accademia Toscana di scienze e lettere "La Colombaria," Studi*, 30)
WILLIAM OCKHAM, *Quaestiones in librum quartum Sententiarum (Reportatio)*. Ed. R. Wood & G. Gál. St. Bonaventure, New York, 1984. (*Opera Theologica*, 7)
—, *Quodlibeta septem*. Ed. J. C. Wey. St. Bonaventure, New York, 1980. (*Opera Theologica*, 9)
—, *Scriptum in librum primum Sententiarum. Ordinatio*. Ed. G. Gál, S. Brown, G. Etzkorn & F. Kelley. St. Bonaventure, New York, 1967–1979. (*Opera Theologica*, 1–4)
—, *Summa logicae*. Ed. Ph. Boehner, G. Gál & S. Brown. St. Bonaventure, New York, 1974. (*Opera Philosophica*, 1)

3. Modern Works

AERTSEN, J. A. & A. SPEER (EDS.), *Was ist Philosophie im Mittelalter?* Berlin & New York, 1998. (*Miscellanea Mediaevalia*, 26)
BAKKER, P. J. J. M., *La raison et le miracle. Les doctrines eucharistiques (c. 1250 – c. 1400). Contribution à l'étude des rapports entre philosophie et théologie*. 2 vols. Nijmegen, 1999.
BAYER, O., R. W. JENSEN & S. KNUUTTILA (EDS.), *Caritas Dei. Beiträge zum Verständnis Luthers und der gegenwärtigen Ökumene*. Helsinki, 1997. (*Schriften der Luther-Agricola-Gesellschaft*, 39)

BECKMAN, J. P., L. HONNEFELDER & G. JÜSSEN (EDS.), *Sprache und Erkenntnis im Mittelalter*. 2 vols. Berlin & New York, 1981. (*Miscellanea Mediaevalia*, 13)

BERNARD, J. (ED.), *Logical Semiotics*. Wien, 1991. (*European Journal for Semiotic Studies*, vol. 3. no. 4)

BIANCHI, L., *Censure et liberté intellectuelle à l'université de Paris (XIIIe à XIVe siècles*. Paris, 1999. (*L'âne d'or*, 9)

—, *L'errore di Aristotele. La polemica contro l'eternità del mondo nel XIII secolo*. Firenze, 1984.

BIANCHI, L. (ED.), *Filosofia e teologia nel Trecento. Studi in ricordo di Eugenio Randi*. Louvain-la-Neuve, 1994. (*Textes et études du Moyen Âge*, 1)

BIANCHI, L. & E. RANDI, *Le verità dissonanti. Aristotele alla fine del Medioevo*. Roma & Bari, 1990.

BIARD, J., "L'idée de nature dans la Physique de Jean Buridan." In Vescovini (ed.), *Filosofia, scienza classica*, pp. 97–113.

—, *Logique et théorie du signe au XIVe siècle*. Paris, 1989. (*Études de philosophie médiévale*, 64)

—, "Le statut du mouvement dans la philosophie naturelle buridanienne." In Caroti e.a. (eds.), *La nouvelle physique*, pp. 141–159.

—, "Le système des causes dans la philosophie naturelle de Jean Buridan." In Hasnawi e.a. (eds.), *Perspectives arabes et médiévales*, pp. 491–504.

BOGEN, J. & J. E. MCGUIRE (EDS.), *How Things Are. Studies in Predication and the History of Philosophy and Sience*. Dordrecht, Boston & Lancaster, 1985. (*Philosophical Studies Series in Philosophy*, 29)

BOS, E. P. & H. A. KROP (EDS.), *John Buridan: A Master of Arts. Some Aspects of his Philosophy*. Nijmegen, 1993. (*Artistarium Supplementa*, 8)

—, *Ockham and Ockhamists*. Nijmegen, 1987. (*Artistarium Supplementa*, 4)

BRAAKHUIS, H. A. G., *De 13de eeuwse tractaten over syncategorematische termen*. 2 vols. Meppel, 1979.

—, "John Buridan and the 'Parisian school' on the Possibility of Returning as Numerically the Same. A Note on a Chapter in the History of the Relationship between Faith and Natural Science." In Caroti e.a. (eds.), *La nouvelle physique*, pp. 111–140.

BRAAKHUIS, H. A. G., C. H. KNEEPKENS & L. M. DE RIJK (EDS.), *English Logic and Semantics. From the End of the Twelfth Century to the Time of Ockham and Burleigh*. Nijmegen, 1981. (*Artistarium Supplementa*, 1)

BROWN, S. F., "A Modern Prologue to Ockham's Natural Philosophy." In Beckman e.a. (eds.), *Sprache und Erkenntnis im Mittelalter*, pp. 107–129.

CAROTI, S. (ED.), *Studies in Medieval Natural Philosophy*. Firenze, 1989. (*Biblioteca di Nuncius, Studi e testi*, 1)

CAROTI, S. & P. SOUFFRIN (EDS.), *La nouvelle physique du XIVe siècle*. Firenze, 1997. (*Biblioteca di Nuncius, Studi e testi*, 24)

CARROLL, W. E. (ED.), *Nature and Motion in the Middle Ages*. Washington D.C., 1985.

CHARLTON, W., "Aristotle's Potential Infinites." In Judson (ed.), *Aristotle's Physics*, pp. 129–151.

CHENEVAL, F., R. IMBACH & TH. RICKLIN, *Albert le Grand et sa réception au moyen âge. Hommage à Zénon Kaluza*. Fribourg, 1998. (Special issue of *Freiburger Zeitschrift für Philosophie und Theologie*, 45 (1998), 1–2)

CLAGETT, M., *The Science of Mechanics in the Middle Ages*. 2nd edition. Madison, Wisc., 1961.

CORSI, G., C. MANGIONE & M. MUGNAI (EDS.), *Atti del convegno internazionale di storia della logica: le theorie delle modalità*. Bologna, 1989.

COURTENAY, W. J., "Inquiry and Inquisition: Academic Freedom in Medieval Universities." *Church History*, 58 (1989), 168–181.

CRAEMER-RÜGENBERG, I. & TH. SPEER (EDS.), *Scientia und Ars im Hoch- und Spätmittelalter*. 2 vols. Berlin & New York, 1984. (*Miscellanea Mediaevalia*, 22)

DALES, R. C., *Medieval Discussions of the Eternity of the World*. Leiden, 1990.

DAVENPORT, A. A., *Measure of a Different Greatness: The Instensive Infinite, 1250–1650*. Leiden, 1999.

DUERING, I. (ED.), *Naturphilosophie bei Aristoteles und Theophrast*. Heidelberg, 1969. (*Symposium Aristotelicum*, 4)

DUHEM, P., *Études sur Leonard de Vinci*. 3 vols. Paris, 1906–1913.

EBBESEN, S., "The Way Fallacies Were Treated in Scholastic Logic." *Cahiers de l'Institut du Moyen Âge Grec et Latin*, 55 (1987), 107–34.

—, "Proof and its Limits according to Buridan." In Kaluza e.a. (eds.), *Preuve et raisons*, pp. 97–110.

EBBESEN, S. & R. L. FRIEDMAN (EDS.), *Medieval Analyses in Language and Cognition*. København, 1999. (*Historisk-filosofiske Meddelelser*, 77)

FEYERABEND, P., *Realism, Rationalism and Scientific Method*. Vol. 1. Cambridge, 1981.

FRIED, J., *Die Abendländische Freiheit vom 10. zum 14. Jahrhundert. Der Wirkungszusammenhang von Idee und Wirklichkeit im europäischen Vergleich*. Sigmaringen, 1991.

FRIEDMAN, R. L., *In principio erat verbum: The Incorporation of Philosophical Psychology into Trinitarian Theology 1250–1325*. Ph.D. diss. University of Iowa, 1997.

GARCÍA Y GARCÍA, A. (ED.), *Constitutiones Concilii quarti Lateranensis una cum Commentariis glossatorum*. Città del Vaticano, 1981. (*Monumenta iuris canonici*, A/2)

GHISALBERTI, A., *Giovanni Buridano dalla metafisica alla fisica*. Milano, 1975. (*Scienze filosofiche. Università del Sacro Cuore*, 13)

GOERING, J., "The Invention of Transubstantiation." *Traditio*, 46 (1991), 147–170.

GRACIA, J. J. E. (ED.), *Individuation in Scholasticism: the Later Middle Ages and the Counter-Reformation, 1150–1650*. Albany & New York, 1994. (*SUNY series in philosophy*)

GRANT, E., "The Condemnation of 1277, God's Absolute Power, and Physical Thought in the Late Middle Ages." *Viator*, 10 (1979), 211–244.

—, "Science and the Medieval University." In Kittelson e.a. (eds.), *Rebirth, Reform and Resilience*, pp. 68–102.

—, *A Source Book in Medieval Science*. Cambridge, Mass., 1974.

HARRÉ, R., *Laws of Nature*. London, 1993.

HARRÉ, R. & E. H. MADDEN, *Causal Powers: A Theory of Natural Necessity*. Oxford, 1975.
HASNAWI, A., A. ELAMRANI-JAMAL & M. AOUAD (EDS.), *Perspectives arabes et médiévales sur la tradition scientifique et philosophique grecque*. Turnhout, 1997.
HEGEL, G. W. F., *Vorlesungen über die Geschichte der Philosophie*. Ed. H. Glockner. Stuttgart, 1928. (*Sämtliche Werke*, 17-19)
HEIDEGGER, M., *Einführung in die Metaphysik*. Frankfurt am Main, 1983. (*Gesamtausgabe*, 40)
—, *Der Satz vom Grund*. Frankfurt am Main, 1997. (*Gesamtausgabe*, 10)
—, *Wegmarken*. Frankfurt am Main, 1976. (*Gesamtausgabe*, 9)
—, "Vom Wesen und Begriff der Physis. Aristoteles, Physik B, 1." In Heidegger, *Wegmarken*, pp. 239-303.
HENRY, D. P., *Medieval Mereology*. Amsterdam & Philadelphia, 1991. (*Bochumer Studien zur Philosophie*, 16)
HINTIKKA, J., "Aristotelian Infinity." In Hintikka, *Time and Necessity*, pp. 321-330.
—, *Time and Necessity. Studies in Aristotle's Theory of Modality*. Oxford, 1973.
HISSETTE, R., *Enquête sur les 219 articles condamnés à Paris le 7 Mars 1277*. Louvain & Paris, 1977. (*Philosophes médiévaux*, 22)
HÖDL, L., "Der Transsubstantiationsbegriff in der scholastischen Theologie des 12. Jahrhunderts." *Recherches de théologie ancienne et médiévale*, 31 (1964), 230-259.
HOENEN, M. J. F. M., *Marsilius of Inghen. Divine Knowledge in Late Medieval Thought*. Leiden, 1993. (*Studies in the History of Christian Thought*, 50)
—, "Marsilius von Inghen in der Geistesgeschichte des ausgehenden Mittelalters." In Hoenen e.a., *Philosophie und Theologie*, pp. 21-45.
HOENEN, M. J. F. M. & P. J. J. M. BAKKER (EDS.), *Philosophie und Theologie des ausgehenden Mittelalters. Marsilius von Inghen und das Denken seiner Zeit*. Leiden, 2000.
HONNEFELDER, L., R. WOOD & M. DREYER (EDS.), *John Duns Scotus: Metaphysics and Ethics*. Leiden, New York & Köln, 1996. (*Studien und Texte zur Geistesgeschichte des Mittelalters*, 53)
HUGHES, G. E, "The Modal Logic of John Buridan." In Corsi e.a. (eds.), *Atti*, pp. 93-113.
IMBACH, R., "Metaphysik, Theologie und Politik. Zur Diskussion zwischen Nikolaus von Straßburg und Dietrich von Freiberg über die Abtrennbarkeit der Akzidentien." *Theologie und Philosophie*, 61 (1986), 359-395.
—, "Philosophie und Eucharistie bei Wilhelm von Ockham. Ein vorläufiger Entwurf." In Bos e.a. (eds.), *Ockham and Ockhamists*, pp. 43-51.
—, "Pourquoi Thierry de Freiberg a-t-il critiqué Thomas d'Aquin? Remarques sur le *De accidentibus*." In Chenevale.a. (eds.), *Albert le Grand*, pp. 116-129.
—, "Le traité de l'eucharistie de Thomas d'Aquin et les averroïstes." *Revue des sciences philosophiques et théologiques*, 77 (1993), 175-194.
IMBACH, R. & A. MAIERÙ (EDS.), *Gli studi di filosofia medievale fra Otto e Novecento. Contributo a un bilancio storiografico*. Roma, 1991. (*Storia e Letteratura*, 179)

JANIS, A. I. & T. HOROWITZ (EDS.), *Scientific Failure*. S. l., 1993.
JANSEN, F., "Eucharistiques (accidents)." In *Dictionnaire de théologie catholique*, 5/2, Paris, 1924, 1368–1452.
JORDAN, M., *The Alleged Aristotelianism of Thomas Aquinas*. Toronto, 1992. (*Etienne Gilson Lecture Series, The Pontifical Institute of Mediaeval Studies*)
JORISSEN, H., *Die Entfaltung der Transsubstantiationslehre bis zum Beginn der Hochscholastik*. Münster, 1965. (*Münsterische Beiträge zur Theologie*, 28/1)
JUDSON, L. (ED.), *Aristotle's Physics. A Collection of Essays*. Oxford, 1991.
KALUZA, Z. & P. VIGNAUX (EDS.), *Preuve et raisons à l'université de Paris. Logique, ontologie et théologie au XIVe siècle*. Paris, 1984. (*Études de philosophie médiévale, hors série*)
KEITH, G., "Deconstructing Star Trek." *Wired*, 6 (1998).
KENNY, A., *Descartes: A Study of His Philosophy*. New York, 1968.
KING, P., "Jean Buridan." In Gracia (ed.), *Individuation in Scholasticism*, pp. 397–430.
—, "Jean Buridan's Philosophy of Science." *Studies in History and Philosophy of Science*, 18 (1987), 109–132.
—, "Scholasticism and the Philosophy of Mind: The Failure of Aristotelian Psychology." In Janis e.a. (eds.), *Scientific Failure*, pp. 109–138.
KITTELSON, J. M. & P. J. TRANSUE (EDS.), *Rebirth, Reform and Resilience: Universities in Transition, 1300–1700*. Columbus, Ohio, 1984.
KLIMA, G., "Buridan's Logic and the Ontology of Modes." In Ebbesen e.a. (eds.), *Medieval Analyses*, pp. 458–479.
—, "Ockham's Semantics and Metaphysics of the Categories." In Spade (ed.), *The Cambridge Companion to Ockham*, pp. 118–142.
—, "Ontological Alternatives vs. Alternative Semantics in Medieval Philosophy." In Bernard (ed.), *Logical Semiotics*, pp. 587–618.
KNUUTTILA, S., "Duns Scotus and the Foundations of Logical Modalities." In Honnefelder e.a. (eds.), *John Duns Scotus*, pp. 127–143.
—, *Modalities in Medieval Philosophy*. London & New York, 1993. (*Topics in Medieval Philosophy*, 8)
—, "Natural Necessity in John Buridan." In Caroti (ed.), *Studies*, pp. 155–176.
—, "*Positio impossibilis* in Medieval Discussions of the Trinity." In Marmo, *Vestigia, Imagines, Verba*, pp. 277–288.
KNUUTTILA, S. & R. SAARINEN, "Innertrinitarische Theologie in der Scholastik und bei Luther." In Bayer e.a. (eds.), *Caritas Dei*, pp. 243–264.
KRETZMANN, N., "*Sensus compositus, sensus divisus*, and Propositional Attitudes." *Mediaevo*, 7 (1981), 195–229.
—, "*Syncategoremata, exponibilia, sophismata*." In Kretzmann e.a. (eds.), *The Cambridge History of Later Medieval Philosophy*, pp. 211–246.
KRETZMANN, N. (ED.), *Infinity and Continuity in Ancient and Medieval Thought*. Ithaca & New York, 1982.
KRETZMANN, N., A. KENNY & J. PINBORG (EDS.), *The Cambridge History of Later Medieval Philosophy*.
KRIEGER, G., *Subjekt und Metaphysik. Die Transformation der Metaphysik im Denken des Johannes Buridan*. (forthcoming)

KUHN, TH., *The Structure of Scientific Revolutions.* Chicago, 1970.
LAGERLUND, H., *Modal Syllogistics in the Middle Ages.* Ph.D. diss. University of Uppsala, 1999.
LEAR, J., "Aristotelian Infinity." *Proceedings of the Aristotelian Society, New series,* 80 (1980), 187-210.
LECQ, R. VAN DER, "Buridan on Modal Propositions." In Braakhuis e.a. (eds.), *English Logic and Semantics,* pp. 427-442.
—, "Confused Individuals and Moving Trees: John Buridan on the Knowledge of Particulars." In Bos e.a. (eds.), *John Buridan,* pp. 1-21.
—, "Paul of Venice on Composite and Divided Sense." In Maierù (ed.), *English Logic and Semantics in Italy,* pp. 114-134.
LIBERA, A. DE, *Albert le Grand et la philosophie.* Paris, 1990.
LIBERA, A. DE, A. ELAMRANI-JAMAL & A. GALONNIER (EDS.), *Langages et philosophie. Hommage à Jean Joliveted.* Paris, 1997. (*Études de philosophie médiévale,* 74)
LINDBERG, D. C. (ED.), *Science in the Middle Ages.* Chicago & London, 1978.
LIVESEY, S., "The Oxford *Calculatores,* Quantification of Qualities, and Aristotle's Prohibition of *Metabasis.*" *Vivarium,* 24 (1986), 50-69.
—, "William of Ockham, the Subalternate Sciences, and Aristotle's Theory of Metabasis." *British Journal for the History of Science,* 18 (1985), 127-145.
MAIER, A., *An der Grenze von Scholastik und Naturwissenschaft.* 2nd edition. Roma, 1952. (*Storia e letteratura,* 41)
—, *Ausgehendes Mittelalter. Gesammelte Aufsätze zur Geistesgeschichte des 14. Jahrhunderts.* 3 vols. Roma, 1964-1977. (*Storia e letteratura,* 97, 105 & 138)
—, "Les commentaires sur la Physique d'Aristote attribués à Siger de Brabant." *Revue philosophique de Louvain,* 47 (1949), 334-350. Repr. Maier, *Ausgehendes Mittelalter,* pp. 189-206.
—, "Diskussionen über das aktuell Unendliche in der ersten Hälfte des 14. Jahrhunderts." In Maier, *Ausgehendes Mittelalter,* pp. 41-87.
—, *Die Vorlaüfer Galileis im 14. Jahrhundert.* Roma, 1949. (*Storia e letteratura,* 22)
—, *Metaphysische Hintergründe der spätscholastischen Naturphilosophie.* Roma, 1955. (*Storia e letteratura,* 52)
—, "Das Problem der Evidenz in der Philosophie des 14. Jahrhunderts." In Maier, *Ausgehendes Mittelalter,* pp. 367-418.
—, *Zwei Grundprobleme der Scholastischen Naturphilosophie.* Third enlarged edition. Roma, 1968. *Storia e letteratura,* 37)
MAIERÙ, A. (ED.), *English Logic and Semantics in Italy in the 14th and 15th Centuries.* Napoli, 1982. (*History of Logic,* 1)
MAIERÙ, A., & A. PARAVICINI BAGLIANI (EDS.), *Studi sul XIV secolo in memoria di Anneliese Maier.* Roma, 1981. (*Storia e letteratura,* 151)
MANDONNET, P., *Siger de Brabant et l'averroïsme latin au XIIIe siècle.* Louvain, 1908-1911.
MARKOWSKI, M., "Les *Questiones* de Marsile d'Inghen sur la Métaphysique d'Aristote. Notes sur les manuscrits et le contenu des questions." *Mediaevalia philosophica Polonorum,* 13 (1968), 8-32.
MARMO, C. (ED.), *Vestigia, Imagines, Verba. Semiotics and Logic in Medieval*

*Theological Texts (XII*ᵗʰ*–XIV*ᵗʰ *Century)*. Turnhout, 1997. (*Semiotic and Cognitive Studies*, 4)
MARSHALL, P. C. , "Parisian Psychology in the Mid-Fourteenth Century." *Archives d'histoire doctrinale et littéraire du Moyen Âge*, 50 (1983), 101–193.
MARTIN, C. J., "Obligations and Liars." In Read, *Sophisms*, pp. 358–381.
MCLAUGHLIN, M. M., *Intellectual Freedom and its Limitations in the University of Paris in the Thirteenth and Fourteenth Centuries*. New York, 1977.
MICHAEL, B., *Johannes Buridan: Studien zu seinem Leben, seinen Werken und zur Rezeption seiner Theorien im Europa des späten Mittelalters*. 2 vols. Ph.D. diss. Freie Universität Berlin, 1985.
MICHALSKI, K., "La physique nouvelle et les différents courants philosophiques du XIVᵉ siècle." *Bulletin international de l'Académie polonaise des sciences et des lettres*, (1928), 1–71.
MIETHKE, J., "Bildungsstand und Freiheitsforderung (12. bis 14. Jahrhundert)." In Fried, *Die Abendländische Freiheit*, pp. 221–247.
MILLER, R., "Buridan on Singular Concepts." *Franciscan Studies*, 45 (1985), 57–72.
MOJSISCH, B. & O.PLUTA (EDS.), *Historia philosophiae medii aevi. Studien zur Geschichte der Philosophie des Mittelalters*. Festschrift für Kurt Flasch 2 vols. Amsterdam & Philadelphia, 1991–1992.
MOORE, G. E., *Philosophical Papers*. London, 1959.
MURDOCH, J. E., "The Analytic Character of Late Medieval Learning. Natural Philosophy without Nature." In Roberts (ed.), *Approaches to Nature*, pp. 171–213.
—, "From Social to Intellectual Factors: An Aspect of the Unitary Character of Late Medieval Learning." In Murdoch e.a. (eds.), *The Cultural Context*, pp. 271–339.
—, "Henry of Harclay and the Infinite." In Maierù e.a. (eds.), *Studi*, pp. 219–261.
—, "Infinity and Continuity." In Kretzmann e.a. (eds.), *The Cambridge History of Later Medieval Philosophy*, pp. 564–593.
—, "Pierre Duhem and the History of Late Medieval Science and Philosophy in the Latin West." In Imbach e.a. (eds.), *Gli studi di filosofia medievale*, pp. 253–302.
—, "*Scientia mediantibus vocibus*. Metalinguistic Analysis in Late Medieval Natural Philosophy." In Beckman e.a. (eds.), *Sprache und Erkenntnis im Mittelalter*, pp. 73–106.
—, "William of Ockham and the Logic of Infinity and Continuity." In Kretzmann (ed.), *Infinity and Continuity*, pp. 165–207.
MURDOCH, J. E. & E. D. SYLLA, "The Science of Motion." In Lindberg, *Science in the Middle Ages*, pp. 206–264.
MURDOCH, J. E. & E. D. SYLLA (EDS.), *The Cultural Context of Medieval Learning*. Dordrecht, 1975. (*Boston Studies in the Philosophy of Science*, 26)
MURDOCH, J. E. & J. M. M. H. THIJSSEN (EDS.), *Medieval Science*. Leiden, 1993. (Special issue *Vivarium*, 31 (1993), 1–191)
NARDI, B. (ED.), *La filosofia della natura nel medioevo. Atti del terzo congresso internazionale di filosofia medioevale*. Milano, 1966.

NIELSEN, L. O., "Dictates of Faith versus Dictates of Reason: Peter Aureole on Divine Power, Creation, and Human Rationality." *Documenti e studi sulla tradizione filosofica medievale*, 7 (1996), 213–241.
NORMORE, C., "Buridan's Ontology." In Bogen e.a. (eds.), *How Things Are*, pp. 189–203.
NUCHELMANS, G., *Late-Scholastic and Humanist Theories of the Proposition*. Amsterdam, 1980.
PANACCIO, C., "Connotative Terms in Ockham's Mental Language." *Cahiers d'épistémologie*, 9016 (1990), 1–22.
PARFIT, D., *Reasons and Persons*. Oxford, 1984.
PATTIN, A., *Pour l'histoire du sens agent. La controverse entre Barthélemy de Bruges et Jean de Jandun, ses antécédents et son évolution*. Louvain, 1988. (*Ancient and Medieval Philosophy*, 6)
PERREIAH, A. R., "Buridan and the Definite Description." *Journal of the History of Philosophy*, 10 (1972), 153–160.
PLUTA, O., "Der Alexandrismus an den Universitäten im späten Mittelalter." *Bochumer Philosophisches Jahrbuch für Antike und Mittelalter*, 1 (1996), 81–109.
—, "Die Diskussion der Unsterblichkeitsfrage bei Marsilius von Inghen." In Wielgus (ed.), *Marsilius von Inghen*, pp. 119–164.
—, "Einige Bemerkungen zur Deutung der Unsterblichkeitsdiskussion bei Johannes Buridan." In Bos e.a. (eds.), *John Buridan*, pp. 107–119.
—, "Ewigkeit der Welt, Sterblichkeit der Seele, Diesseitigkeit des Glücks – Elemente einer materialistischen Philosophie bei Johannes Buridanus." In Mojsisch e.a. (eds.), *Historia philosophiae medii aevi*, pp. 847–872.
—, "Die Frage nach der *felicitas humana* bei Marsilius von Inghen." *Studia Mediewistyczne*, 34 (1999), 175–190.
—, "*Homo sequens rationem naturalem* – die Entwicklung einer eigenständigen Anthropologie in der Philosophie des späten Mittelalters." In Zimmermann e.a. (eds.), *Mensch und Natur im Mittelalter*, pp. 752–763.
PORRO, P., *Forme e modelli di durata nel pensiero medievale. L'aevum, il tempo discreto, la categoria "quando"*. Louvain, 1996. (*Ancient and Medieval Philosophy*, 1, 16)
PUTALLAZ, F.-X. & R. IMBACH, *Profession: philosophe. Siger de Brabant*. Paris, 1997. (*Initiations au Moyen Âge*)
PUTNAM, H., *Realism with a Human Face*. Cambridge, Mass., 1990.
RANDI, E., *Il sovrano e l'orologiaio. Due immagini di Dio nel dibattito sulla "potentia absoluta" fra XIII e XIV secolo*. Firenze, 1987.
READ, S. (ED.), *Sophisms in Medieval Logic and Grammar*. Dordrecht, Boston & London, 1993. (*Nijhoff International Philosophy Series*, 48)
REINA, M. E., "*Comprehensio veritatis*. Una questione di Marsilio di Inghen sulla Metafisica." In Bianchi (ed.), *Filosofia e teologia nel Trecento*, pp. 283–335.
—, "L'ipotesi del 'casus supernaturaliter possibilis' in Giovanni Buridano." In Nardi (ed.), *La filosofia della natura*, pp. 683–690.
RIJK, L. M. DE, "On Buridan's View of Accidental Being." In Bos e.a. (eds.), *John Buridan*, pp. 41–51.

—, "Foi chrétienne et savoir humain. La lutte de Buridan contre les theologizantes." In De Libera e.a. (eds.), *Langages et philosophie*, pp. 393–409.
—, *Jean Buridan (c. 1292 – c. 1360). Eerbiedig ondermijner van het aristotelisch substantie-denken*. Amsterdam, 1994. (*Koninklijke Nederlandse Akademie van Wetenschappen. Mededelingen van de Afdeling Letterkunde. Nieuwe Reeks*, 57/1)
—, "John Buridan on Man's Capability of Grasping the Truth." In Craemer-Rügenberg e.a. (eds.), *Scientia und Ars*, pp. 281–303.
—, "John Buridan on Universals." *Revue de métaphysique et de morale*, 97 (1992), 35–59.
ROBERTS, L. D. (ED.), *Approaches to Nature in the Middle Ages*. Binghamton & New York, 1982. (*Medieval and Renaissance Texts and Studies*, 16)
SAARINEN, R., "John Buridan and Donald Davidson on "Akrasia"." *Synthese*, 96 (1993), 133–154.
—, *Weakness of the Will in Medieval Thought. From Augustine to Buridan*. Leiden & New York, 1994. (*Studien und Texte zur Geistesgeschichte des Mittelalters*, 44)
SCHÖNBERGER, R., "*Causa causalitatis*. Zur Funktion der aristotelischen Ursachenlehre in der Scholastik." In Craemer-Rügenberg e.a. (eds.), *Scientia und Ars*, pp. 421–439.
—, "Eigenrecht und Relativität des Natürlichen bei Johannes Buridanus." In Zimmermann e.a. (eds.), *Mensch und Natur im Mittelalter*, pp. 216–233.
—, "Evidenz und Erkenntnis. Zu mittelalterlichen Diskussionen um das erste Prinzip." *Philosophisches Jahrbuch*, 102 (1995), 4–19.
—, "Das gleichzeitige Auftreten von Nominalismus und Mystik." In Speer (ed.), *Die Bibliotheca Amploniana*, pp. 409–433.
—, *Relation als Vergleich. Die Relationstheorie des Johannes Buridan im Kontext seines Denkens und der Scholastik*. Leiden, 1994. (*Studien und Texte zur Geistesgeschichte des Mittelalters*, 43)
SHAH, M. H., *The General Principles of Avicenna's Canon of Medicine*. Karachi, 1966.
SORABJI, R. R. K., *Time, Creation and the Continuum. Theories in Antiquity and the Early Middle Ages*. London, 1983.
SOSA, E. & M. TOOLEY (EDS.), *Causation*. Oxford, 1993.
SPADE, P. V., "Synonymy and Equivocation in Ockham's Mental Language." *The Journal of the History of Philosophy*, 1980 (18), 9–22.
SPADE, P. V. (ED.), *The Cambridge Companion to Ockham*. Cambridge, 1999. (*Cambridge Companions to Philosophy*)
SPEER, A. (ED.), *Die Bibliotheca Amploniana. Ihre Bedeutung im Spannungsfeld von Aristotelismus, Nominalismus und Humanismus*. Berlin & New York, 1995. (*Miscellanea Mediaevalia*, 23)
STEENBERGHEN, F. VAN, *Maître Siger de Brabant*. Louvain & Paris, 1977. (*Philosophes médiévaux*, 21)
SYLLA, E. D., "Aristotelian Commentaries and Scientific Change: The Parisian Nominalists on the Cause of the Natural Motion of Inanimate Bodies." In Murdoch e.a. (eds.), *Medieval Science*, pp. 37–83.
—, "Autonomous and Handmaiden Science: St. Thomas Aquinas and Wil-

liam of Ockham on the Physics of the Eucharist." In Murdoch e.a. (eds.), *The Cultural Context*, pp. 349–391.
—, "William Heytesbury on the Sophism *Infinita sunt finita*." In Beckman e.a. (eds.), *Sprache und Erkenntnis im Mittelalter*, pp. 628–636.
THIJSSEN, J. M. M. H., *Johannes Buridanus over het oneindige. Een onderzoek naar zijn theorie over het oneindige in het kader van zijn wetenschaps- en natuurfilosofie.* 2 vols. Nijmegen, 1988.
—, "John Buridan and Nicholas of Autrecourt on Causality and Induction." *Traditio*, 43 (1987), 237–255.
—, "Late-Medieval Natural Philosophy: Some Recent Trends in Scholarship." *Recherches de théologie et philosophie médiévale*, 67 (2000), 158–190.
TRIFOGLI, C., "La dottrina del tempo in Egidio Romano." *Documenti e studi sulla tradizione filosofica medievale*, 1 (1990), 247–276.
—, "Il problema dello statuto ontologico del tempo nelle *Quaestiones super Physicam* di Thomas Wylton e di Giovanni di Jandun." *Documenti e studi sulla tradizione filosofica medievale*, 2 (1990), 491–548.
TURKLE, S., *Life on the Screen: Identity in the Age of the Internet.* New York, 1995. Repr. London, 1997.
VERBEKE, G., "L'argument du livre VII de la Physique: Une impasse philosophique." In Duering (ed.), *Naturphilosophie*, pp. 250–267. Repr. in Verbeke, *D'Aristote à Thomas d'Aquin*, pp. 147–165.
—, *D'Aristote à Thomas d'Aquin. Antécédents de la pensée moderne.* Louvain, 1990. (*Ancient and Medieval Philosophy*, 8)
FEDERICI VESCOVINI, G., "La concezione della natura di Giovanni Buridano." In Nardi (ed.), *La filosofia della natura*, pp. 616–624.
FEDERICI VESCOVINI, G. (ED.), *Filosofia, scienza classica: filosofia, scienza arabo-latina medievale et l'età moderna.* Louvain-la-Neuve, 1999.
VLASTOS, G., "On Heraclitus." *American Journal of Philology*, 76 (1955), 337–368. Repr. Vlastos, *Studies in Greek Philosophy*, vol 1, pp. 127–150.
—, *Studies in Greek Philosophy.* Princeton, New Jersey, 1995.
WALSH, J. J., "Buridan on the Connection of the Virtues." *Journal of the History of Philosophy*, 24 (1986), 453–482.
WARDY, R., *The Chain of Change: A Study of Aristotle's Physics VII.* Cambridge, 1990.
WEISHEIPL, J. A., "The Principle *Omne quod movetur ab alio movetur* in Medieval Physics." *Isis*, 56 (1965), 26–45. Repr. in Carroll (ed.), *Nature and Motion*, pp. 75–99.
WIELGUS, S. (ED.), *Marsilius von Inghen. Werk und Wirkung. Akten des Zweiten Internationalen Marsilius-von-Inghen-Kongresses.* Lublin, 1993.
WILSON, M. D., *Descartes.* London & Boston, 1978. (*The Arguments of the Philosophers*)
WINKELMANN, E. (ED.), *Urkundenbuch der Universität Heidelberg.* 2 vols. Heidelberg, 1886.
WITTGENSTEIN, L., *On Certainty.* Ed. G. E. M. Anscombe & G. H. von Wright. Translated by Denis Paul & G. E. M. Anscombe. Oxford, 1969.
YRJÖNSUURI, M., *Obligationes. Fourteenth-Century Logic of Disputational Duties.* Ph.D. diss. University of Helsinki, 1994.

ZIMMERMANN, A., *Verzeichnis ungedruckter Kommentare zur Metaphysik und Physik des Aristoteles aus der Zeit von etwa 1250–1350*. Leiden & Köln, 1971. (*Studien und Texte zur Geistesgeschichte des Mittelalters*, 9)

ZIMMERMANN, A. & A. SPEER (EDS.), *Mensch und Natur im Mittelalter*. 2 vols. Berlin & New York, 1991. (*Miscellanea Mediaevalia*, 21)

ZUPKO, J., "What Is the Science of the Soul? A Case Study in the Evolution of Late Medieval Natural Philosophy." *Synthese*, 110 (2) (1997), 297–334.

—, "Buridan and Skepticism." *Journal of the History of Philosophy*, 31 (1993), 91–221.

—, "Freedom of Choice in Buridan's Moral Psychology." *Mediaeval Studies*, 57 (1995), 75–99.

—, "How Are Souls Related to Bodies? A Study of John Buridan." *Review of Metaphysics*, 46 (1993), 575–601.

—, "Sacred Doctrine, Secular Practice: Theology and Philosophy in the Faculty of Arts at Paris, 1325–1400." In Aertsen e.a., *Was ist Philosophie im Mittelalter?*, pp. 656–666.

INDEX OF NAMES

Albert of Saxony: 131
Albert The Great: xiii, 78, 85, 86, 186, 187, 194, 212
Alexander of Aphrodisias: 147
Alfarabi: 81, 83
al-Ghazali: 74
Anselm of Canterbury: 272
Aristotle: : ix, xi, xii, xiv, xv, xvi, 19, 39, 55, 57, 65, 66, 67, 69, 70, 71, 73, 77, 81, 82, 88, 116, 117, 122, 127, 128, 145, 147, 148, 152, 156, 157, 161, 171, 182, 183, 184, 185, 188, 189, 190, 194, 195, 196, 197, 198, 201, 222, 223, 226, 228, 231, 235, 236, 237, 239, 240, 243, 247, 251, 252, 254, 255, 256, 258, 260, 262, 268, 269, 270, 271, 272, 273, 274, 275, 277, 278, 279, 281.
Augustine: 209, 216, 272
Avempace: 274
Averroes: xvi, 66, 74, 86, 129, 184, 186, 187, 223, 227, 232, 269, 274, 278, 279
Avicenna: xvii, 81, 82, 83, 86, 193, 195, 269, 281

Bakker, P.J.J.M.: xvi, xvii
Bernard of Arezzo: 167
Biard, J.: xiii
Blaise Pascal: 77, 81
Boethius of Dacia: xvi, 223, 224, 237, 244
Bonaventure: 78

Chalcidius: 173
Clagett, M.: x

Dante Alighieri: 200, 276

Dekker, D.-J.: xiv
Descartes, R.: 167, 168, 172
Duhem, P.: x, 224, 238, 244

Feyerabend, P.: x

Galileo: x
Galen: 195
Godfrey of Fontaines: 68
Gregory of Rimini: 131

Harré, R.: 73, 74, 75, 76
Henry of Ghent: 67
Heraclitus: 50, 55, 56
Hippocrates: 92
Hume, D.: 73, 74

John Buridan: passim
John Duns Scotus: xvii, 59, 71, 76, 83, 199, 200, 208, 209, 210, 263, 265, 269, 270, 272
John of Jandun: 263

King, P.: xi
Klima, G.: xi
Knuuttila, S.: xii, 90
Krieger, G.: xiii
Kuhn, Th.: x

Macrobius: 272
Madden, E.H.: 73, 74, 75, 76
Maier, A.: x, 129, 152
Malebranche: 77
Marsilius of Inghen: xvii, 247, 249, 257, 258, 260, 261, 262, 263, 264
Meister Eckhart: 265, 270
Moore, G.E.: 166
Murdoch, J.E.: xiii

Newton, I.: 224, 244

Nicholas of Autrecourt: 122, 166, 167, 169, 179, 182
Nicole Oresme: 222
Normore, C.: xiii

Pelagius: 216
Peter Aureoli: 70
Peter Lombard: 152, 221, 250
Pinborg, J.: x
Pironet, F.: xv, xvi
Plato: 3, 15, 16, 17, 20, 21, 22, 24, 50, 78, 89, 115, 171, 172, 173, 272
Pluta, O.: xii
Plutarch: 50
Putnam, H.: 73, 74

Randi, E.: 83
Roger Bacon: 189, 191

Saarinen, R.: 201, 206, 212

Schönberger, R.: xvii
Seneca: 56
Siger of Brabant: 78, 218
Sobol, P.G.: xv
Sylla, E.D.: xvi

Thijssen, J.M.M.H.: : xiii, 23
Thomas Aquinas: : xi, xvii, 13, 43, 44, 45, 59, 67, 78, 200, 211, 212, 222, 223, 265, 268, 269
Thomas Bradwardine: 131

William Ockham: xiv, 2, 6, 7, 59, 68, 76, 81, 84, 129, 134, 149, 153, 200, 223, 266, 270

Zupko, J.: xiv, 199, 200, 205, 219

INDEX OF MANUSCRIPTS

Carpentras, Bibliothèque Municipale, 292: p. 251, n.9-10-11; p. 252, n.12-13; p. 253, n.14-15; p. 254, n.16; p. 255, n.18; p. 256, n.19-20; p. 257, n.21

Città del Vaticano, Biblioteca Apostolica Vaticana, Cod. Vat. lat. 2163: p. 53, n.13

—, Biblioteca Apostolica Vaticana, Cod. Chigi E VI 199: p. 53, n.13

Frankfurt am Main, Stadt- und Universitätsbibliothek, Cod. Praed. 52: p. 53, n.13

Paris, Bibliothèque nationale de France, lat. 14.716: p. 251, n.9-10-11; p. 252, n.12-13; p. 253, n.14-15; p. 254, n.16; p. 255, n.18; p. 256, n.19-20; p. 257, n. 21

København, Kongelige Bibliotek, Cod. Ny kgl. Saml. 1801 fol.: p. 53, n.13; p. 145, n.45; p. 148, n.54; p. 153, n.12

Kraków, Biblioteka Jagiellońska, Cod. 709: p. 257, n.24; p. 258, n.25-26; p. 259, n.27; p. 260, n.28-29; p. 261, n.30-31-32; p. 262, n.33

Wien, Österreichische Nationalbibliothek, CVP 5297: p. 257, n.24; p. 258, n.25-26; p. 259, n.27; p. 260, n.28-29; p. 261, n.30-31-32; p. 262, n.33

CONTRIBUTORS

Paul J.J.M. Bakker, *University of Nijmegen*, Nijmegen, The Netherlands
Joël Biard, *University of Tours*, Tours, France
Dirk-Jan Dekker, *University of Nijmegen*, Nijmegen, The Netherlands
Peter King, *The Ohio State University*, Columbus, Ohio, U.S.A.
Gyula Klima, *Fordham University*, New York, U.S.A.
Simo Knuuttila, *University of Helsinki*, Helsinki, Finland
Gerhard Krieger, *University of Trier*, Trier, Germany
John E. Murdoch, *Harvard University*, Cambridge, Mass., U.S.A.
Olaf Pluta, *University of Nijmegen*, Nijmegen, The Netherlands
Fabienne Pironet, *University of Montreal*, Montreal, Canada
Rolf Schönberger, *University of Regensburg*, Regensburg, Germany
Peter G. Sobol, Madison, Wisconsin, U.S.A.
Edith D. Sylla, *University of North Carolina*, Raleigh, North Carolina, U.S.A.
Johannes M.M.H. Thijssen, *University of Nijmegen*, Nijmegen, The Netherlands
Jack Zupko, *Emory University*, Atlanta, Georgia, U.S.A.